Bhanu Prasad and S.R. Mahadeva Prasanna (Eds.)

Speech, Audio, Image and Biomedical Signal Processing using Neural Networks

Studies in Computational Intelligence, Volume 83

Editor-in-chief

Prof. Janusz Kacprzyk
Systems Research Institute
Polish Academy of Sciences
ul. Newelska 6
01-447 Warsaw
Poland
E-mail: kacprzyk@ibspan.waw.pl

Bhanu Prasad
S.R. Mahadeva Prasanna
(Eds.)

Speech, Audio, Image and Biomedical Signal Processing using Neural Networks

With 157 Figures and 81 Tables

 Springer

Bhanu Prasad
Department of Computer and Information Sciences
Florida A&M University
Tallahassee, FL 32307
USA

S.R. Mahadeva Prasanna
Department of Electronics and Communication Engineering
Indian Institute of Technology Guwahati
Guwahati
India

ISBN 978-3-540-75397-1   e-ISBN 978-3-540-75398-8

Studies in Computational Intelligence ISSN 1860-949X

Library of Congress Control Number: 2007938407

Cover Design: Deblik, Berlin, Germany

Printed on acid-free paper

9 8 7 6 5 4 3 2 1

springer.com

Foreword

Welcome to the book "Speech, Audio, Image and Biomedical Signal Processing using Neural Networks". Neural networks play an important role in several scientific and industrial applications. This peer reviewed and edited book presents some recent advances on the application of neural networks in the areas of speech, audio, image and biomedical signal processing. Each chapter in the book is reviewed by at least two independent experts in the topical area.

The first chapter, Information Theoretic Feature Selection and Projection, authored by Deniz Erdogmus, Umut Ozertem and Tian Lan demonstrates some algorithms for the purpose of learning optimal feature dimensionality reduction solutions in the context of pattern recognition. The techniques and algorithms can be applied to the classification of multichannel electroencephalogram (EEG) signals for brain computer interface design, as well as sonar imagery for target detection. The results are compared with some widely used benchmark alternatives.

The next chapter, Recognition of Tones in Yoruba Speech: Experiments with Artificial Neural Networks, authored by Odetunji Ajadi Odejobi provides an artificial neural networks approach for the recognition of Standard Yoruba (SY) language tones. Yoruba is widely used in the African continent and also in some other parts of the world. Multi-layered perceptron and recurrent neural network models are used for training and recognition purposes. Some performance results are also presented.

Stefan Scherer, Friedhelm Schwenker and Guunther Palm provide a detailed overview of related work on the emotion recognition from speech. In the chapter, Emotion Recognition from Speech using Multi-Classifier Systems and RBF-Ensembles, the authors provide a comparison of possible data acquisition methods. A complete automatic emotion recognizer scenario comprising a possible way of collecting emotional data, a human perception experiment for data quality benchmarking, the extraction of commonly used features, and recognition experiments using multi-classifier systems and RBF ensembles, is also provided.

The next chapter in the book, Modeling Supra-segmental Features of Syllables using Neural Networks, is authored by K. Sreenivasa Rao. The author discusses modeling of supra-segmental features of syllables, and suggests some applications of these models. A four layer feedforward neural network is used for this purpose. The performance of the prosody models is evaluated for applications such as speaker recognition, language identification and text-to-speech synthesis.

In the chapter, Objective Speech Quality Evaluation Using an Adaptive Neuro-Fuzzy Network, the authors Guo Chen and Vijay Parsa, investigate a novel non-intrusive speech quality metric based on adaptive neuro-fuzzy network techniques. A first-order Sugeno type fuzzy inference system is applied for objectively estimating the speech quality. The performance of the method was demonstrated through comparisons with the state-of-the-art non-intrusive quality evaluation standard, the ITU-T P.563.

The next chapter in the book, A Novel Approach to Language Identification Using Modified Polynomial Networks, is authored by Hemant A. Patil and T. K. Basu. The authors present a modified polynomial networks for the language identification problem. The method consists of building up language models for each language using the normalized mean of the training feature vectors for all the speakers in a particular language class. Some experimental details are provided by considering four major Indian languages.

Enrique Alexandre, Lucas Cuadra, Manuel Rosa-Zurera and Francisco López-Ferreras explore the feasibility of using some kind of tailored neural networks to automatically classify sounds into either speech or non-speech in hearing aids. In their chapter, Speech/Non-speech Classification in Hearing Aids Driven by Tailored Neural Networks, the authors emphasized on the designing the size and complexity of the multilayer perceptron constructed by a growing method. The number of simple operations is evaluated.

Audio Signal Processing authored by Preeti Rao is the next chapter of the book. The author reviews the basic methods for audio signal processing, mainly from the point of view of audio classification. General properties of audio signals are discussed followed by a description of time-frequency representations for audio. Features useful for classification are reviewed. In addition, a discussion on prominent examples of audio classification systems with a particular emphasis on feature extraction is provided.

In the chapter, Democratic Liquid State Machines for Music Recognition, the authors Leo Pape, Jornt de Gruijl and Marco Wiering present a democratic liquid state machine (LSM) that uses majority voting by combining two dimensions. The results show that the democratic LSM significantly outperforms the basic LSM and other methods on two music composer classification tasks.

Arvind Nayak, Subhasis Chaudhuri and Shilpa Inamdar present two different methods to address the problem of accommodating the images from the same scene with varying illumination conditions. In the chapter, Color

Transfer and its Applications, the authors also discuss the issues related to color correction.

The next chapter of the book, A Neural Approach to Unsupervised Change Detection of Remote-sensing Images, is authored by Susmita Ghosh, Swarnajyoti Patra and Ashish Ghosh. The authors present two unsupervised context-sensitive change detection techniques for remote sensing images. Experimental results are also provided to determine the effectiveness of the proposed approaches.

Manjunath Aradhya V N, Hemantha Kumar G and Noushath S present a new scheme of image recognition. It is based on the Fisher Linear Discriminant (FLD) analysis and Generalized Regression Neural Networks (GRNN). In the chapter, Fisher Linear Discriminant Analysis and Connectionist Model for Efficient Image Recognition, the authors also presented some experiments conducted on an image database and demonstrated the effectiveness and feasibility of the presented method.

Detection and Recognition of Human Faces and Facial Features, authored by Milos Oravec, Gregor Rozinaj and Marian Beszedes, is the next chapter of the book. The authors discuss linear and nonlinear feature extraction methods from face images. Higher-order and nonlinear methods of feature extraction as well as feature extraction by kernel methods are also considered. The methods are illustarted by the results of several face recognition systems. Several methods for facial feature detection are also discussed.

The next chapter is Classification of Satellite Images with Regularized AdaBoosting of RBF Neural Networks. The authors Gustavo Camps-Valls and Antonio Rodrigo-González present the soft margin regularized AdaBoost method for the classification of images acquired by satellite sensors. The method is compared with the state-of-the-art support vector machines.

The chapter, Convolutional Neural Networks for Image Processing with Applications in Mobile Robotics, by Matthew Browne, Saeed Shiry Ghidary and Norbert Michael Mayer presents a description of the Convolutional Neural Network (CNN) architecture and reports some work on applying CNNs to theoretical as well as real-world image processing problems.

SVM based Adaptive Biometric Image Enhancement using Quality Assessment is the next chapter of the book. The authors Mayank Vatsa, Richa Singh and Afzel Noore discuss the influence of input data quality on the performance of a biometric system. The proposed quality assessment algorithm associates the quantitative quality score of the image that has a specific type of irregularity such as noise, blur, and illumination. The performance algorithm is validated by considering face biometrics as the case study.

The next chapter, Segmentation and Classification of Leukocytes using Neural Networks - A Generalization Direction, is authored by Pedro Rodrigues, Manuel Ferreira and Joao Monteiro. The authors initially present an exclusively supervised approach for segmentation and classification of blood white cells images. In addition, another approach formed by two neural networks is also presented to overcome some drawbacks with the initial approach. Some performance results of these two approaches are also presented.

The final chapter of the book, A Closed Loop Neural Scheme to Control Knee Flex-extension Induced by Functional Electrical Stimulation: Simulation Study and Experimental Test on a Paraplegic Subject, is authored by Simona Ferrante, Alessandra Pedrocchi and Giancarlo Ferrigno. The authors present an Error Mapping Controller (EMC) for neuromuscular stimulation of the quadriceps during the knee flex-extension movement. The EMC is composed by a feedforward inverse model and a feedback controller, both implemented using neural networks. The controller was developed and also tested.

Tallahassee, USA   *Bhanu Prasad*
Guwahati, India   *S.R. Mahadeva Prasanna*

Contents

Information Theoretic Feature Selection and Projection

Deniz Erdogmus[1,2], Umut Ozertem[1] and Tian Lan[2]

[1] Department of CSEE, Oregon Health and Science University, Portland, OR, USA
[2] Department of BME, Oregon Health and Science University, Portland, OR, USA

Summary. Pattern recognition systems are trained using a finite number of samples. Feature dimensionality reduction is essential to improve generalization and optimal exploitation of the information content in the feature vector. Dimensionality reduction eliminates redundant dimensions that do not convey reliable statistical information for classification, determines a manifold on which projections of the original high- dimensional feature vector exhibit maximal information about the class label, and reduces the complexity of the classifier to help avoid over-fitting. In this chapter we demonstrate computationally efficient algorithms for estimating and optimizing mutual information, specifically for the purpose of learning optimal feature dimensionality reduction solutions in the context of pattern recognition. These techniques and algorithms will be applied to the classification of multichannel EEG signals for brain computer interface design, as well as sonar imagery for target detection. Results will be compared with widely used benchmark alternatives such as LDA and kernel LDA.

1 Introduction

In pattern recognition, a classifier is trained solve the multiple hypotheses testing problem in which a particular input feature vector's membership to one of the classes is assessed. Given a finite number of training examples, feature dimensionality reduction to improve generalization and optimal exploitation of the information content in the feature vector regarding class labels is essential. Such dimensionality reduction enables the classifier to achieve improved generalization through (1) eliminating redundant dimensions that do not convey reliable statistical information for classification, (2) determining a manifold on which projections of the original high dimensional feature vector exhibit maximal information about the class label, and (3) reducing the complexity of the classifier to help avoid over-fitting. In other words, feature dimensionality reduction through projections with various constraints can exploit salient features and eliminate irrelevant feature fluctuations by representing the discriminative information in a lower dimensional manifold embedded in the original Euclidean feature space.

In principle, the maximally discriminative dimensionality reduction solution is typically nonlinear – in fact, the minimum-risk Bayesian classifier can be interpreted as the optimal nonlinear dimensionality reduction mapping. However, given finite amount of training data, arbitrary robust nonlinear projections are challenging to obtain, thus there is wide interest in the literature in finding regularized nonlinear projections as well as simple linear projections. Furthermore, in some situations, such as real-time brain computer interfaces, it is even desirable to completely eliminate the computational and hardware requirements of evaluating the values of certain features, in which case feature selection – a special case of linear projections constrained to sparse orthonormal matrices – is required. Existing approaches for feature dimensionality reduction can be classified into the so-called *wrapper* and *filter* categories.

The wrapper approach aims to identify the optimal feature dimensionality reduction solution for a particular classifier, therefore involves repeatedly adjusting the projection network based on the cross validation performance of a corresponding trained classifier with the particular topology. This approach is theoretically the optimal procedure to find the optimal feature dimensionality reduction given a particular training set, a specific classifier topology, and a particular projection family. However, repeated training of the classifier and iterative learning of the projection network under this framework is computationally very expensive and could easily become unfeasible when the original feature dimensionality is very high and the number of training samples is large. The complexity is further increased by the cross-validation procedure and repeated training of the classifier to ensure global optimization, if particular topologies that are prone to local optima, such as neural networks, are selected. It is also widely accepted, and is intuitive, that some classification algorithms, such as decision tree, multi-layer perceptron neural networks have inherent ability to focus on relevant features and ignore irrelevant ones, when properly trained [1].

The filter approach provides a more flexible and computationally attractive approach at the expense of not identifying the optimal feature-classifier combination. This technique relies on training a feature projection network through the optimization of a suitable optimality criterion that is relevant to classification error/risk. Since the filter approach decouples the feature projection optimization from the following classifier-training step, this approach enables the designer to conveniently compare the performance of various classifier topologies on the reduced dimensionality features obtained through the filter approach. In this chapter, we will propose linear and nonlinear feature projection network training methodologies based on the filter approach. These projections will be optimized to approximately maximize the mutual information between the reduced dimensionality feature vector and the class labels. The selection of mutual information is motivated by information theoretic bounds relating Bayes probability of error for a particular feature vector and its mutual information with the class labels. Specifically, Fano's lower bound [2–4] provides a performance bound on the classifiers and more importantly the Hellman–Raviv bound [2–4], expressed as $p_e \leq (H(C) - I(\mathbf{x}; C))/2$ where C are the class labels corresponding to feature vectors $\mathbf{x}$, p_e denotes the minimum probability of error, which is obtained with a Bayes classifier, and H and I denote Shannon's entropy and mutual information, respectively. The entropy of the class labels depends only on the class priors. Consequently, maximizing the mutual information between the (projected) features and the class labels results in a dimensionality reduction that minimizes this tight bound on Bayes error. The maximum mutual information principle outlined here can be utilized to solve the following three general problems: determining optimal (1) feature ranking and selection, (2) linear feature projections, and (3) nonlinear feature projections that contain maximal discriminative information, minimal redundancy, and irrelevant statistical variations. Earlier work on utilizing mutual information to select input features for a classifier includes [5–8].

Many feature selection and projection methods have been developed in the past years [9–18]. Guyon and Elisseeff also reviewed several approaches used in the context of machine learning [19]. The possibility of learning the optimal feature projections sequentially decreases the computational requirements making the filter approach especially attractive. Perhaps, historically the first dimensionality reduction technique is linear principle components analysis (PCA) [9,10]. Although this technique is widely used, its shortcomings for pattern recognition are well known. A generalization to nonlinear projections, Kernel PCA [20], still exhibits the same shortcoming; the projected features are not necessarily useful for classification. Another unsupervised (i.e., ignorant of class labels) projection method is independent component analysis (ICA), a modification of the uncorrelatedness condition in PCA to independence, in order to account for higher order statistical dependencies in non-Gaussian distributions [21]. Besides statistical independence, source sparsity and nonnegativity is also utilized as a statistical assumption in achieving dimensionality reduction through sparse bases, a technique called nonnegative matrix factorization (NMF) [22]. These methods, however, are linear and restricted in their ability to generate versatile projections for curved data distributions. Local linear projections is an obvious method to achieve globally nonlinear yet locally linear dimensionality reduction. One such method that aims to achieve dimensionality reduction while preserving neighborhood topologies is local linear embedding (LLE) [23]. Extensions of this approach to supervised local linear embeddings that consider class label information also exist [24]. Linear Discriminant Analysis (LDA) attempts to eliminate the shortcoming of PCA by finding linear projections that maximize class separability under the Gaussian distribution assumption [11]. The LDA projections are optimized based on the means and covariance matrix of each class, which are not descriptive of an arbitrary probability density function (pdf). In addition, only linear projections are considered. Kernel LDA [25], generalizes this principle to finding nonlinear projections under the assumption that the kernel function induces a nonlinear transformation (dependent on the eigenfunctions of the kernel) that first projects the data to a hypothetical high dimensional space where the Gaussianity assumption is satisfied. However, the kernel functions used in practice do not necessarily guarantee the validity of this assumption. Nevertheless, empirical evidence suggests that robust nonparametric solutions to nonlinear problems in pattern recognition can be obtained by first projecting the data into a higher dimensional space (possibly infinite) determined by the eigenfunctions of the selected interpolation kernel. The regularization of the solution is achieved by the proper selection of the kernel. Torkkola [14] proposes utilizing quadratic density distance measures to evaluate an approximate mutual information measure to find linear and parametric nonlinear projections, based on the early work on information theoretic learning [26], and Hild et al. [18] propose optimizing similar projections using a discriminability measure based on Renyi's entropy. The latter two proposals are based on utilizing the nonparametric kernel density estimation (KDE) technique (also referred to as Parzen windowing) [27].

Estimating mutual information requires assuming a pdf estimate explicitly or implicitly. Since the data pdf might take complex forms, in many applications determining a suitable parametric family becomes a nontrivial task. Therefore, mutual information is typically estimated more accurately using nonparametric techniques [28, 29]. Although this is a challenging problem for two continuous-valued random vectors, in the feature transformation setting the class labels are discrete-valued. This reduces the problem to simply estimating multidimensional entropies of continuous random vectors. The entropy can be estimated nonparametrically using a number of techniques. Entropy estimators based on sample spacing, such as the minimum spanning tree, are not differentiable making them unsuitable for adaptive learning of feature projections [29–33]. On the other hand, entropy estimators based on kernel density estimation (KDE) provide a differentiable alternative [28, 33, 34].

In this chapter we derive and demonstrate computationally efficient algorithms for estimating and optimizing mutual information, specifically for the purpose of learning optimal feature dimensionality reduction solutions in the context of pattern recognition. Nevertheless, the estimators could be utilized in other contexts. These techniques and algorithms will be applied to the classification of multichannel EEG signals for brain computer interface design, as well as sonar imagery for target detection. Results will be compared with widely used benchmark alternatives such as LDA and kernel LDA.

2 Nonparametric Estimators for Entropy and Mutual Information with a Discrete Variable

The probability distribution of a feature vector $\mathbf{x} \in \Re^n$ is a mixture of class distributions conditioned on the class label c:$p(\mathbf{x}) = \sum_c p_c p(\mathbf{x}|c)$. Maximizing mutual information between the projected features and the class labels requires implicitly or explicitly estimating this quantity. Shannon's mutual information between the continuous feature vector and the discrete class label can be expressed in terms of the overall Shannon entropy of the features and their average class conditional entropy: $I(\mathbf{x}; c) = H(c) - \sum_c p_c H(\mathbf{x}|c)$. Specifically, the Shannon joint entropy for a random vector x and this vector conditioned on a discrete label c are defined as:

$$H(\mathbf{x}) = -\int p(\mathbf{x})log(p(\mathbf{x}))d\mathbf{x}, \quad H(\mathbf{x}|c) = -\int p(\mathbf{x}|c)log(p(\mathbf{x}|c))d\mathbf{x} \quad (1)$$

This demonstrates that estimating the mutual information between a continuous valued random feature vector and a discrete valued class label is relatively easy since only multiple joint entropy estimates (including conditionals) with the dimensionality of the feature vector of interest need to be obtained. Given a finite labeled training data set $\{(\mathbf{x}_1, c_1), \ldots, (\mathbf{x}_N, c_N)\}$, where c_i takes one of the class labels as its value, the multidimensional entropy terms in (1) can be estimated nonparametrically with most convenience, although parametric and semiparametric estimates are also possible to obtain. The parametric approach assumes a family of distributions for the class conditional densities and utilizes Bayesian model fitting techniques, such as maximum likelihood. For iterative optimization of mutual information, this procedure is computationally very expensive, therefore not always feasible. Semiparametric approaches utilize a suitable truncated series expansion to approximate these distributions around a reference density (for instance, Hermite polynomial expansion around a multivariate Gaussian is commonly utilized in the independent component analysis literature and leads to the well known kurtosis criterion in that context) [21]. These estimates are accurate provided that the reference density is selected appropriately and the series converges fast, so that low-order truncations yield accurate approximations.

In this chapter, we will place the most emphasis on nonparametric estimators of entropy and mutual information due to their computational simplicity and versatile approximation capabilities. Various approaches that depend on pairwise sample distances and order statistics (such as ranked samples and minimum spanning trees). All of these approaches can in fact be explained as special cases of kernel density estimator based plug-in entropy estimation. Specifically, variable kernel size selection results in highly accurate density representations, therefore entropy estimates. For a sample set $\{\mathbf{x}_1, \ldots, \mathbf{x}_N\}$ the kernel density estimate with variable kernel size is:

$$p(\mathbf{x}) \approx N^{-1} \sum_{k=1}^{N} K_{\Sigma_k}(\mathbf{x} - \mathbf{x}_k) \quad (2)$$

where typical kernel functions in the literature are uniform (leads to K-nearest-neighbor sliding histogram density estimates) and Gaussian. Especially important is the latter, since the Gaussian kernel is commonly utilized in many kernel machine solutions to nonlinear regression, projection, and classification problems in machine learning:

$$G_\Sigma(\xi) = exp(-\xi^T \Sigma^{-1} \xi/2)/\sqrt{(2\pi)^n |\Sigma|} \tag{3}$$

where n is the dimensionality of the kernel function.

In this chapter, parametric and semiparametric approaches will be briefly reviewed and sufficient detail will be provided for the reader to understand the basic formulations. In addition, detailed expressions for various nonparametric estimators will be provided, their theoretical properties will be presented, and their application to the determination of maximally informative and discriminative feature projections will be illustrated with many real datasets including EEG classification in the context of brain interfaces, neuron spike detection from microelectrode recordings, and target detection in synthetic aperture radar imagery.

2.1 Parametric Entropy Estimation

The parametric approach relies on assuming a family of distributions (such as the Gaussian, Beta, Exponential, mixture of Gaussians, etc.) that parametrically describes each candidate distribution. The *optimal* pdf estimate for the data, is then determined using maximum likelihood (ML) or maximum *a posteriori* (MAP) statistical model fitting using appropriate regularization through model order selection criteria. For example, the ML estimate yields $p(x; \Theta_{ML})$ as the density estimate by solving

$$\Theta_{ML} = \arg\max_{\Theta} \sum_{k=1}^{N} log\ p(\mathbf{x}_k; \Theta) \tag{4}$$

where $p(\mathbf{x}_k; \Theta)$ is the selected parametric family of distributions. The modeling capability of the parametric approach can be enhanced by allowing mixture models (such as mixture of Gaussians), in which case, the family of distributions is in the form:

$$p(\mathbf{x}; \{\alpha_k, \Theta_k\}) = \sum_{k=1}^{M} \alpha_k\ G(\mathbf{x}, \Theta_k) \tag{5}$$

Once the density estimate is optimized as described, it is plugged-in to the entropy expression to yield the sample estimator for entropy, which is obtained in two-stages, since analytical expressions for the entropy of most parametric families is not available. Using the sample mean approximation for expectation, we obtain:

$$H(\mathbf{x}) \approx -\frac{1}{N} \sum_{j=1}^{N} logp(\mathbf{x}_j; \Theta_{ML}) \tag{6}$$

The difficulty with parametric methods in learning optimal feature projections is that it requires solving an optimization problem (namely ML or other model fitting procedure) within the main projection optimization problem. Another drawback of the parametric approach is the insufficiency of parametric models for general-purpose data modeling tasks. As the features in an arbitrary pattern recognition problem might exhibit very complicated structures, the selected parametric family for modeling might remain too simplistic to be able to accurately model all possible variations of the data distribution during learning. It can be shown that the ML density estimate asymptotically converges to the member of the parametric family that minimizes the KLD with the true underlying density [35].

An alternative approach to parametric estimation involves exploiting the maximum entropy principle. Consider the following constrained optimization problem that seeks the maximum entropy distribution given some nonlinear moments:

$$\max_{p(\mathbf{x})} = -\int p(\mathbf{x}) log\ p(\mathbf{x}) d\mathbf{x}\ sub.\ to \int p(\mathbf{x}) f_k(x) d\mathbf{x} = \alpha_k\ k = 0, 1, \ldots, m \quad (7)$$

where $f_0(\mathbf{x}) = 1$ and $\alpha_0 = 1$ 1 in order to guarantee normality. The solution to this maximum entropy density problem is in the form of an exponential: $p(\mathbf{x}) = exp(\lambda_0 + \sum_{k=1}^m \lambda_k f_k(\mathbf{x}))$, where the Lagrange multipliers can be approximated well by solving a linear system of equations if the true data distribution is close to the maximum entropy density with the same moments [36]. Further refinement can be obtained using these estimates as initial condition in a fixed-point equation solver for the constraints in (7). In this form, given the Lagrange multipliers, the entropy is easy to calculate:

$$H(\mathbf{x}) = -\int p(\mathbf{x})\lambda_0 + \sum_{k=1}^m \lambda_k f_k(\mathbf{x}) d\mathbf{x} = -\lambda_0 - \sum_{k=1}^m \lambda_k \alpha_k \quad (8)$$

The nonlinear moment functions can be selected to be various polynomial series, such as Taylor, Legendre, or Hermite (the first one is the usual polynomials and the latter two are explained below).

2.2 Semiparametric Entropy Estimation

The semiparametric approach provides some additional flexibility over the parametric approaches as they are based on using a parametric density model as the reference point (in the Taylor series expansion) and additional flexibility is introduced in terms of additional series coefficients. Below, we present a few series expansion models for univariate density estimation including Legendre and Gram–Charlier series. Multidimensional versions become computationally infeasible due to the combinatorial expansion of cross terms and their associated coefficients [21].

Legendre Series Expansion: Consider the Legendre polynomials defined recursively as follows using the initial polynomials $P_0(x) = 1$ and $P_1(x) = x$:

$$P_k(x) = \frac{1}{k}[(2k-1)x \, P_{k-1}(x) - (k-1)P(k-2)(x)] \quad k \geq 2 \tag{9}$$

Any pdf (satisfying certain continuity and smoothness conditions) can expressed in terms of the Legendre polynomials and in terms of polynomial statistics of the data.

$$q(x) = \sum_{k=0}^{\infty}(k + \frac{1}{2})E[P_k(X)]P_k(x) \tag{10}$$

The expectations can be approximated by sample mean approximations to obtain an approximation from a finite sample set.

Gram–Charlier Series Expansion: The characteristic function of a distribution $q(x)$, denoted by $\Phi_q(w)$, can be expressed in terms of characteristic function $\Phi_r(w)$ of an *arbitrary* reference distribution $r(x)$ as

$$\Phi_q(w) = exp\left(\sum_{k=1}^{\infty}(c_{q,k} - c_{r,k})\frac{(jw)^k}{k!}\right)\Phi_r(w) \tag{11}$$

where $c_{q,k}$ and $c_{r,k}$ cumulants of $q(x)$ and $r(x)$, respectively. The cumulants are expressed in terms of the moments using the Taylor series expansion of the cumulant generating function, defined as $\Psi(w) = log\Phi(w)$.

For the special case of a zero-mean and unit-variance Gaussian distribution as the reference pdf (denoted by $G(x)$ below), expanding the series and collecting the coefficients of same order derivatives together leads to the following expression in terms of the Hermite polynomials and polynomial moments as before.

$$q(x) = G(x)\sum_{k=0}^{\infty}\frac{E[H_k(X)]}{k!}H_k(x) \tag{12}$$

The Hermite polynomials are obtained using the initial polynomials $H_0(x) = 1$, $H_1(x) = x$, and the recursion

$$H_k(x) = x \, H_{k-1}(x) - (k-1)H_{k-2}(x) \quad k \geq 2 \tag{13}$$

Expansion on Complete Bases: All pdfs (satisfying the usual continuity and smoothness conditions) can be approximated by a truncated linear combination of basis functions (preferably orthonormal). Given infinitely many orthonormal bases $\{b_1(x), b_2(x), \ldots\}$, it is possible to express an arbitrary pdf in terms of the following linear combination.

$$q(x) = \sum_{k=1}^{\infty} E[b_k(X)]b_k(x) \tag{14}$$

An abundance of orthonormal bases for Hilbert spaces can be found in the function approximation literature (the eigenfunctions of any reproducing

kernel forms a bases for the pdf space). A drawback of the series approximations is that the truncation leads to pdf estimates that do not satisfy the two
basic conditions for a function to be a probability distribution: nonnegativity
and integration to unity [37].

2.3 Nonparametric Entropy Estimates

In the remainder of this chapter we will focus on nonparametric estimates,
especially the following two that emerge naturally from plug-in entropy
estimation utilizing a variable-size kernel density estimate; namely, sample-
spacing estimate and Parzen estimate. For illustration purposes, the former will be presented for univariate entropy estimation and extensions to
multidimensional cases will be discussed.

The most straightforward nonparametric approach in entropy estimation,
usually leading to poor estimates, yet surprisingly utilized frequently in the, is
to consider a histogram approximation for the underlying distribution. Fixed-
bin histograms lack the flexibility of sliding histograms, where the windows are
placed on every sample. A generalization of sliding histograms is obtained by
relaxing the rectangular window and assuming smoother forms that are continuous and differentiable (and preferably symmetric and unimodal) pdfs. This
generalization is referred to as kernel density estimation (KDE) and is shown
in (2). Another generalization of histograms is obtained by letting the bin-
size vary in accordance with local data distribution. In the case of rectangular
windows, this corresponds to nearest neighbor density estimation [38], and
for KDE this means variable kernel size [38, 39]. The corresponding entropy
estimates are presented below.

Entropy Estimation Based on Sample Spacing: Suppose that the ordered
samples $\{x_1 < x_2 < \ldots < x_N\}$ drawn from $q(x)$ are provided. We assume
that the distribution is piecewise constant in m-neighborhoods of samples [31],
leading to the following approximation:

$$p(x) = (N+1)^{-1}(x_{i+1} - x_i), \ i = 0, \ldots, N \tag{15}$$

Denoting the corresponding empirical cdf by $P(x)$, for ordered statistics, it is
known that

$$E[P(x_{i+m} - P(x_i))] = \frac{m}{N+m}, \ i = 1, \ldots, N-m \tag{16}$$

where the expectation is evaluated with respect to the joint data distribution
$q(x_1), \ldots, q(x_N)$, assuming iid samples. Substituting this in entropy, we obtain
the m-spacing estimator as

$$H(x) \approx -\frac{1}{N-m} \sum_{i=1}^{N-m} log((N+1)(x_{i+m} - x_i)/m) \tag{17}$$

The spacing interval m is chosen to be a slower-than-linear increasing function of N in order to guarantee asymptotic consistency and efficiency. Typically, $m = N^{1/2}$ is preferred in practice due to its simplicity, but other roots are viable. A difficulty with the sample spacing approach is its generalization to higher dimensionalities. Perhaps the most popular extension of sample-spacing estimators to multidimensional random vectors is the one based on the minimum spanning tree recently popularized in the signal processing community by Hero [30]. This estimator relies on the fact that the integral in Renyi's definition [40] of entropy is related to the sum of the lengths of the edges in the minimum spanning tree with useful asymptotic convergence guarantees. One drawback of this approach is that it only applies to entropy orders of $0 < \alpha < 1$ (Shannon entropy is the limiting case as a approaches 1). Another drawback is that finding the minimum spanning tree itself is a computationally cumbersome task that is also prone to local minima due to the heuristic selection of a neighborhood search radius by the user in order to speed-up.

Another generalization of sample spacing estimates to multi-dimensional entropy estimation has relied on the L1-norm as the distance measure between the samples instead of the usual Euclidean norm [39]. This technique can, in principle, be generalized to arbitrary norm definitions. The drawback of this method is its nondifferentiability, which renders it next to useless for traditional iterative gradient-based adaptation, but could be useful for feature ranking. This approach essentially corresponds to extending (15) to the multidimensional case as data-dependent variable-volume hyperrectangles. One could easily make this latter approach differentiable through the use of smooth kernels rather than rectangular volumes. Such modification will also form the connection between the sample-spacing methods and kernel based methods described next.

Parzen Windowing Based Entropy Estimation: Kernel density estimation is a well-understood and useful nonparametric technique that can be employed for entropy estimation in the plug-in estimation framework [33]. For a given set of iid samples $\{\mathbf{x}_1, \ldots, \mathbf{x}_N\}$ the variable size KDE is given in (2). To simplify computational requirements, fized size isotropic kernels could be utilized by assuming $\Sigma_k = \sigma^2 \mathbf{I}$ for all samples. The kernel function and its size can be optimized in accordance with the ML principle [42, 43] or other rules-of-thumb could be employed to obtain approximate optimal parameter selections [39, 41]. Given a suitable kernel function and size, the corresponding plug-in entropy estimate is easily obtained to be:

$$H(\mathbf{x}) = -\frac{1}{N} \sum_{j=1}^{N} log \frac{1}{N} \sum_{i=1}^{N} K_{\Sigma_i}(\mathbf{x}_j - \mathbf{x}_i) \tag{18}$$

Next, we demonstrate how these estimators can be utilized to design maximally discriminative linear and nonlinear feature projections via maximization of mutual information based on nonparametric estimators.

3 Linear and Nonlinear Feature Projection Design via Maximum Mutual Information

In this section we present two different techniques for determining linear and nonlinear projections respectively. For finding optimal linear projections, including feature selection, we will employ ICA decomposition in conjunction with the m-spacing estimator given in (17). For determining optimal nonlinear projections, we will employ the more general KDE-based plug-in estimate given in (18). In both cases, we assume that independent and identically distributed (iid) training data of the form $\{(\mathbf{x}_1, c_1), \ldots, (\mathbf{x}_N, c_N)\}$, where $x_i \in \Re^n$ is available.

3.1 Linear Feature Projections

Given the training data, we seek to determine a maximally informative linear feature projection $\mathbf{y} = \mathbf{A}\mathbf{x}$ from n to m dimensions that is characterized by a projection matrix $A \in \Re^{m \times n}$. Recently, methods based on optimizing this matrix via direct maximization of an approximation to the mutual information $I(\mathbf{y}, c)$ [14, 18]. These methods are based on slow iterative updates of the matrix due to the fact that at every update the gradient or other suitable update for the matrix must be computed using double-sum pairwise averages of samples due to the form in (18) that arises from the plug-in formalism. Stochastic gradient updates are a feasible tool to improve speed by reducing the computational load at each iteration, yet they may still not be sufficiently fast for very high dimensional scenarios.

We have recently proposed an approximation to this procedure by assuming that the class-mixture and class-conditional distributions of the given features obey the linear-ICA generative statistical model. Under this assumption, each class distribution as well as the mixture data density can be linearly brought to a separable form consisting of *independent* feature coordinates. This assumption is realistic for circularly symmetric class distributions, as well as elliptically symmetric class distributions where the independent axes of different classes are aligned, such that all classes can be separated into independent components simultaneously. In other cases, the assumption will fail and result in a biased estimate of the mutual information and therefore the optimal projection.

Under the assumption of linear separability of class distributions (note that this is not traditional linear separability of data points), we can obtain an independent linearly transformed feature coordinate system: $\mathbf{y} = \mathbf{W}\mathbf{x}$, where $\mathbf{W}$ is optimized using a suitable linear ICA algorithm on the training data $\{\mathbf{x}_1, \ldots, \mathbf{x}_N\}$ [21]. Under the assumption of overall and class-conditional independence, the mutual information of $\mathbf{y}$ with c, can be decomposed into the sum of mutual informations between each marginal of $\mathbf{y}$ and c:

$$I(\mathbf{y}; c) \approx \sum_{d=1}^{n} I(y_d; c) \tag{19}$$

Each *independent* projection can now be ranked in terms of its individual contribution to the total discriminative information by estimating $I(y_d; c)$. From the training data, one obtains: $\mathbf{y}_i = \mathbf{W}\mathbf{x}_i$ and the samples y_d are $\{y_{d1}, \ldots, y_{dN}\}$. Employing the m-spacing estimator in (17), we obtain:

$$H(y_d) \approx -\frac{1}{N-m} \sum_{i=1}^{N-m} log((N+1)(y_{d(i+m)} - y_{di})/m)$$

$$H(y_d|c) \approx -\frac{1}{N_c-m_c} \sum_{i=1}^{N_c-m_c} log((N_c+1)(y_{d(i+m_c)} - y_{di})/m_c) \tag{20}$$

$$I(y_d; c) = H(y_d) - \sum_c p_c H(y_d|c)$$

Suppose that the following ranking is obtained: $I(y(1); c) > I(y(2); c) > \ldots > I(y(n); c)$. Then the rows of $\mathbf{W}$ corresponding to the top m marginals $y(1), \ldots, y(m)$ are retained as the linear projection matrix.

3.2 Feature Subset Selection

Consider the mixture density of the features: $p(\mathbf{z}) = \sum_c p_c p(\mathbf{z}|c)$. Assuming that $\mathbf{W}$ and $\mathbf{W}^c$ are the linear ICA solutions (separation matrices) for $p(\mathbf{z})$ and $p(\mathbf{z}|c)$ respectively, let $\mathbf{y} = \mathbf{W}\mathbf{z}$ and $\mathbf{y}^c = \mathbf{W}^c \mathbf{z}^c$, where z^c is a random vector distributed according to $p(\mathbf{z}|c)$ and $\mathbf{z}$ is a random vector drawn from $p(\mathbf{z})$. It can be shown that the following identities hold:

$$H(\mathbf{z}) = \sum_{d=1}^{m} H(y_d) - log|\mathbf{W}| - I(\mathbf{y}) \quad H(\mathbf{z}|c) = \sum_{d=1}^{m} H(y_d^c) - log\|\mathbf{W}^c\| - I(\mathbf{y}^c) \tag{21}$$

The residual mutual information due to imperfect ICA solutions are denoted by $I(\mathbf{y})$ and $I(\mathbf{y}^c)$. Under the assumption of linear separability (in the ICA sense mentioned above rather than in the traditional meaning), these residual mutual information terms will (be assumed to) become zero. Since mutual information is decomposed into class-conditionals entropies as $I(\mathbf{z}; c) = H(\mathbf{z}) - \sum_c p_c H(\mathbf{z}|c)$, given linear ICA solutions $\mathbf{W}$ and $\mathbf{W}_c$, we obtain the following decomposition of mutual information:

$$I(\mathbf{z}; c) = \sum_{d=1}^{m} \left(H(y_d) - \sum_c p_c H(y_d^c) \right) + \left(log|\mathbf{W}| - \sum_c p_c log|\mathbf{W}^c| \right)$$

$$+ \left(I(y) - \sum_c p_c I(\mathbf{y}^c) \right) \tag{22}$$

Given any subset of features selected from the components (marginals) of $\mathbf{x}$, the high-dimensional feature vectors that needs to be reduced by selection, (22) can be utilized to estimate the mutual information of this subset, denoted by $\mathbf{z}$, with the class labels, by assuming that linear ICA solutions $\mathbf{W}$ and $\mathbf{W}^c$ is obtained using a suitable algorithm [21] and the training samples corresponding to each class (or the whole set). Further, the bias is assumed to be zero:

$I(\mathbf{y}) - \sum_c p_c I(\mathbf{y}^c) = 0$.[1] The feature subset can be performed by evaluating the mutual information between all possible subsets (there are 2^n of them) or utilizing a heuristic/greedy ranking algorithm to avoid the combinatorial explosion of subsets to consider. For ranking, forward (additive), backward (subtractive), or forward/backward strategies can be employed. In the purely forward strategy we perform the following additive ranking iteration:

Initialization: Let the Unranked Feature Set (UFS) be $\{x_1, \ldots, x_n\}$ and the *Ranked Feature Set* be empty. *Ranking iterations*: For d from 1 to n perform the following. Let *Candidate Set i* (CSi) be the union of RFS and x_i, evaluate the MI $I(CSi; C)$ between the features in the candidate set and the class labels for every x_i in UFS. Label the feature x_i with the highest $I(CSi, C)$ as $x_{(d)}$, redefine RFS as the union of RFS and $x_{(d)}$, and remove the corresponding feature from UFS.

Alternatively, a purely backward strategy would iteratively remove the least informative features from UFS and include in the RFS as the worst features that should be eliminated. A better alternative to both approaches is to initialize as above and allow both additive and subtractive (replacing) operations to the RFS such that earlier ranking mistakes can be potentially corrected at future iterations.

The main drawback of the linear ICA approach presented in these sections is the high possibility of the feature distributions not following the main underlying assumption of linear separability. To address this issue, linear ICA can be replaced with nonlinear ICA, especially local linear ICA, which has the flexibility of nonlinear ICA in modelling independent nonlinear coordinates in a piecewise linear manner, and the simplicity of linear ICA in solving for optimal separation solutions. The application of local linear ICA essentially follows the same procedures, except the whole process is initialized by a suitable partitioning of the data space using some vector quantization or clustering algorithm (for instance k-means or mean shift clustering could be utilized).

Due to the additivity of Shannon entropy, if the data space is partitioned into P nonoverlapping bur complementary regions, the entropies and mutual information become additive over these regions:

$$H(\mathbf{z}) = \sum_p q_p H(z^{(p)})$$
$$H(\mathbf{z}|c) = \sum_p q_p H(\mathbf{z}^{(p)}|c) \tag{23}$$
$$I(\mathbf{z}; c) = \sum_p q_p I(z^{(p)}; c)$$

where q_p is the probability mass of partition p (number of samples in partition p divided by the total number of samples). Within each partition, we still have

---

[1] Optionally, computational complexity can be further reduced by assuming that all class-conditional and class mixture densities can be linearly separated by the same ICA solution $\mathbf{W}$, such that $\mathbf{W} = \mathbf{W}^{(c)}$ for all c. This additional assumption would also eliminate need to include the correction terms depending on the separation matrices.

$I(\mathbf{z}^{(p)}; c) = H(\mathbf{z}^{(p)}) - \sum_c q_{pc} H(\mathbf{z}^{(p)}|c)$, thus (21) and (22) can be employed on each partition separately and summed up to determine the total mutual information.

3.3 Smooth Nonlinear Feature Projections

In order to derive a nonparametric nonlinear feature projection, consider the following equivalent definition of Shannon's mutual information and the KDE plug-in estimate with some positive semidefinite kernel $K(.)$:

$$I_S(\mathbf{z}; c) = \sum_c p_c E_{\mathbf{z}|c} \left[log \frac{p_{\mathbf{z}|c}(\mathbf{z}|c)}{p_\mathbf{z}(\mathbf{z})} \right] \approx \sum_c \frac{p_c}{N_c} \sum_{j=1}^{N_c} log \frac{(1/N_c) \sum_{i=1}^{N} K(\mathbf{z}_j^c - \mathbf{z}_i^c)}{(1/N_c) \sum_{i=1}^{N} K(\mathbf{z}_j^c - \mathbf{z}_i)} \tag{24}$$

where $\mathbf{z}_1, \ldots, \mathbf{z}_N$ is the training set and its subset corresponding to class c is $\mathbf{z}_1^c, \ldots, \mathbf{z}_N^c$. According to the theory of reproducing kernels for Hilbert spaces (RKHS), the eigenfunctions $\overline{\varphi}_1(\mathbf{z}), \overline{\varphi}_2(\mathbf{z}), \ldots$ collected in vector notation as $\overline{\varphi}(\mathbf{z})$, of a kernel function K that satisfy the Mercer conditions [44] form a basis for the Hilbert space of square-integrable continuous and differentiable nonlinear functions [45,46]. Therefore, every smooth nonlinear transformation $g_d(\mathbf{x})$ in this Hilbert space can be expressed as a linear combination of these bases:

$$y_d = g_d(\mathbf{z}) = \mathbf{v}_d^T \overline{\varphi}(\mathbf{z}) \tag{25}$$

where y_d is the dth component of the projection vector $\mathbf{y}$. For a symmetric positive semidefinite, translation invariant, and nonnegative (since we will establish a connection to KDE) kernel, we can write

$$K(\mathbf{z} - \mathbf{z}') = \sum_{k=1}^{\infty} \overline{\lambda}_k \overline{\varphi}_k(\mathbf{z}) \overline{\varphi}_k(\mathbf{z}') = \overline{\varphi}^T(\mathbf{z}) \overline{\Lambda} \overline{\varphi}^T(\mathbf{z}') \geq 0 \tag{26}$$

Notice that for a nonnegative kernel, kernel induced feature space (KIFS) defined by the $\overline{\varphi}(\mathbf{z})$ transformation maps all the data points into the same half of this hyper-sphere; i.e., the *angles* between all transformed data pairs are less than p radians. This is a crucial observation for the proper geometrical interpretation of what follows. substituting (26) into (24), we get:

$$I_S(\mathbf{z}; c) \approx \sum_c \frac{p_c}{N_c} \sum_{j=1}^{N_c} log \left[\frac{N \overline{\varphi}^T(\mathbf{z}_j^c) \overline{\Lambda} \boldsymbol{\Phi}_\mathbf{z} \mathbf{m}_c}{N \overline{\varphi}^T(\mathbf{z}_j^c) \overline{\Lambda} \boldsymbol{\Phi}_\mathbf{z} \mathbf{1}} \right] \tag{27}$$

where $\mathbf{m}_c i = 1$ *if* $c_i = c$, 0 *otherwise*, $\mathbf{1}$ is the vector of ones, $N = N_1 + \ldots + N_C$, and $p_c = N_c/N$. The matrix $\boldsymbol{\Phi}_\mathbf{z} = [\overline{\varphi}(\mathbf{z}_1) \cdots \overline{\varphi}(\mathbf{z}_N)]$. The class-average vectors in the KIFS are $\overline{\mu}_c = (1/N_c) \overline{\boldsymbol{\Phi}}_\mathbf{z} \mathbf{m}_c$ and for the whole data it is $\overline{\mu} = (1/N_c) \overline{\boldsymbol{\Phi}}_\mathbf{z} \mathbf{1}$. Substituting these:

$$I_S(\mathbf{z}; c) \approx \sum_c \frac{p_c}{N_c} \sum_{j=1}^{N_c} log \left[\frac{N \overline{\varphi}^T(\mathbf{z}_j^c) \overline{\Lambda} \mu}{N \overline{\varphi}^T(\mathbf{z}_j^c) \overline{\Lambda} \mu} \right] \tag{28}$$

Consider a projection dimensionality of m; we have $\mathbf{y} = \mathbf{V}_T \overline{\varphi}(\mathbf{x})$, where $\mathbf{V} = [\mathbf{v}_1, \ldots, \mathbf{v}_m]$ consists of orthonormal columns $\mathbf{v}_d$. Note that the orthonormality constraint of any linear projection is reasonable because any full rank linear transformation can be written as the product of an orthonormal matrix times an invertible arbitrary linear transformation (due to the existence of singular value decomposition with nonzero eigenvalues). The arbitrary transformation does not change the information content of the projection, thus can be omitted. The back-projection of y to the KIFS is

$$\widetilde{\varphi}(\mathbf{z}) = \mathbf{V}\mathbf{V}^T \overline{\varphi}(\mathbf{z}) \tag{29}$$

The eigenfunctions of the kernel are not explicitly known in practice typically, therefore, we employ the common Nystrom approximation [47], $\overline{\varphi}(\mathbf{z}) \approx \varphi(\mathbf{z}) = \sqrt{N} \boldsymbol{\Lambda}^{-1} \boldsymbol{\Phi}_\mathbf{z} \mathbf{k}(\mathbf{z})$, where $\mathbf{k}(\mathbf{z}) = [K(\mathbf{z} - \mathbf{z}_1), \ldots, K(\mathbf{z} - \mathbf{z}_N)]^T$ and the eigendecomposition $K = \boldsymbol{\Phi}_\mathbf{z}^T \boldsymbol{\Lambda} \boldsymbol{\Phi}_\mathbf{z}$ of the data affinity matrix whose entries are $K_{ij} = K(\mathbf{z}_i - \mathbf{z}_j)$ provide the other necessary terms. Combining (29) with this approximation and substituting in (28) leads to the following cost function that needs to be maximized by optimizing an orthonormal $\mathbf{V} \in \Re^{N \times m}$:

$$J(\mathbf{V}) = \sum_c \frac{p_c}{N_c} \sum_{j=1}^{N_c} log \left[\frac{\varphi^T(\mathbf{x}_j) \mathbf{V}\mathbf{V}^T \boldsymbol{\Lambda} \mathbf{V}\mathbf{V}^T \boldsymbol{\mu}_c}{\varphi^T(\mathbf{x}_j) \mathbf{V}\mathbf{V}^T \boldsymbol{\Lambda} \mathbf{V}\mathbf{V}^T \boldsymbol{\mu}} \right] \tag{30}$$

where $\boldsymbol{\mu}_c = (1/N_c) \boldsymbol{\Phi}_\mathbf{z} \mathbf{m}_c$ and $\boldsymbol{\mu} = (1/N) \boldsymbol{\Phi}_\mathbf{z} \mathbf{1}$ are the class and overall mean vectors of the data in the $\boldsymbol{\Phi}$-space. Note that $\boldsymbol{\mu} = p_1 \boldsymbol{\mu}_1 + \ldots + p_C \boldsymbol{\mu}_C$ and with the approximation, we have $\mathbf{y} = \mathbf{V}^T \varphi(\mathbf{x})$.

By observation, (30) is seen to be maximized by any orthonormal matrix $\mathbf{V}$ whose columns span the intersection of the subspace orthogonal to $\boldsymbol{\mu}$ and $span(\boldsymbol{\mu}_1, \boldsymbol{\mu}_2, \ldots, \boldsymbol{\mu}_C)$ [48]. This also points out the fact that any projection that leads to a reduced dimensionality of more than $C\text{-}1$, where C is the number of classes, is redundant. It is also possible to find the analytical solution for the optimal projection to $C\text{-}1$ dimensions or less. Let $\mathbf{M} = [\boldsymbol{\mu}_1 \ldots \boldsymbol{\mu}_C]$, note that $\mathbf{M}^T \mathbf{M} = \mathbf{P}^{-1}$ with $\mathbf{p} = [\mathbf{p}_1, \ldots, \mathbf{p}_C]$ and $\mathbf{P} = diag(\mathbf{p})$. We observe that $\boldsymbol{\mu} = \mathbf{M}\mathbf{p}$ is unit-norm and

$$\mathbf{V} = \mathbf{M} - \boldsymbol{\mu}(\boldsymbol{\mu}^T \mathbf{M}) = \mathbf{M} - \boldsymbol{\mu}(\mathbf{p}^T \mathbf{M}^T \mathbf{M}) = \mathbf{M} - \boldsymbol{\mu} \mathbf{1}^T \tag{31}$$

for a $C\text{-}1$ dimensional projection. For lower dimensional projections, deflation can be utilized and the procedure can be found in [48].

Special case of 2-classes: We illustrate the analytical solution for the case of projection to a single dimension in the case of two classes. We parameterize the projection vector as $\mathbf{v} = \mathbf{M}\mathbf{P}^{-1/2}\boldsymbol{\alpha}$, where $\boldsymbol{\alpha}^T \boldsymbol{\alpha} = 1$ (so that $\mathbf{v}^T \mathbf{v} = 1$). It can be found that the optimal solution is provided by $\boldsymbol{\alpha} = [-p_2^{1/2}, p_1^{1/2}]^T$ (and its negative yields a projection equivalent in discriminability where the two class projections are flipped in sign). For the projection, the natural threshold is zero since the data mean is projected to this point (due to the fact that v is orthogonal to $\boldsymbol{\mu}$).

4 Experimental Evaluation and Comparisons

In this section, we present the experimental evaluation for the methods explained above and we also provide comparison with LDA and kernel LDA. The first example is an illustrative toy example, and the preceding two experiments are challenging problems performed on real data.

4.1 Synthetic Dataset

This dataset consists of four features: x_i $i = 1, \ldots, 4$, where x_1 and x_2 are nonlinearly related, x_3 and x_4 are independent from the first two features and are linearly correlated Gaussian-distributed with different mean and variance. There are two classes in this dataset represented with different markers in Figs. 1a and 1b. Forming an almost separable distribution with a nonlinear separation boundary in the x_1 and x_2 plane, and overlapping in the x_3 and x_4 plane, this dataset forms a good example to compare linear and nonlinear methods.

For ICA feature projection and selection, we use Support Vector Machine (SVM) to classify them. For the SVM, we use Chang and Lin's library toolbox. Based on the experiment results, we select the parameter of SVM as: penalty parameter $c = 10$, and kernel size $g = 10$. We apply ICA-MI feature projection and ICA-MI feature selection, nonlinear MI projection methods on the dataset, as well as LDA and kernel LDA. Each class contains 500 samples and we divide the dataset into five equal parts, four of which are used as training simples, one of which is used as testing samples. Figure 2a shows the classification accuracy vs. number of n best features, the Fig. 2b presents a comparison of five above mentioned methods. To present a fair comparison, while comparing with other methods, we consider one dimensional projections for ICA based methods.

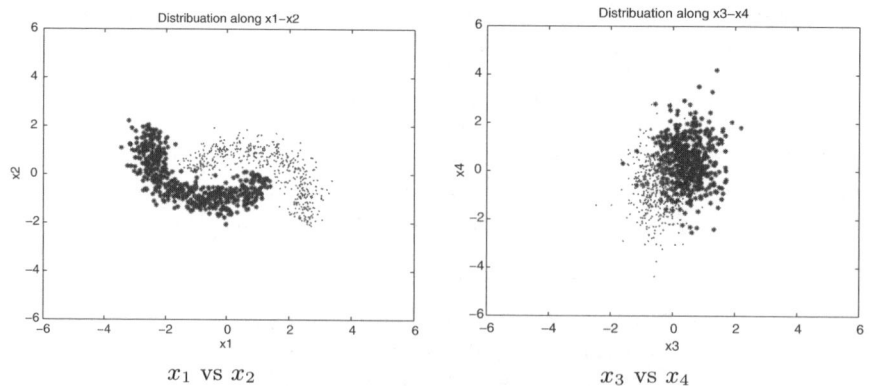

x_1 vs x_2 $\qquad\qquad$ x_3 vs x_4

Fig. 1. The synthetic dataset

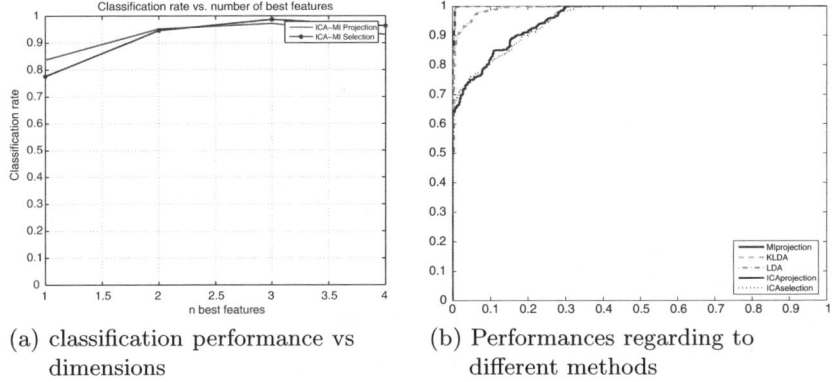

(a) classification performance vs dimensions

(b) Performances regarding to different methods

Fig. 2. The performance evaluation on synthetic dataset

4.2 Brain Computer Interface Dataset

In this experiment, we apply the same five methods on Brain Computer Interfaces Competition III dataset V. This dataset contains human brain EEG data from 3 subjects during 4 non-feedback sessions. The subject is sitting on a chair, relaxed arms resting on their legs and executed one of the three tasks: imagination of left hand movements; imagination of right hand movements; generation of words beginning with the same random letter. The EEG data were collected during the sessions. The data of all 4 sessions of a given subject were collected on the same day, each lasting 4 m with 5–10 m breaks in between. We want to classify one of the three tasks from the EEG data. The raw EEG data contains 32 channels at 512 Hz sampling rate. The raw EEG potentials were first spatially filtered by means of a surface Laplacian. Then, every 62.5 ms, the power spectral density (PSD) in the band 8–30 Hz was estimated over the last second of data with a frequency resolution of 2 Hz for the 8 centro-parietal channels C3, Cz, C4, CP1, CP2, P3, Pz, and P4. As a result, an EEG sample is a 96-dimensional vector (8 channels times 12 frequency components).

To be able to present the results in a ROC curve we only use the first the classes in the dataset. We also mix the data from all sessions together, then and use five-fold cross validation as in the previous experiment. The classification performance vs. number of selected features, and comparison of one dimensional projections by different methods are presented in Figs. 3a, and 3b, respectively.

4.3 Sonar Mine Detection

The mine detection dataset consists of sonar signals bounced off either a metal cylinder or a roughly cylindrical rock. Each sonar reflection is represented by a 60-dimensional vector, where each dimension represents the energy that

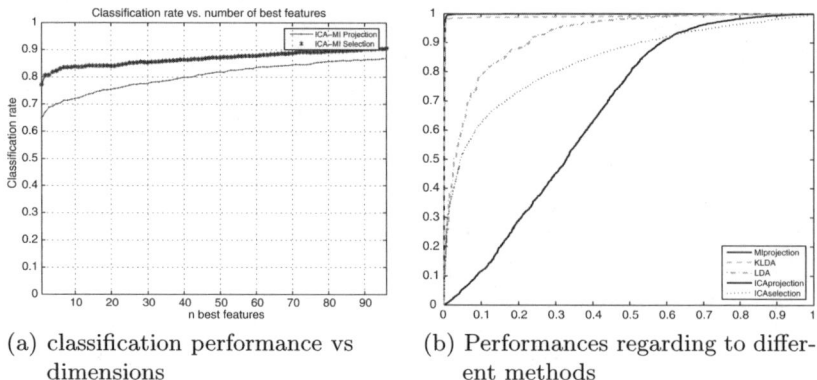

(a) classification performance vs dimensions

(b) Performances regarding to different methods

Fig. 3. The performance evaluation on BCI dataset

falls within a particular energy band, integrated over a certain period of time. There are 208 60-dimensional sonar signals in this dataset, 111 of them belongs to mines and 97 of them obtained by bouncing sonar signals from cylindrical rocks under similar conditions. These sonar signals are collected from a variety of different aspect angles, and this dataset was originally used by Gorman and Sejnowski in their study of sonar signal classification [49]. The dataset is available in UCI machine learning repository [50].

As in the previous experiments, here we compare five different methods: MI Projections, ICA feature selection, ICA feature projection, LDA, and Kernel LDA. As in the previous experiments, for all these methods, we present the results of five-fold cross validation, where the class a priori probabilities in the bins are selected according to the a priori probabilities in the dataset. The results for projections into a single dimension is presented with a ROC curve, whereas the performance increase of the ICA based linear methods for different dimensions are presented separately. Due to the nonlinear structure of the optimal class separation boundary, nonlinear methods show superior performance in this experiment. Figure 4a presents the performance of ICA based methods for different number of dimensions, and a comparison of one dimensional projections with all five methods is presented in Fig. 4b.

5 Conclusions

We presented and compared several information theoretic feature selection an projection methods. Selection and projection methods based on ICA are either linear or locally linear methods, which are simply analyzable. As seen from the original feature space, the mutual information projection method is nonlinear and not easy to analyze. Although it is hard in the original input space, this method is also simple to analyze in the KIFS, where the MI projection becomes a linear method with the use of kernel trick.

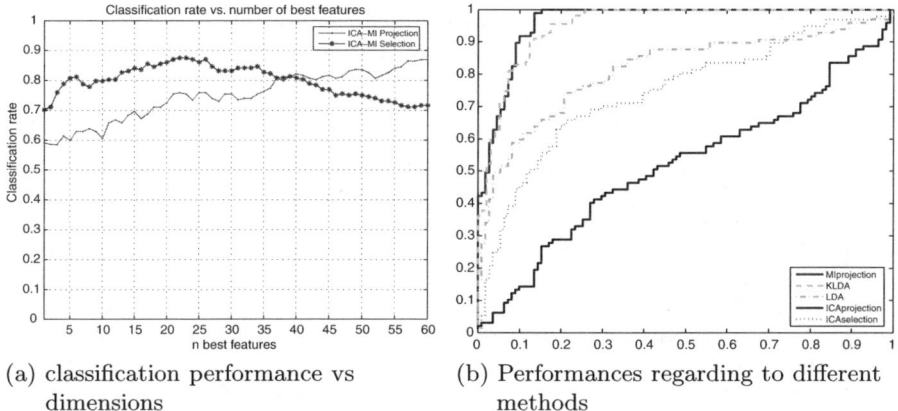

(a) classification performance vs
dimensions

(b) Performances regarding to different
methods

Fig. 4. The performance evaluation on sonar mine detection dataset

MI projection method also provided similar -sometimes slightly better-
performance as compared to widely used KLDA method. At this point, note
that KLDA is known to be numerically unstable if there are not enough data
points, and MI projection method provides same/better performance with an
analytical solution. Among the information theoretic methods presented here,
MI projection methods provided the best performance, and it also does not
suffer from the input data dimensionality, whereas methods based on ICA
transformation severely lose accuracy with increasing input dimensionality in
the estimation of the inverse mixing matrix.

Acknowledgements

This work is partially supported by NSF grants ECS-0524835, and ECS-
0622239.

References

1. Duch W, Wieczorek T, Biesiada J, Blachnik M (2004) Comparison of Feature
   Ranking Methods Based on Information Entropy, Proceedings of International
   Joint Conference on Neural Networks 1415–1420
2. Erdogmus D, Principe J C (2004) Lower and Upper Bounds for Misclassification
   Probability Based on Renyi's Information, Journal of VLSI Signal Processing
   Systems 37:305–317
3. Fano R M (1961) Transmission of Information: A Statistical Theory of Com-
   munications, Wiley, New York
4. Hellman M E, Raviv J (1970) Probability of Error, Equivocation and the
   Chernoff Bound, IEEE Transactions on Information Theory 16:368–372

5. Koller D, Sahami M (1996) Toward Optimal Feature Selection, Proceedings of the International Conference on Machine Learning 284–292
6. Battiti R (1994) Using Mutual Information for Selecting Features in Supervised Neural Net Learning, Neural Networks 5:537–550
7. Bonnlander B V, Weigend A S (1994) Selecting Input Variables Using Mutual Information and Nonparametric Density Estimation, Proceedings of International Symposium on Artificial Neural Networks 42–50
8. Yang H, Moody J (2000) Data Visualization and Feature Selection: New Algorithms for Nongaussian Data, Advances in Neural Information Processing Systems 687–693
9. Oja E (1983) Subspace Methods of Pattern Recognition, Wiley, New York
10. Devijver P A, Kittler J (1982) Pattern Recognition: A Statistical Approach, Prentice Hall, London
11. Fukunaga K (1990) Introduction to Statistical Pattern Recognition, Academic, New York
12. Everson R, Roberts S (2003) Independent Component Analysis: A Flexible Nonlinearity and Decorrelating Manifold Approach, Neural Computation 11:1957–1983
13. Hyvärinen A, Oja E, Hoyer P, Hurri J (1998) Image Feature Extraction by Sparse coding and Independent Component Analysis, Proceedings of ICPR 1268–1273
14. Torkkola K (2003) Feature Extraction by Non-Parametric Mutual Information Maximization, Journal of Machine Learning Research 3:1415–1438
15. Battiti R (1994) Using Mutual Information for Selecting Features in Supervised Neural Net Training, IEEE Transaction Neural Networks 5:537–550
16. Kira K, Rendell L (1992) The Feature Selection Problem: Traditional Methods and a New Algorithm, Proceedings of Conference on Artificial Intelligence 129–134
17. John G H, Kohavi R, Pfleger K (1994) Irrelevant Features and the Subset Selection Problem, Proceedings of Conference on Machine Learning 121–129
18. Hild II K E, Erdogmus D, Torkkola K, Principe J C (2006) Feature Extraction Using Information-Theoretic Learning, IEEE Transactions on Pattern Analysis and Machine Intelligence 28:1385–1392
19. Guyon I, Elisseeff A (2003) An Introduction to Variable and Feature Selection, Journal of Machine Learning Research 3:1157–1182 (Special Issue on Variable and Feature Selection)
20. Scholkopf B, Smola A, Muller K R (1998) Nonlinear Component Analysis as a Kernel Eigenvalue Problem, Neural Computation 10:1299–1319
21. Hyvarinen A, Karhunen J, Oja E (2001) Independent Component Analysis, Wiley, New York
22. Lee D D, Seung H S (1999) Learning the parts of objects by non-negative matrix factorization, Nature 401:788–791
23. Roweis S, Saul L (2000) Nonlinear Dimensionality Reduction by Locally Linear Embedding, Science 290:2323–2326
24. Costa J, Hero A O (2005) Classification Constrained Dimensionality Reduction, Proceedings of ICASSP 5:1077–1080
25. Baudat G, Anouar F (2000) Generalized Discriminant Analysis Using a Kernel Approach, Neural Computation 12:2385–2404
26. Principe J C, Fisher J W, Xu D (2000) Information Theoretic Learning, In Haykin S (Ed.) Unsupervised Adaptive Filtering, Wiley, New York, 265–319

27. Parzen E (1967) On Estimation of a Probability Density Function and Mode, Time Series Analysis Papers, Holden-Day, San Diego, California
28. Erdogmus D (2002) Information Theoretic Learning: Renyi's Entropy and its Applications to Adaptive System Training, PhD Dissertation, University of Florida, Gainesville, Florida
29. Kraskov A, Stoegbauer H, Grassberger P (2004) Estimating Mutual Information, Physical Review E 69:066138
30. Learned-Miller E G, Fisher J W (2003) ICA Using Spacings Estimates of Entropy, Journal of Machine Learning Research 4:1271–1295
31. Vasicek O, (1976) A Test for Normality Based on Sample Entropy, Journal of the Royal Statistical Society B 38:54–59
32. Hero A O, Ma B, Michel O J J, Gorman J (2002) Applications of Entropic Spanning Graphs, IEEE Signal Processing Magazine 19:85–95
33. Beirlant J, Dudewicz E J, Gyorfi L, Van Der Meulen E C (1997) Nonparametric Entropy Estimation: An Overview, International Journal of Mathematical and Statistical Sciences 6:17–39
34. Erdogmus D, Principe J C (2002) An Error-Entropy Minimization Algorithm for Supervised Training of Nonlinear Adaptive Systems, IEEE Transactions on Signal Processing 50:1780–1786
35. Erdogmus D, Principe J C (2006) From Linear Adaptive Filtering to Nonlinear Information Processing, IEEE Signal Processing Magazine 23(6):14–33
36. Erdogmus D, Hild II K E, Rao Y N, Principe J C (2004) Minimax Mutual Information Approach for Independent Components Analysis, Neural Computation 16:1235–1252
37. Girolami M (2002) Orthogonal Series Density Estimation and the Kernel Eigenvalue Problem, Neural Computation, MIT Press 14:669–688
38. Duda R O, Hart P E, Stork D G (2000) Pattern Classification (2nd ed.), Wiley, New York
39. Devroye L, Lugosi G (2001) Combinatorial Methods in Density Estimation, Springer, Berlin Heidelberg New York
40. Renyi A (1970) Probability Theory, North-Holland, Amsterdam
41. Silverman B W, (1986) Density Estimation for Statistics and Data Analysis, Chapman and Hall, London
42. Duin R P W (1976) On the Choice of the Smoothing Parameters for Parzen Estimators of Probability Density Functions, IEEE Transactions on Computers 25:1175–1179
43. Schraudolph N (2004) Gradient-Based Manipulation of Nonparametric Entropy Estimates, IEEE Transactions on Neural Networks 15:828–837
44. Mercer J (1909) Functions of Positive and Negative Type, and Their Connection with the Theory of Integral Equations, Transactions of the London Philosophical Society A 209:415–446
45. Wahba G (1990) Spline Models for Observational Data, SIAM, Philedelphia, Pennsylvania
46. Weinert H (ed.) (1982) Reproducing Kernel Hilbert Spaces: Applications in Statistical Signal Processing, Hutchinson Ross Publisher Co., Stroudsburg, Pennsylvania
47. Fowlkes C, Belongie S, Chung F, Malik J (2004) Spectral Grouping Using the Nystrom Method, IEEE Transactions on Pattern Analysis and Machine Intelligence 23:298–305

48. Ozertem U, Erdogmus D, Jenssen R (2006) Spectral Feature Projections That Maximize Shannon Mutual Information with Class Labels, Pattern Recognition 39:1241–1252
49. Gorman R P, Sejnowski T J (1988) Analysis of Hidden Units in a Layered Network Trained to Classify Sonar Targets, Neural Networks 1:75–79
50. http://www.ics.uci.edu/mlearn/MLRepository.html

Recognition of Tones in YorÙbÁ Speech: Experiments With Artificial Neural Networks

Ọdẹ́túnjí Àjàdí ỌDélọBí

Ọbáfẹ́mi Awólọ́wọ̀ University, Ilé-Ifẹ̀, Nigeria,
oodejobi@oauife.edu.ng, oodejobi@yahoo.com

Summary. The speech recognition technology has been applied with success to many Western and Asian languages. Work on African languages still remains very limited in this area. Here a study into automatic recognition of the Standard Yoruba (SY) language tones is described. The models used fundamental frequency profile of SY syllables to characterize and discriminate the three Yoruba tones. Tonal parameters were selected carefully based on linguistic knowledge of tones and observation of acoustic data. We experimented with Multi-layered Perceptron (MLP) and Recurrent Neural Network (RNN) models by training them to classify feature parameters corresponding to tonal patterns. The results obtained exhibited good performances for the two tone recognition models, although the RNN achieved accuracy rates which are higher than that of the MLP model. For example, the outside tests for the H tone, produced a recognition accuracy of 71.00 and 76.00% for the MLP and the RNN models, respectively. In conclusion, this study has demonstrated a basic approach to tone recognition for Yoruba using Artificial Neural Networks (ANN). The proposed model can be easily extended to other African tone languages.

1 Introduction

The application of Artificial Neural Networks (ANN) to the processing of speech has been demonstrated for most Western and Asian languages. However, there is limited work reported on African languages. In this chapter we provide the background to the application of Artificial Neural Networks (ANN) to the processing of African language speech. We use the problem of tone recognition for the Standard Yorùbá (SY) language as a case study. This presentation has two aims (1) to provide background materials for research in tone language speech recognition, (2) to motivate and provide arguments for the application of ANN to the recognition of African languages. To achieve the stated aims, we demonstrate the design and implementation of Multilayer Perceptron (MLP) and Recurrent Neural Network (RNN) in the recognition of SY tones.

Ọdẹ́túnjí Àjàdí ỌDélọBí: *Recognition of Tones in YorÙbÁ Speech: Experiments With Artificial Neural Networks*, Studies in Computational Intelligence (SCI) **83**, 23–47 (2008)
www.springerlink.com
© Springer-Verlag Berlin Heidelberg 2008

Tone languages, such as Yorùbá and Mandarin, differ from non-tone languages, such as English and French [20, 71]. In non-tone language, lexical items are distinguished by the stress pattern on the syllables that constitute an utterance. For example, the English words *re*cord (verb) and re*cord* (noun) differ in syntactic class and meaning because of the stress pattern on their component syllables. In the verb *re*cord the first syllable is stressed. In the noun re*cord* the second syllable is stressed.

In tone languages, tone, rather than stress, is used to distinguish lexical items. The tones are associated with the individual syllables in an utterance. For example, the following mono-syllabic Yorùbá words: *bí* (H) [to give birth], *bi* (M) [to ask], *bì* (L) [to vomit] differ in meaning because of the tone associated with each syllable. A high tone is associated with the first monosyllabic word, i.e. *bí*. The second and third words carry the mid and low tones, respectively. Most tone languages have distinguishable tones and the number of tones vary for different languages, e.g. two for Hausa and Igbo, four for Mandarin (plus a neural tone), five for Thai and nine for Cantonese [35]. Standard Yorùbá has three phonological tones [1].

Two important features of tone languages make them an interesting subject of research in the area of speech recognition. First, tones are associated with syllables which are unambiguous speech units. Second, each tone, in isolated utterance, has a unique fundamental frequency (f_0) curve. Although the f_0 curves are affected by their context in continuous speech, they can still be easily recognised in speeches articulated at moderate or slow speaking rates. The complexities of speech recognition for tone languages can be reduced considerably if these two features are exploited in the design of speech recognition system. It is important to note that the recognition of tones is a major step in the recognition of speech in tone languages [15].

In speech signal, the timing, intensity and the fundamental frequency (f_0) dimensions contribute, in one way or the other, to the recognition and perception of tones. However, the fundamental frequency (f_0) curve has been shown to be the most influential acoustic correlate of tone in tone languages [54]. The possible application of tone recognition system such as the one presented here include the following:

- Recognition of SY monosyllabic utterances
- A component in a system for the automatic segmentation of continuous speech into syllables
- A component in a system for the automatic syllable level speech database annotation
- Application in automatic stylisation of tone f_0 curves

In Sect. 2 we give a brief description of the Standard Yorùbá language. Section 3 presents background to Automatic Speech Recognition (ASR) technology. The Hidden Markov Model (HMM) is the most popular method applied in ASR, hence we provide a detailed review of work on HMM in Sect. 4. Section 5 contains a literature review on the application of ANN to speech and tone recognition. The data used for developing the models presented in this chapter is presented in Sect. 6. The tone recognition framework developed in this work is presented in Sect. 7. Experiments, results and discussion on the developed tone recognition models are presented in Sect. 8. Section 9 concludes this Chapter.

2 A Brief Description of the Standard Yorùbá Language

Yorùbá is one of the four major languages spoken in Africa and it has a speaker population of more than 30 million in West Africa alone [17, 62]. There are many dialects of the language, but all speakers can communicate effectively using Standard Yorùbá (SY). SY is used in language education, mass media and everyday communication. The present study is based on the SY language.

The SY alphabet has 25 letters which is made up of 18 consonants (represented by the graphemes: *b, d, f, g, gb, h, j, k, l, m, n, p, r, s, ṣ, t, w, y*) and seven vowels (*a, e, ẹ, i, o, ọ, u*). Note that the consonant *gb* is a diagraph, i.e. a consonant written in two letters. There are five nasalised vowels in the language (*an, en, in, ọn, un*) and two pure syllabic nasals (*m, n*). SY has three phonologically contrastive tones: High (H), Mid (M) and Low (L). Phonetically, however, there are two additional allotones or tone variants, namely; rising (R) and falling (F) [2, 14]. A rising tone occurs when an L tone is followed by an H tone, while a falling tone occurs when an H tone is followed by an L tone. This situation normally occurs during assimilation, elision or deletion of phonological object as a result of co-articulation phenomenon in fluent speech.

A valid SY syllable can be formed from any combination of a consonant and a vowel as well as a consonant and a nasalised vowel. When each of the eighteen consonants is combined with a simple vowel, we will have a total of 126 *CV* type syllables. When each consonant is combined with a nasalised vowel, we have a total of 90 *CVn* type syllables. SY also has two syllabic nasals *n* and *m*. Table 1 shows the distribution of the components of the phonological structure of SY syllables.

It should be noted that although a *CVn* syllable ends with a consonant, the consonant and its preceding vowels are the orthographic equivalent of a

Table 1. Phonological structure of SY syllables

Tone syllables (690)[1]				
Base syllables (230)				Tones (3)
ONSET (18)	*RYHME* (14)			
		Nucleus	Coda	
Consonant	Vocalic	Non-Vocalic		
C	V(7)	N(2)	n(1)	H, M, L

[1] The numbers within a parenthesis indicates the total number of the specified unit.

nasalised vowel. There is no closed syllable and there is no consonant cluster in the SY language.

3 Background

Generally, there are two approaches to the problem of tone recognition: (1) rule base and (2) data driven. In the rule-based approach to tone recognition, the acoustic properties of tones are studied using established theories: e.g. acoustic, phonetics, and/or linguistic theories. The knowledge elicited from such study is coded into rules. Each rule is represented in the form **IF** {*premise*} **THEN** {*consequence*}. The *premise* specifies a set of conditions that must be true for the {*consequence*} to be fired.

The strength of this approach to tone recognition is that the description of the properties of speech are defined within the context of established theories. This facilitates the extension and generalisation of the resulting tone recogniser. The resulting model is also easy to understand making it useful as a tool for the explanation of phenomena underlying tone perception particularly in tone languages, e.g. the tone *sandhi* [46]) phenomenon.

A major weakness of the rule-based approach to speech recognition is that its development requires a collaboration between a number of experts, e.g. linguists, phoneticians, etc. These experts are not readily available for most African languages. The information available in the literature is not sufficient as many phenomena that occur in speech are yet to be described definitively [3, 29]. The development of practical speech recogniser requires the coding of speech information into rules. Such rules usually results in a large rule-base. It is well known that, the maintenance and organisation of such a large rule-base can be very complicated [50].

Another limitation of the rules-driven approach is that it is difficult to generate and represent all the possible ways in which rules are interdependent. Therefore, it is inevitable that rules compete with each other to explain the same phenomenon while others are in direct contradiction [39]. Although tools for managing large databases are available, such tools are not primarily designed for speech databases. It is also very difficult to extend such tools to speech database manipulation. These weaknesses are responsible for the poor performances of rule-based speech recognisers and motivated the application of the data-driven approach to speech recognition [19, 61, 70].

In data-driven (also called machine learning) approach to tone recognition, the aim is to develop a model that can learn tonal patterns from speech data. In this approach, a well designed statistical or computational model is trained on carefully selected data. The most commonly used models include; the Classification and Regression Trees (CART) [8, 34, 58, 59], the Hidden Markov Model (HMM) [7, 31, 56, 69] and the Artificial Neural Networks (ANN) [4, 26, 42, 53]. The aim of training data-driven models is to "store" the pattern to be recognised into their memories. The stored pattern is then used to recognise new patterns after the training phases. A few works have applied the CART in speech recognition. In the following section we review work on the Hidden Markov Model (HMM) and the ANN in more details as they have become more popular in tone language speech recognition.

4 The Hidden Markov Model (HMM) and Artificial Neural Networks (ANN)

HMMs and their various forms (discrete, continuous, and semi-continuous) have been applied to speech recognition problems in general and tone recognition in particular [10,19,38,69]. For example, Liu et al. [43], and later Wang et al. [66], demonstrated a left-to-right continuous HMM with two transitions in the recognition of Mandarin continuous speech with very large vocabulary. They are able to achieve a recognition accuracy of 92.20%.

McKenna [45] applied the HMM approach to the recognition of 4 Mandarin tones. In that work, various HMM systems were trained using a single speaker, and isolated monosyllabic words. f_0 and energy related data were the features used for developing the HMM. McKenna [45] reported an accuracy of about 98.00%. In Peng and Wang [54] an approach to tone and duration modelling was developed within the framework of HMMs. Their experiment showed that the HMM approach reduced the relative word error by as much as 52.90 and 46.00% for Mandarin and Cantonese digit string recognition tasks, respectively. HMM are particularly good in the recognition of non-stationary continuous speech signals.

A major weakness of HMMs, which limits their application in the context of tone recognition systems, is that HMM tries to modify the data to fit the model. This weakness of the HMM limits its effectiveness at modelling tone. In fact, Bird [5] has shown that HMM are poor at the modelling of tone in the context of tone languages. A more appropriate approach is to use a model which incorporates the concept of modelling segments of speech, rather than individual frames as done by HMM. Artificial Neural Networks, such as the MLP, provide a potential solution to this problem. This is because unlike HMMs, ANNs do not need to treat features as independent, as they can incorporate multiple constraints and find optimal combinations of constraints for classification [28].

Modern speech recognition systems uses a hybrid of ANN and HMM models to achieve better performance [60]. In such cases, the ANN is used for estimating the state-dependence observation probability for the HMM. The process of generating such a model is complex and the efforts required may not justify its application to the recognition of SY tones. In this chapter, therefore, we address the problem of ANN for SY tone recognition.

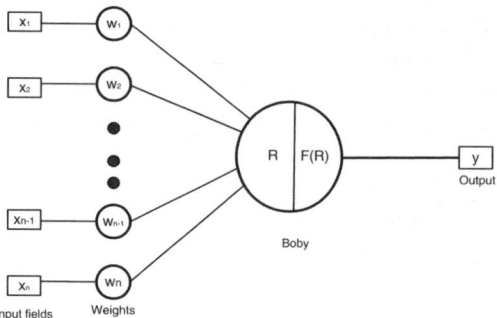

Fig. 1. A model of ANN processing unit

4.1 Introduction to Neural Networks

Artificial Neural Network (ANN), also called *connectionist* model, *neural net*, or *parallel distributed processing* (PDP) model are interconnection of simple, non-linear computational elements. These elements are called nodes or neurons (see Fig. 1). A neuron typically consists of three components: (1) a group of weights, (2) a summation function, and (3) a nonlinear activation function $f(R)$. An ANN can contain a large number of neurons. These neurons are connected by links with variable weights.

The computational behaviour of ANN models is determined by the values of each of the weight in it. To derive the behaviour of an ANN, it is usually trained on sample data. Given an input vector X with a set of n fields represented as $X = \{x_1, x_2, \ldots, x_n\}$, the operations of the processing unit consist of a number of steps [24]. First, each input field x_i is multiplied by a corresponding weight w_i, $w_i \in W = \{w_1, w_2, \ldots, w_n\}$. The product of each input field and its corresponding weight are then summed to produce the cumulative weighted combination R, as shown in (1)

$$R = w_1 x_1 + w_2 x_2, \ldots, w_n x_n = \sum_{1}^{n} w_i x_i = \mathbf{W}.\mathbf{X} \tag{1}$$

The summation is further processed by an activation function $\mathbf{f(R)}$ to produce only one output signal y. We can express the computation of y mathematically as:

$$y = f(\sum_{1}^{n} w_i x_i) = f(\mathbf{W}.\mathbf{X}) \tag{2}$$

The function $f(.)$ can take many forms, e.g. linear, sigmoid, exponential, etc. In this work we used the sigmoid function. The computed value of y can serve as input to other neurons or as an output of the network. Each node is responsible for a small portion of the ANN processing task.

The tone recognition problem being addressed in this work is an example of supervised classification. In this type of problem, decision making process requires the ANN to identify the class to which an input pattern belongs. The ANN meant to achieve such task is first adapted to the classification task through a learning process using training data. Training is equivalent to finding the proper weight for all the connections in an ANN such that a desired output is generated for a corresponding input [32]. Several neural network models have been developed and used for speech recognition. They include: (1) the Kohonen's Self-organising Map (SOM) model, (2) the Hopefield Model, and (3) the multilayer perceptron [18]. In the next subsection, we review the literature on the application of ANN to tone language speech recognition in general and tone recognition in particular.

5 ANN in Speech and Tone Recognition

ANNs have a number of interesting applications including spatiotemporal pattern classification, control, optimization, forecasting and generalization of pattern sequences [55]. ANNs have also been applied in speech recognition and synthesis as well as in prosody modelling [9, 16, 22, 41, 57, 63, 65].

5.1 Multilayered Perceptron in Tone and Speech Recognition

The main difference between various models of ANN is how the neurons are connected to each other. Perhaps the most popular model is the multilayer perceptron trained using the backward propagation (back propagation) algorithm. A number of work has been reported on the application of MLP to Asian tone languages recognition. Cheng et al. [13] implemented a speaker-independent system for Mandarin using a Multilayer Perceptron (MLP). They used ten input features which included: energies, f_0 slopes, normalised f_0 curves as well as the duration of the voiced part of the syllables. They are able to get a recognition accuracy above 80.00%. Chang et al. [11] reported the application of MLP to the problem of recognition of four Mandarin isolated syllables tones. To achieve the tone recognition task, ten features extracted from the fundamental frequency and energy contours of monosyllables are used as the recognition features. In that work, the back-propagation algorithm was used to train the MLP. They are able to achieve a recognition rate of 93.80% in a test data set. The work also confirmed that the MLP outperforms a Gaussian classifier which has a recognition rate of 90.60%.

Similarly, Thubthong and Kijsirikul [64] described the application of a 3 layer MLP to the recognition of Thai tones. The MLP has an input layer of 6 units, a hidden layer of 10 units, and an output layer of 5. Each of the 5 units correspond to the 5 Thai tones. The MLP was trained by the back-propagation algorithm for a maximum of 1000 epochs. They obtained the recognition rates of 95.48 and 87.21% for the data and test sets respectively.

5.2 RNN in Tone and Speech Recognition

The Recurrent Neural Networks (RNN) is another type of ANN model. Its configuration is similar to that of the MLP except that the output of some hidden or output layers are feed back to the input layer. A number of researchers have used the RNN in the recognition of tones and speech. Hunt [30] described a novel syllabification technique using a series of RNNs. Several optimisation for the RNN training algorithms are also employed. The technique was developed and tested using the TIMIT database, an American, speaker-independent, continuous-speech, read-sentence database. It was reported that the system places the start and end points with an accuracy within $20\,msec$ of the desired point. This is a very high accuracy considering the fact that this results will allow a system to find syllable at 94.00% accuracy.

In Hane, et al. [27] an RNN was used for acoustic-to-phoneme mapping. This problem is similar to speech recognition in that acoustic data are used to classify phonemes. The RNN was trained using standard back-propagation method. They are able to obtain 90.00% accuracy in consonant recognition and 98.00% for vowel recognition.

RNN has also been applied to the modelling of Mandarin speech prosody recognition [68]. In that work, the RNN is trained to learn the relationship between the input prosodic feature of the training utterance and the output word-boundary information of the associated text. Accuracy rate of 71.90% was reported for word-tag detection and the character accuracy rate of speech-to-text conversion increased from 73.60 to 74.70%.

The results obtained from the above studies revealed that the MLP and RNN are powerful tools for the processing and modeling of speech data. The results also demonstrated that, when carefully constructed using the appropriate tools, the ANN can provide a robust solution to speech recognition for tone languages. However, RNNs can perform highly non-linear dynamic mappings and thus have temporally extended applications, whereas multi-layer perceptrons are confined to performing static mappings [40].

The ANN approach to tone recognition is particulary useful for our purpose for three reasons. First, the tone recognition problem involves discriminating short input speech patterns of a few, about 100, frames in length. It has been shown that the ANN approach are good at handling this type of problem [21, 26, 27]. Second, ANN techniques provide a framework that makes speech recognition systems easier to design and implement. Third, and most importantly, the availability of free software and tools [6] for modeling and implementing ANN. This makes it a feasible approach to African language speech processing applications for economic reasons.

6 Data

The data for this study was collected as part of a project to construct a language resource for Yorùbá speech technology. There are five types of syllable configurations in SY [52]. These are CV, CVn, V, N and Vn. Based on a careful analysis of SY newspapers and language education textbooks, we collected 150 syllables. Some of the criteria used for selecting the syllables include: ease of articulation, frequency of occurrence in the text as well as the consistency of the recorded speech sound pattern. In addition, the syllables were collected with a view to have a proper representation and distribution of phonetic items in the database.

6.1 Speech Data Recording

Four native male speakers of SY read the syllables aloud. They are undergraduate Yorùbá students of the Ọbáfẹ́mi Awólọ́wọ̀ University, Ilé-Ifẹ̀. Their ages are between 21 and 26 years. Their voices were recorded using the *Wavesurfer* software on a Pentium IV computer running the *Window XP* operating system. The speech sound was captured in a quiet office environment using the head mounted *Stereo-Headphone* by *LightWave*.

The parameter for recording the speech sound is listed in Table 2.

In order to achieve good quality recording, recorded speech corresponding to a syllable was examined for the following defects:

- distortion arising from clippings,
- aliasing via spectral analysis,
- noise effects arising from low signal amplitude-resulting in evidence of quantization noise or poor signal-to-noise ratio (SNR),
- large amplitude variation,
- transient contamination (due to extraneous noise).

Recorded speech samples that have one or more of the above listed defects are discarded and the recording repeated until the resulting speech signal is satisfactory. To prepare the recorded waveform for further processing, we replayed each speech sound to ensure that they are correctly pronounced. Artefacts introduced at the end and beginning of each recording was edited out.

All the speech files are hand labelled. Figure 2 shows the syllable files annotations together with the spectrograph of the syllable Bá. The voiced

Table 2. Speech Sound Recording Parameter

Ser. No.	Parameter	Specification
1	Sampling rate	16 kHz
2	Frame size	10 ms
3	Window type	Hanning
4	Window length	32 ms (512 samples)
5	Window overlap	22 ms (352 sample)
6	Analysis	Short time spectrum
7	Set no. of channel	1
8	Waveform format	.wav (Microsoft)

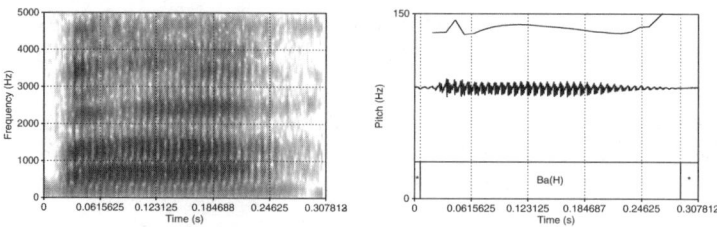

Fig. 2. SY syllable Bá (get to)

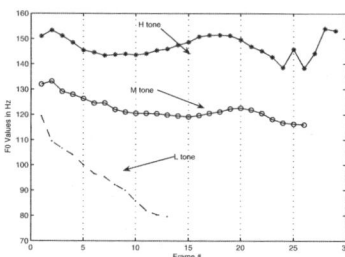

Fig. 3. f_0 curves for H, M & L tone on syllable ẹ, spoken in isolation

part of each speech file is annotated with the corresponding orthography of the syllable. The silence at the beginning and ending of each syllable speech file are annotated with the asterisk, i.e. $*$. The tone data of each syllable is the f_0 curve associated with the RHYTHM part of the speech waveform.

The *Praat* software was used to load and extract the numerical data corresponding to each syllable's f_0 curve. The data were then exported into an ASCII text file, formatted and exported as a *MATLAB mfile*. Third degree polynomials were interpolated using tools available in the *MatLab* environment, such as *polyfit*.

6.2 ANN Modelling Data Preparation

In most work on syllable and tone recognition, the following acoustic features are commonly used: (1) the fundamental frequency (f_0) of the voiced part, (2) energy, (3) zero crossing rate, (4) (linear Predictive Coefficient) LPC coefficient, (5) cepstrum, and delta cepstrum [19, 48, 54, 67]. In this work we are using only the f_0 curve because it provides enough discrimination of the SY tones due to the simplicity of the f_0 profiles. For example, Fig. 3 shows the f_0 curves for the SY High (H), Mid (M) and Low (L) tones over the syllable ẹ. The excursions of these f_0 curves are distinct in that, while the low tone f_0 is low and falling, that of the high tone is high while the f_0 of the mid tone occupies a region between the high and low tone f_0 curves.

To represent the f_0 profile of each syllable we interpolate a third degree polynomial into the f_0 data over the voiced portion of the syllable. Ọdéjọbí,

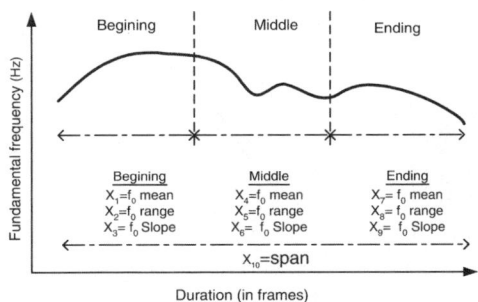

Fig. 4. f_0 feature parameters for the ANN

et al., [51] have shown that the third degree polynomials adequately approximate the f_0 data of SY syllables. To generate the data for training the ANNs, the interpolated f_0 curve for each syllable is partitioned into three equal sections: beginning, middle and ending (see Fig. 4). This partitioning is achieved by dividing the total duration of the f_0 curve into three equal parts. This approach is informed by the findings that the nonlinear pattern of the f_0 curve of each tone type has unique turning points [33, 67]. The location in time and the amplitude of the turning points provide a good discriminating parameter for each of the tones [36].

For each of the three parts, three data values were computed: f_0 range, f_0 slope and f_0 mean. The f_0 range is computed as the difference between the minimum and maximum f_0 value for that part. The f_0 slope is computed as the f_0 range divided by the time span, i.e. change of f_0 with time (i.e. number of frames). The computations here is similar to that used by Lee et al. [36] in the recognition of Cantonese tones. A similar approach was also used by Hanes et al. [27] for generating formant data for acoustic-to-phoneme mapping. Our process produced three data values for each of the three parts of the f_0 curve. These data together with the total duration of the f_0 curve produced a total of 10 data values for each syllable. It is important to note that the beginning and ending sections of the f_0 curve have the tendency of being affected by the phonetic structure of the syllable with which the tone is associated [25]. The middle section of an f_0 curve, on the other hand, is more stable.

The ten data values are mapped into a $3 - bit$ binary pattern representing the three tones. Assume the pattern is represented as $b_1 b_2 b_3$, the most significant bit, b_1 indicates the status of the High tone. The least significant bit, b_3, represents the status of the Low tone while the middle bit, b_2, indicates the status of the Mid tone. When a data set is for a tone, the bit for that tone is set to 1 and that of the others is set to 0. For example, 100 is the output pattern for a High tone data. Sample data generated and used for the modelling are shown in Fig. 5.

No.	Beginning (f_0 Hz)			Middle (f_0 Hz)			End (f_0 Hz)			X_{10}	Output fields		
	X_1	X_2	X_3	X_4	X_5	X_6	X_7	X_8	X_9	Frames	H	M	L
1	125.09	48.41	02.25	131.11	31.24	01.24	141.11	24.09	00.93	45.00	1	0	0
2	105.30	10.58	00.60	104.33	03.09	00.31	104.91	03.11	00.29	35.04	0	1	0
3	119.72	57.01	−0.92	112.73	34.01	−0.71	112.13	27.31	−0.60	23.06	0	0	1
4	128.14	36.17	00.38	129.17	29.12	00.24	130.41	30.09	00.23	38.75	1	0	0
5	118.70	9.18	−0.37	119.31	01.17	−0.32	118.18	02.68	−0.30	33.00	0	1	0
6	126.46	62.01	−2.23	116.58	41.17	−1.98	117.87	39.11	−2.13	23.60	0	0	1
7	122.45	42.01	00.05	128.15	44.73	00.11	126.33	34.17	00.13	42.00	1	0	0
8	113.41	8.01	−0.06	103.41	06.33	−0.12	100.17	01.31	−0.11	34.54	0	1	0
9	121.69	49.11	−2.05	121.77	40.93	−2.11	118.38	32.01	−2.06	23.60	0	0	1
10	124.49	39.71	01.81	124.47	28.11	01.37	128.13	29.37	01.11	44.75	1	0	0

Fig. 5. Sample data for the ANN modelling

Table 3. Syllable statistics for the training set

Tone	Syllable types					% of Total
	CV	V	Vn	CVn	N	
H	59 (36%)	36 (22%)	29 (18%)	28 (17%)	12 (7%)	164 (46%)
M	20 (34%)	16 (27%)	8 (13%)	11 (19%)	4 (7%)	59 (17%)
L	43 (34%)	31 (25%)	22 (17%)	23 (18%)	8 (6%)	127 (36%)
Total	122	83	59	62	24	350

Table 4. Syllable statistics for the test set

Tone	Syllable types					% of Total
	CV	V	Vn	CVn	N	
H	18 (31%)	13 (22%)	10 (17%)	12 (21%)	5 (9%)	58 (46%)
M	10 (37%)	8 (30%)	2 (22%)	6 (22%)	1 (4%)	27 (22%)
L	14 (35%)	9 (22%)	5 (13%)	9 (22%)	3 (8%)	40 (32%)
Total	42	30	17	27	9	125

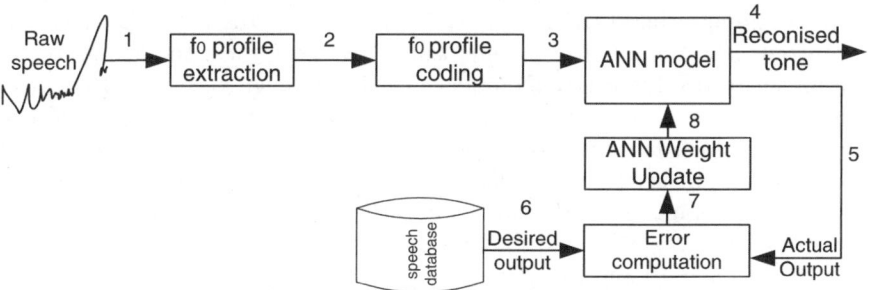

Fig. 6. Overview of the tone recognition system

A total of 475 data items were generated from the computed f_0 data. These data were divided into two disjoint sets: training (350) and test set (125). The distribution of the data is shown in Tables 3 and 4 for the training and test set respectively.

7 Overview of the Tone Recognition System

An overview of the architecture of the proposed SY tone recognition system is shown in Fig. 6. The system is composed of five main parts: f_0 profile extraction, f_0 profile coding, the ANN model, weight update and feedback system, and the speech database. The raw speech signal is applied to the model through the f_0 profile extraction component. The f_0 profile extraction component extracts the f_0 profile from the voiced portion of the speech signal. The f_0 profile coding component computes the data that are used to model the ANN. To achieve this, a third degree polynomial is first interpolated into the f_0 data. The resulting f_0 curve is divided into three equal parts and the 10 parameters discussed in Sect. 6.2. The extracted feature is then fed into the ANN for the training and evaluation processes. The standard back-propagation algorithm is used for supervised training of the two ANN models [4, 23, 60].

The system operates in two phases: *train phase* and *operation phase*. In the train phase, the ANN is presented with the training data set. After training, it is assumed that the ANN had "learnt" the behaviour embedded in the data. During the operation phase, the ANN is presented with test data and required to produce the corresponding output. The ANN performance is determined by its ability to correctly classify or predict the test data.

The ANN is used to learn the f_0 features corresponding to SY tones. During the training phase of the ANN, the path 1, 2, 3, 5, 6, 7 and 8 is traversed repeatedly. During this training process, the network uses the tone features to compute the connection weights for each node in the network. The feedback error is computed based on the desired output (i.e. output from the database, line 6) and the actual output of the ANN (i.e. output from line 5). During the operation phrase, the path 1, 2, 3 and 4 is traversed. By activating the output corresponding to a tone, the ANNs recognises the lexical tone of the input features. For the ANN models, the input layer consists of 10 neurons. Each of these neurons is responsible for one of the input variables. Various number of hidden layer neurons were experimented with in the range 20–65 units. The output layer consisted of 3 units corresponding to each of the 3 SY tones. The mapping vectors from the input space $\mathfrak{R}^n$ to the output space $\mathfrak{R}^m$ of the ANN can be expressed mathematically as $MLP : \mathfrak{R}^n \to \mathfrak{R}^m$. In these ANN models $n = 10$ and $m = 3$.

7.1 The MLP Architecture

The bahaviour of an ANN is determined by its structure. The structure of an ANN model is defined by its architecture and the activation functions. A representation of the architecture of the MLP used in this work is shown in Fig. 7. The MLP is composed of three layers: input, hidden and output. The output from each neuron in the input layer is connected to the input of every neuron in the hidden layer. Also, the output of each neuron in the hidden layer is connected to the input of the output layer. The sigmoid activation function was used in implementing the model.

This represents a feed-forward neural network architecture. It is important to note that there is no interconnection between neurons in the same layer. To train the neural network, all the weights in the network are assigned small initial values. These values were generated randomly. In our model we used random values in the range 0.0 to 1.0. After the weight initialisation process, the following steps are taken iteratively:

1. Apply the next training data to nodes in the input layer.

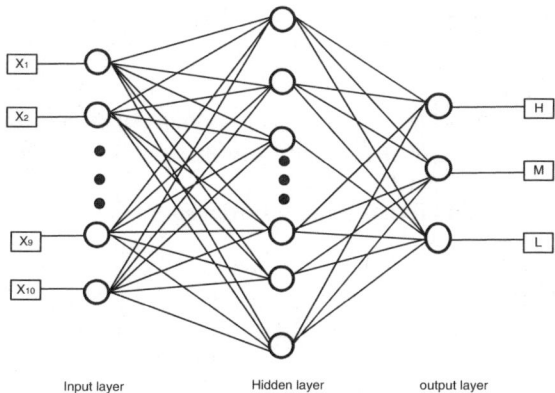

Fig. 7. Architecture of the MLP

2. Propagate the signal computed by each input node forward through the nodes in the others layers of the MLP.
3. Compute the actual output of the network.
4. Compute the error for each processing unit in the output layer by comparing the actual output to the desired output.
5. Propagate the errors backwards from the output layer to the hidden and then the input layer.
6. Modify each weight based on the computed error.

The above process is repeated until all the data set had been processed or the training converges. The MLP training has converged when the error computed is *negligible*. In such cases, the error propagated backwards will not affect the weights in the MLP significantly. The algorithm to implement the process described above is shown in Table 5.

7.2 The RNN Model Architecture

The recognition capability of a feed-forward network is affected by the number of hidden layers and increasing the number of hidden layers increases the complexity of the decision regions that can be modelled. This problem is addressed in the RNN through the delay in the Elman network, identified as Z in Fig. 8, which creates an explicit memory of one time lag [21]. By delaying the output, the network has access to both prior and following context for recognising each tone pattern. An added advantage of including the delay is that the recognition capability of the network is enhanced. This is because the number of hidden layers between any input features and its corresponding output tones is increased. The delay, however, has the disadvantage of making the training process slower requiring more iteration steps than is required in the standard MLP training.

Table 5. Back-propagation algorithm

Begin
  Initialisation
      Initialise all weights w_{ij} to small random values with w_{ij}
      being the value of the weight connecting node j to another
      node i in the next lower layer.
  While(MoreTrainingData() and !Converged())
  {
    ApplyData
      Apply the input from class m, i.e. $X_i^m = \{x_1^m, x_2^m, \ldots, x_n^m\}$, to the
      input layer. Apply the desired output corresponding to the input
      class n i.e. $Y_i^l = \{y_1^l, y_2^l, \ldots, y_n^l\}$ to the output layer.
      Where l is the number of layers.
      In our model, $m = 12$, $n = 3$ and we set the desired output
      to 0 for all the output nodes except the mth node, which is
      set to 1
    ComputeOutput
      Compute actual output of the jth node in layer n,
      where $n = 1, 2, \ldots, l$ using the following expression
      $y_j^n = F\left(\sum_j x_j^{n-1} W_{ij}^n\right)$
      where w_{ij}^n is the weight from node j in the
      $(n-1)^{th}$ layer to node i in the nth layer.
      $F(.)$ is the activation function described in 2.
      The set of output at the output layer can then be represented
      as $X_p^l = \{x_1^l, x_2^l, \ldots, x_n^l\}$.
    ComputeErrorTerm
      Compute an error term, δ_j, for all the nodes. If d_j
      and y_j are the desired and actual output, respectively, then
      for an output node:
      $\delta_j = (d_j - y_j)y_j(1.0 - y_j)$
      and the hidden layer node,
      $\delta_j = y_j(1.0 - y_j)\sum_k^N \delta_k w_{jk}$
    AdjustWeights
      Adjust weight by
      $w_{ij}(n+1) = w_{ij}(n) + \alpha\delta_j y_i + \zeta(w_{ij}(n) - w_{ij}(n-1))$

  } **EndWhile**
  End

We used an Elman [21] model for our RNN based tone recognition system. An Elman RNN is a network which, in principle, is set up as a regular feed-forward network. Figure 8 shows the architecture of the Elman RNN used for our experiment. It is a three-layered network with all the outputs of the neurons in the hidden layer delayed and fed back into the input layer as additional input. Similar to a regular feed-forward neural network, the strength

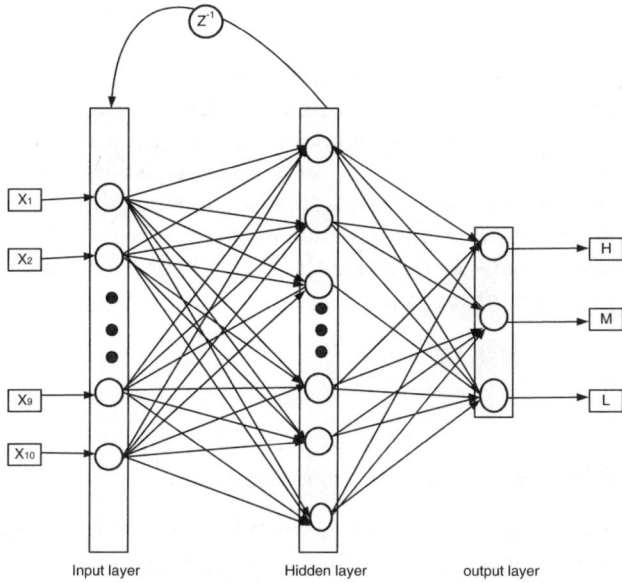

Fig. 8. Architecture of the RNN

of all connections between neurons are indicated with a weight. Initially, all weight values are chosen randomly and are optimized during the training process.

Let $O_k^{(i)}(n)$ denote the output of the kth neuron in the ith layer and the connection weight from the jth in the i_1^{th} layer to the kth neuron in the i_2^{th} layer be denoted by $W_{k,j}^{i_2,i_1}$, where n is the time index. Then,

$$O_k^{(3)} = f(\sum_j W_{k,j}^{(3,2)} O_j^{(2)}(n)) \tag{3}$$

$$= f(\sum_j W_{k,j}^{(3,2)} f(net_j^{(2)}(n))) \tag{4}$$

$$= f(net_k^{(3)}(n)) \tag{5}$$

$$net_j^{(2)}(n) = \sum_l W_{j,l}^{(2,1)} I_l(n) + \sum_{l'} W_{j,l'}^{(2,2)} O_l'^{(2)}(n-1) \tag{6}$$

Where $I_l(n)$ represents the input signals and $f()$ is the sigmoid function. An RNN of this type has been shown to possess a good ability of learning the complex relationship [68]. The model was trained on same data used for the MLP.

7.3 Model Implementation

The two ANN described above were simulated using the MATLAB 7.0 software package with neural networks toolbox [49]. This problem requires that the networks read 10 single value input signal and output three binary digits, at each time step. Therefore, the networks must have ten input element, and three output neuron. The input fields are stored as two dimensional array IP and the output are stored in the one dimensional array OP. The input vector to the input layer is same as those for the MLP. There are generally five steps in the ANN implementation process:

- Assemble the training data.
- Create the network object.
- Initialise the network.
- Train the network.
- Simulate the network response to new inputs.

For the MLP simulation, the network reads 10 single value input data and output three binary digits, at each time step. The network was created using *newf()* command. The network was trained using the *train()* command and simulated using the *sim()* command. We experimented using various number of neurons for the hidden layers. Thirty seven hidden layer neurons, i.e. $n = 37$, produced the best result.

For the RNN network, we want the input fields and target outputs to be considered a sequence, so we make the conversion from the matrix format using the command *con2seq()*. These data are used to train the Elman network. The network were created using the *newelm()* command. It was trained using the *train()* command and simulated using the *sim()* command. We experimented using various number of neurons for the hidden layer. Thirty two hidden layer neurons, i.e. $n = 32$, produced the best result.

The recognition performances of the MLP and RNN using training data set (inside data) and test set (outside test) was obtained. The reason for evaluating with the inside test is to determine how well the model represent the data. The outside test was used to determine how well the model can extrapolate to unknown data. In the RNN, for example the 350 data items (from three speakers) were used for training and the 125 data items from forth speaker, were used for testing. Training of the RNN was done in 1500 epochs since the values of mean square error (MSE) converged to small constants at that value. However, the MLP converged in 1200 epochs. Thus, the convergence rate of the RNN presented in this study was found to be lower than that of the MLP.

8 Results and Discussions

We have experimented with MLP and RNN for SY tone recognition. The experiments show that the adequate functioning of neural networks depends on the sizes of the training set and test set. For the evaluation, we used percent recognition rate. This is computed, for a tone type, as the ratio of the total number of correctly recognised tone to that of the total number of tones of that type multiplied by 100. For example, if there are 30 H tone in the test sample and 15 H tones are recognised correctly, then the recognition rate for the H tone is $(15.00/30.00) \times 100.00 = 50.00\%$. We are using this approach because the occurrence of the three tone types is not equal. The data for the experiments were divided into two disjoint set: the *training set* and the *test set*. The training set is used to develop the model. The test set is used to obtain an objective evaluation the model.

The results for the two ANN models are shown in Table 6. Generally, the inside test produced a higher accuracy than the inside test. This implies that although the ANN models the data relatively well, they do not extrapolate to new data at the same level of accuracy. For example, while the MLP produced an accuracy of 87.50% for the inside test for H tone, it produces 71.30% for the outside test. Similarly, the RNN model produced an inside test of 89.50% for the H tone while it produces 76.10% for outside test. Another thing that the results in Table 6 indicates is that the RNN produces a better recognition accuracy than the MLP. However, a student *t-test* revealed that there is no statistically significant ($p > 0.05$) difference between the accuracy of recognition for the two models.

The tone confusion matrix of the ANN models were generated as shown in Table 7 and 8 for the MLP and RNN respectively. This result showed that tones H and M are more easily confused by the two models. For example, the

Table 6. Tone recognition results

Models		Tone recognition Rate %			
		H	M	L	Mean
MLP	Inside test	87.50	92.71	73.50	82.30
	Outside test	71.30	85.50	81.20	75.32
RNN	Inside test	89.50	97.07	85.50	87.52
	Outside test	76.10	86.15	83.42	79.11

Table 7. Confusion matrix for MLP

Tones	H	M	L
H	78.00	17.00	5.00
M	12.00	82.00	6.00
L	6.00	17.00	77.00

Table 8. Confusion matrix for RNN

Tones	H	M	L
H	91.00	7.00	2.00
M	2.00	98.00	0.00
L	2.00	3.00	95.00

MLP recognises the H tone as such 73.00% of the time. It also recognises the M and L tone as such in 82.00 and 75.00% of the time. The RNN on the other hand recognises the H tone as such 91.00% of the time. It also recognises the M and L tone as such in 98.00% and 95.00% of the time. These recognition results are encouraging despite the simplicity of the input parameters used for the modelling.

The general conclusion from these results is that the M tone has the best recognition rate. Our results agree with those reported in the literature. For example, the f_0 curve of Mandarin Tone 1 is similar to that of the SY Mid tone (cf. [12]). The results of Mandarin tone recognition showed that Tone 1 is the least difficult to recognise with recognition accuracy as high as 98.80% while the neural tone (Tone 5) has recognition accuracy of 86.40%. A similar result was obtained by Cao et al. [10] although the HMM was used in that work.

Also Tone 1 and Tone 7 in Cantonese have similar f_0 curve as SY Mid tone although at difference f_0 ranges. Lee [35, 37] implemented a MLP for Cantonese. He found that the overall recognition accuracy for training and test data are 96.60 and 89.00% respectively. Tone 1 and 7, which have remarkably high pitch level, show the best recognition rates. We speculate that, a reason for the high recognition accuracy of the SY Mid tone is related to the relatively stable excursion of its f_0 curve when compared with those of the High and Low tones.

We did not use a linguistic model, such the Finite State Automaton (FSA), to further clarify the SY tone data. This because linguistic categories are not required for the recognition of isolated tones as will be the case in continuous speech [10, 44, 69]. Moreover the SY tones have simple f_0 signatures when compared with other tone languages such as Mandarin, Thai and Mandarin. This simple signature does not require that the f_0 curves of syllables be model linguistically before accurate recognition can be obtained. The fewer number of SY tones also reduces the complexities of the classification problem.

9 Conclusion

We have presented the Multi-layered Perceptron (MLP) and the Recurrent Neural Network (RNN) models for Standard Yorùbá (SY) isolated tone recognition. The neural networks were trained on SY tone data extracted from recorded speech files. Our results led to three major conclusions:

1. SY tone recognition problem can be implemented with MLP and RNN;
2. The RNN training converges slower than the MLP using the same training data;
3. The accuracy rates achieved using the RNN was found to be higher (although not significantly) than that of the MLP on the inside and outside test data sets;
4. the SY Mid tone has highest recognition accuracy.

However, the efforts required for building the RNN is relatively more than those required for building the MLP. An extension of this work could be to apply the ANN models to continuous SY speech recognition. However, it is well known that ANNs have difficulty in modelling the time-sequential nature of speech. They have therefore not been very successful at recognising continuous utterances. In addition, the training algorithm is not guaranteed to find the global minimum of the error function since gradient descent may get stuck in local minima. Therefore, the ANN approach does not generalise to connected speech or to any task which requires finding the best explanation of an input pattern in terms of a sequence of output classes.

To address this problem, there has been an interest in approaches which are a hybrid of HMMs and neural networks to produce hybrid systems. The aim of this type of approach is to combine the connectionist capability provided by neural networks with the ability of HMM to model the time-sequential nature of speech sound [7]. One approach has been to use MLPs to compute HMM emission probabilities [47] with better discriminant properties and without any hypotheses about the statistical distribution of the data. An alternative approach [4] is to use a neural network as a post-processing stage to an *N-best* HMM system [31, 39, 47, 47, 60].

References

1. L. O. Adéwọlé. *The categorical status and the function of the Yorùbá auxiliary verb with some structural analysis in GPSG*. PhD thesis, University of Edinburgh, Edinburgh, 1988.
2. A. Akinlabí. Underspecification and phonology of Yorùbá /r/. *Linguistic Inquiry*, 24(1):139–160, 1993.
3. A. M. A. Ali, J. Spiegel, and P. Mueller. Acoustic-phonetic features for the automatic classifcation of stop consonants. *IEEE Transactions on Speech and Audio Processing*, 9(8):833–841, 2001.
4. S. Austin, G. Zavaliagkos, J. Makhoul, and R. Schwartz. Continuous speech recognition using segmental neural nets. In *IEEE ICASSP*, 625–628, San Francisco, 2006.
5. S. Bird. Automated tone transcription. http://www.idc.upenn.edu/sb/home/papers/9410022/941002.pdf, May 1994. Visited: Apr 2004.
6. P. Boersma and D. Weenink. *Praat, doing phonetics by computer*. http://www.fon.hum.uva.nl/praat/, Mar 2004. Visited: Mar 2004.
7. H. Bourlard, N. Morgan, and S. Renals. Neural nets and hidden Markov models: Review and generalizations. *Speech Communication*, 11:237–246, 1992.
8. L. Breiman, J. H. Friedman, R. A. Olshen, and C. J. Stone. *Classification and Regression Tree*. Wadworth, CA, 1984.
9. T.-L. Burrows. *Trainable Speech Synthesis*. PhD thesis, Cambridge, Mar 1996.
10. Y. Cao, S. Zhang, T. Huang, and B. Xu. Tone modeling for continuous Mandarin speech recognition. *International Journal of Speech Technology*, 7:115–128, 2004.
11. P. C. Chang, S. W. Sue, and S. H. Chen. Mandarin tone recognition by multi-layer perceptron. In *Proceedings on IEEE International Conference on Acoustics, Speech and Signal Processing (ICASSP)*, 517–520, 1990.

12. S.-H. Chen, S.-H. Hwang, and Y.-R. Wang. An RNN-based prosodic information synthesiser for Mandarin text-to-speech. *IEEE Transactions on Speech and Audio Processing*, 6(3):226–239, 1998.

13. P.-C. Cheng, S.-W. Sun, and S. Chen. Mandarin tone recognition by multi-layer perceptron. In *IEEE 1990 Internertional Conference on Acoustics, Speech and Signal Processing, ICASSP-90*, 517–520, 1990.

14. B. Connell and D. R. Ladd. Aspect of pitch realisation in Yorùbá. *Phonology*, 7:1–29, 1990.

15. B. A. Connell, J. T. Hogan, and A. J. Rozsypal. Experimental evidence of interaction between tone and intonation in Mandarin Chinese. *Journal of Phonetics*, 11:337–351, 1983.

16. R. Córdoba, J. M. Montero, J. M. Gutiérrez, J. A. Vallejo, E. Enriquez, and J. M. Pardo. Selection of most significant parameters for duration modelling in a Spanish text-to-speech system using neural networks. *Computer Speech and Language*, 16:183–203, 2002.

17. D. H. Crozier and R. M. Blench. *An Index of Nigerian Languages*. Summer Institute of Linguistics, Dallas, 2nd edition, 1976.

18. E. Davalo and P. Naïm. *Neural Networks*. MacMillan, Hong Kong, 1991.

19. T. Demeechai and K. Mäkeläinen. Recognition of syllable in a tone language. *Speech Communication*, 33:241–254, 2001.

20. S. J. Eady. Difference in the f_0 patterns of speech: Tone languages versus stress languages. *Language and Speech*, 25(Part 1):29–41, 1982.

21. J. L. Elman. Finding structure in time. *Cognitive Science*, 12(2):179–211, 1990.

22. J. W. A. Fackrell, H. Vereecken, J. P. Martens, and B. V. Coile. Multilingual prosody modelling using cascades of regression trees and neural networks. *http://chardonnay.elis.rug.ac.be/papers/1999 0001.pdf*, 1999. Visited: Sep 2004.

23. A. K. Fernando, X. Zhang, and P. F. Kinley. Combined sewer overflow forecasting with feed-forward back-propagation artificial neural network. *Transactions On Engineering, Computing And Technology*, 12:58–64, 2006.

24. S. C. Fox and E. K. Ong. A high school project on artificial intelligence in robotics. *Artificial Intelligence in Engineering*, 10:61–70, 1996.

25. J. Gandour, S. Potisuk, and S. Dechnonhkit. Tonal co-articulation in Thai. *Phonetica*, 56:123–134, 1999.

26. N. F. Gülera, E. D. Übeylib, and I. Gülera. Recurrent neural networks employing Lyapunov exponents for EEG signals classification. *Expert Systems with Applications*, 29:506–514, 2005.

27. M. D. Hanes, S. C. Ahalt, and A. K. Krishnamurthy. Acosutic-to-phonetic mapping using recurrent neural networks. *IEEE Transactions on Neural Networks*, 4(5):659–662, 1994.

28. W. Holmes and M. Huckvale. Why have HMMs been so successful for automatic speech recognition and how might they be improved? Technical report, Phonetics, University Colledge London, 1994.

29. M. Huckvale. 10 things engineers have discovered about speech recognition. In *NATO ASI workshop on speech pattern processing*, 1997.

30. A. Hunt. Recurrent neural networks for syllabification. *Speech Communication*, 13:323–332, 1993.

31. B. H. Juang and L. R. Rabiner. Hidden Markov models for speech recognition. *Technometrics*, 33:251–272, 1991.

32. A. Khotanzak and J. H. Lu. Classification of invariant images representations using a neural netwrok. *IEEE Transactions on Speech and Audio Processing*, 38(6):1028–1038, 1990.

33. D. R. Ladd. Tones and turning points: Bruce, pierrehumbert, and the elements of intonation phonology. In M. Horne (ed.) *Prosody: Theory and Experiment – Studies presented to Gösta Bruce*, 37–50, Kluwer Academic Publishers, Dordrecht, 2000.

34. S. Lee and Y.-H. Oh. Tree-based modelling of prosodic phrasing and segmental duration for Korean TTS systems. *Speech Communication*, 28(4):283–300, 1999.

35. T. Lee. *Automatic Recognition Of Isolated Cantonese Syllables Using Neural Networks*. PhD thesis, The Chinese University of Hong Kong, Hong Kong, 1996.

36. T. Lee, P. C. Ching, L. W. Chan, Y. H. Cheng, and B. Mak. Tone recognition of isolated Cantonese syllables. *IEEE Transactions on Speech and Audio Processing*, 3(3):204–209, 1995.

37. W.-S. Lee. The effect of intonation on the citation tones in Cantonese. In *International Symposium on Tonal Aspect of Language*, 28–31, Beijing, Mar 2004.

38. Y. Lee and L.-S. Lee. Continuous hidden Markov models integrating transition and instantaneous features for Mandarin syllable recognition. *Computer Speech and Language*, 7:247–263, 1993.

39. S. E. Levinson. A unified theory of composite pattern analysis for automatic speech recognition. In F. F. and W. A. Woods (eds) *Computer Speech Processing*, Prentice-Hall International, London, 1985.

40. Y.-F. Liao and S.-H. Chen. A modular RNN-based method for continuous Mandarin speech recognition. *IEEE Transactions on Speech and Audio Processing*, 9(3):252–263, 2001.

41. C.-H. Lin, R.-C. Wu, J.-Y. Chang, and S.-F. Liang. A novel prosodic-information synthesizer based on recurrent fuzzy neural networks for Chinese TTS system. *IEEE Transactions on Systems, Man and Cybernetics*, B:1–16, 2003.

42. R. P. Lippman. Review of neural networks for speech recognition. *Neural Computing*, 1:1–38, 1989.

43. F.-H. Liu, Y. Lee, and L.-S. Lee. A direct-concatenation approach to training hidden Markov models to recognize the highly confusing Mandarin syllables with very limited training data. *IEEE Transactions on Speech and Audio Processing*, 1(1):113–119, 1993.

44. L. Liu, H. Yang, H. Wang, and Y. Chang. Tone recognition of polysyllabic words in Mandarin speech. *Computer Speech and Language*, 3:253–264, 1989.

45. J. McKenna. Tone and initial/final recognition for Mandarin Chinese. Master's thesis, University of Edingbrugh, U.K., 1996.

46. N. Minematsu, R. Kita, and K. Hirose. Automatic estimation of accentual attribute values of words for accent sandhi rules of Japanese text-to-speech conversion. *IEICE Transactions on Information and System*, E86-D(3):550–557, Mar 2003.

47. N. Morgan and H. Bourlard. Continuous speech recognition using multilayer perceptrons with Hidden Markov Models. In *Proceedings of IEEE ICASSP*, 413–416, Albuquerque, 1990.

48. R. D. Mori, P. Laface, and Y. Mong. Parallel algorithms for syllable recognition in continuous speech. *IEEE Transactions on Pattern Analysis and Machine Intelligence*, PAMI-7(1):56–69, 1985.

49. Y. Morlec, G. Bailly, and V. Aubergé. Generating prosodic attitudes in French: data, model and evaluation. *Speech Communication*, 33:357–371, 2001.

50. S. M. O'Brien. Knowledge-based systems in speech recognition: A survey. *International Journal of Man-Machine Studies*, 38:71–95, 1993.

51. O. A. Ọdéjọbí, A. J. Beaumont, and S. H. S. Wong. A computational model of intonation for Yorùbá text-to-speech synthesis: Design and analysis. In P. Sojka, I. Kopeček, and K. Pala, (eds) *Lecture Notes in Artificial Intelligence*, Lecture Notes in Computer Science (LNAI 3206), 409–416. Springer, Berlin Heidelberg New York, Sep 2004.

52. O. A. Ọdéjọbí, A. J. Beaumont, and S. H. S. Wong. Experiments on stylisation of standard Yorùbá language tones. Technical Report KEG/2004/003, Aston University, Birmingham, Jul 2004.

53. S. M. Peeling and R. K. Moore. Isolated digit recognition experiments using the multi-layer perceptron. *Speech Communication*, 7:403–409, 1988.

54. G. Peng and W. S.-Y. Wang. Tone recognition of continuous Cantonese speech based on support vector machines. *Speech Communication*, 45:49–62, Sep 2005.

55. A. A. Petrosian, D. V. Prokhorov, W. Lajara-Nanson, and R. B. Schiffer. Recurrent neural network-based approach for early recognition of Alzheimer's disease in EEG. *Clinical Neurophysiology*, 112(8):1378–1387, 2001.

56. L. Rabiner. A tutorial on hidden Markov models and selected applications in speech recognition. In *Proceedings of IEEE*, 77:257–286, 1989.

57. M. Riedi. A neural-network-based model of segmental duration for speech synthesis. In *European Conference on Speech Communication and Technology*, 599–602, 1995.

58. M. D. Riley. Tree-based modelling of segmental durations. In G. Bailly, C. Benoit, and T. R. Sawallis (eds), *Talking Machines: Theories, Models and Designs*, p 265–273. Elsevier, Amsterdam, 1992.

59. S. R. Safavian and D. Landgrebe. A survey of decision tree classifier methodology. *IEEE Transactions on Systems, Man and Cybernetics*, 21:660–674, 1991.

60. P. Salmena. Applying dynamic context into MLP/HMM speech recognition system. *IEEE Transactions on Neural Networks*, 15:233–255, 2001.

61. J. Sirigos, N. Fakotakis, and G. Kokkinakis. A hybrid syllable recognition system based on vowel spotting. *Speech Communication*, 38:427–440, 2002.

62. C. Taylor. Typesetting African languages. http://www.ideography.co.uk/library/afrolingua.html, 2000. Visited: Apr 2004.

63. P. Taylor. Using neural networks to locate pitch accents. In *Proceedings of EuroSpeech '95*, 1345–1348, Madrid, Sep 1995.

64. N. Thubthong and B. Kijsirikul. Tone recognition of continuous Thai speech under tonal assimilation and declination effects using half-tone model. *International Journal of Uncertainty, Fuzziness and Knowledge-Based Systems*, 9(6):815–825, 2001.

65. M. Vainio. *Artificial neural network based prosody models for Finnish text-to-speech synthesis*. PhD thesis, Department of Phonetics, University of Helsinki, Helsinki, 2001.

66. H.-M. Wang, T.-H. Ho, R.-C. Yang, J.-L. Shen, B.-O. Bai, J.-C. Hong, W.-P. Chen, T.-L. Yu, and L.-S. Lee. Complete recognition of continuous Mandarin speech for Chinese language with very large vocabulary using limited training data. *IEEE Transactions on Speech and Audio Processing*, 5(2):195–200, 1997.

67. T.-R. Wang and S.-H. Chen. Tone recognition of continuous Mandarin speech assisted with prosodic information. *Journal of the Acoustical Society of America*, 96(5):2637–2645, 1994.
68. W.-J. Wang, Y.-F. Liao, and S.-H. Chen. RNN-based prosodic modelling for Mandarin speech and its application to speech-to-text conversion. *Speech Communication*, 36:247–265, 2002.
69. Y. R. Wang, J.-M. Shieh, and S.-H. Chen. Tone recognition of continuous Mandarin speech based on hidden Markov model. *International Journal of Pattern Recognition and Artificial Intelligence*, 8(1):233–246, 1994.
70. M. Wester. Pronunciation modeling for ASR knowledge-based and data-driven methods. *Computer Speech and Language*, 38:69–85, 2003.
71. Y. Xu. Understanding tone from the perspective of production and perception. *Language and Lingustics*, 5:757–797, 2005.

Emotion Recognition from Speech Using Multi-Classifier Systems and RBF-Ensembles

Stefan Scherer, Friedhelm Schwenker, and Günther Palm

Institute for Neural Information Processing, University of Ulm, 89069 Ulm, Germany, stefan.scherer@uni-ulm.de, friedhelm.schwenker@uni-ulm.de, guenther.palm@uni-ulm.de

Summary. This work provides a detailed overview of related work on the emotion recognition task. Common definitions for emotions are given and known issues such as cultural dependencies are explained. Furthermore, labeling issues are exemplified, and comparable recognition experiments and data collections are introduced in order to give an overview of the state of the art. A comparison of possible data acquisition methods, such as recording acted emotional material, induced emotional data recorded in Wizard-of-Oz scenarios, as well as real-life emotions, is provided. A complete automatic emotion recognizer scenario comprising a possible way of collecting emotional data, a human perception experiment for data quality benchmarking, the extraction of commonly used features, and recognition experiments using multi-classifier systems and RBF ensembles, is included. Results close to human performance were achieved using RBF ensembles, that are simple to implement and trainable in a fast manner.

1 Introduction

One of the main issues in constructing an automatic emotion recognizer is to specify what exactly an emotion represents. Most people have an informal understanding about what emotion is, but there is also a formal research tradition which tries to define it systematically. This systematic approach, namely *Psychological Tradition*, will be discussed in the first part of this section [1]. The introduction also covers the state of the art in emotion recognition by presenting experiments conducted by various research groups, and output related issues such as labeling methods and emotion representations.

1.1 Psychological Tradition

The *Psychological Tradition* has been developed by several prominent researchers from various disciplines, for instance René Descartes in philosophy, Charles Darwin in biology, and Wilhelm Wundt and William James in psychology. Descartes introduced the idea that a few basic emotions form the

S. Scherer et al.: *Emotion Recognition from Speech Using Multi-Classifier Systems and RBF-Ensembles*, Studies in Computational Intelligence (SCI) **83**, 49–70 (2008)
www.springerlink.com   © Springer-Verlag Berlin Heidelberg 2008

basis for the majority of the emotions. Darwin said that emotions are distinctive action patterns, selected by evolution, because of their survival value [2]. He expressed that emotions were triggered by certain situations in the life of humans and animals, for example, the reaction to danger is mostly experiencing fear and thinking ways for escape, which is a natural reaction. James focused on the direct connection between emotion and somatic arousal [1]. Similarly, Wundt applied the same principles for physiological psychology. According to Wundt, humans are predominantly emotional, involving emotions in all mental and physical activities. However, James and Wundt did not entirely agree in this concept. According to James, humans first respond to a situation, and then experience an emotion, whereas Wundt proposed emotion comes first rather than the mental response [3].

Literature concerning emotions does not specify which of these ideas are correct. However, they indicate that the truth is somewhere in between all of them, producing a standard view. Unfortunately, there are still three main unresolved problems, all of which are relevant to automatic emotion recognizers.

Open Issues

Mainly, there is no general agreement on a set of *basic emotions*. Even if some exist, then it would be difficult to arrive upon a commonly agreed list. The important question to be addressed here is "Does this lack of agreement show there is no natural unit for emotions and has to be relied upon pragmatic decisions on which list has to be chosen?" However, Wundt did not believe that it was possible to come up with a consistent list of basic emotions, for which he defined certain quality dimensions, which are explained in Sect. 1.3.

Secondly, there is a divergence between the thoughts of Darwin, that emotion is a "product" of evolution and the facts, that can not be overseen, which can also be termed as *culturally dependent* emotions. Probably the best known examples are the medieval European concept of *accidie* [4] and the Japanese concept of *amae* [5], translated as "sinful lethargy" and "indulgent dependence". Averill and Harre conclude, that "emotions are transitory social roles – that is, institutionalized ways of interpreting and responding to particular classes of situations" [6]. However, a compromise has to be reached to make a pragmatic decision dealing with cultural differences.

Finally, the basic meaning of *emotion* can be viewed in different perspectives, for example, taking it as a discrete syndrome, it can be compared to the so called *full-blown*[1] emotions, which are characterized as emotions that fully take "control" of the persons actions in a certain moment. *Full-blown* emotions occur for small intervals of time, and the other underlying emotions are somehow overlooked. However, the approaches for determining underlying emotions are the matter of pragmatic decisions.

---

[1] E.g. fear, anger, happiness, ...

1.2 Related Work

In recent times, several approaches toward emotion recognition have been pursued [7–13]. Many of these works differ in the database used, the basic set of emotions, feature extraction and the classification methods. However, the primary goal of all these works was to develop a better performing emotion recognizer.

The utilized database and the set of basic emotions were different and intended for specific applications. For example, in a call center application, separating calm and agitated customers is an important task, as indicated by Petrushin [7] and Yacoub et al. [8]. In some cases it may be necessary to distinguish more than two emotions (believable agent domain), as described in Dellaert et al. [9].

The classification methods vary from k-Nearest-Neighbor (k-NN) classifiers [9, 10] to more sophisticated classifiers such as Support Vector Machines (SVM) [8,11]. This was interpreted as a binary classification problem in [8] and multi class problem in [11]. Hidden Markov Model (HMM) based classification framework was used for emotion recognition task in [12].

However, the type of features used to classify emotions did not vary much, as it is common to use prosodic features based on the fundamental frequency (pitch), which have been important for emotion recognition [9, 10, 13]. Apart from pitch based features, the other commonly used features are energy based and Mel Frequency Cepstral Coefficients (MFCC) [8,12]. Basically the type of features depend upon the classification methods, for example, the MFCC features would be too large to be classified with a multi class SVM algorithm. At the same time, the sequential characteristics of the MFCC features are perfect for HMMs. This shows the complexity of the possible variations that exist. Since there have been a lot of ambiguities regarding basic sets of emotions, features, and algorithms, emotion recognition is a potential research area.

Emotional Corpora

In the context of automatic emotion recognition, a huge variety of emotional corpora exists [14]. They differ from each other based on the recording conditions, language, and basic set of emotions [15,16]. The commonly used corpora are described in the following.

SpeechWorks Dataset: The SpeechWorks Dataset is a collection of several hundred real human machine dialogs taken from a call center application [10]. Since the recordings are taken from a real application over the telephone the sample rate is 8 kHz and not comparable to high end studio recordings. The recordings were labeled with two different labels: *negative* and *non-negative*.

German Database of Emotional Speech: As the name indicates, this database is a collection of German emotional utterances [16]. There are ten different sentences, which are taken from normal life situations, spoken by ten speakers in seven different basic emotions[2] in an anechoic room. The recording atmosphere is artificially setup as the speakers were prepared to say the exact preset sentences in one of the given emotions. The speakers received feedback on the recording, if they overacted the emotion. The recording equipment was professional and the microphone was hanging from the ceiling.

Belfast Naturalistic Emotional Database: The goal of the Belfast Naturalistic Emotional Database was to record realistic audiovisual emotional data [17–19]. In order to achieve this goal, several conversations between a researcher and his colleagues have been conducted. The researcher tried to use his background knowledge about his conversation partner to tease emotions out of him or her. These conversations were extended with several recordings from talk shows on television. Overall the database consists of 298 clips, each varying between 10 and 60 s. Since it is not predictable what emotions will be present beforehand the emotional parts of the clips can not be categorized crisply into a set of basic emotions. That is why a program called "FEEL-TRACE" was implemented in order to label the clips in a continuous manner. The clips are projected into the activation evaluation space, which is explained in Sect. 1.3, providing soft labels for the clips.

Experiments on Emotion Recognition and Results

As mentioned before, emotional data can be classified with many different ways. The following provide an overview of the possible methods, the utilized features, the different applications, and the results of the conducted experiments.

Yacoub, Simske, Lin, and Burns: Since it was not necessary for the targeted application – an interactive voice response system – to be able to distinguish between many emotions, only two emotions namely anger and neutral were categorized [8]. Overall a set of 37 features was used based on pitch, energy, and segments of speech (silence vs speech). As a classification tool the Weka Toolkit [20] was used. SVMs and Neural Networks achieved an accuracy of around 90%.

Dellaert, Polzin, and Waibel: Several experiments were performed using their own dataset containing acted utterances in various emotions such as sadness, anger, happiness, and fear [9]. A human performance test led to an accuracy of 82%. Pitch based features were used for emotion recognition task. A first set of basic pitch features such as maximum, minimum, and mean of pitch in a sequence achieved an accuracy of 64% using a k-NN classifier (k = 19). A more sophisticated set of features led to slightly better results using a k-NN classifier (k = 11, 68% accuracy). Advanced approaches toward feature selection and the combination of subset classifiers improved the performance significantly (80% accuracy).

---

[2] Anger, happiness, fear, boredom, sadness, disgust, and neutral.

Lee, et al.: A different approach using MFCC features is proposed [12]. The MFCC features, obtained with a window of 25 ms sampled every 10 ms were utilized. In order to handle this amount of data and to exploit its sequential characteristics, HMMs were trained. Four different HMMs were trained for each of the targeted emotions anger, happiness, sadness, and fear. For each of the emotion HMMs, five broad phoneme class[3] based HMMs were trained, using the TIMIT database. After training a combination of all the phoneme class based HMMs for each of the emotions reached a classification performance of around 76%.

1.3 Labeling Issues

In this section, different labeling methods for recorded emotional data are discussed. Labeling or classifying emotions is the primary goal of an automatic emotion recognizer. For the same, two possible methods exist discrete and continuous labeling.

Discrete Emotion Labeling

Humans have an enormous vocabulary of distinguishing emotions, as it may be seen from a list in [1]. However, it is impossible for an artificial emotion recognizer to categorize emotions in such a fine grained manner. For this reason, a small fraction of possible labels is chosen to cover most of the *emotional space*. This method of choosing a small amount of different labels results most of the time in a list of four to seven different emotions the so-called *full-blown* emotions, e.g. anger, happiness, boredom, fear, disgust, sadness, and neutral. These seven emotions were taken as labels for the tests in Sects. 4 and 5.

If acted emotional data corpus is chosen, labeling emotions discretely would be the optimum method of choice, as discussed in Sect. 2.1. It is very straightforward to take the targeted emotion as a direct label for the utterance. However, there are possibilities to soften these crisp labels, for example one could take the outputs from the human performance test, as described in 2.3, from some subjects as labels.

Continuous Emotion Labeling

As mentioned above, continuous labels may be obtained by additional experiments from discrete labels. But there are other approaches towards continuously labeled emotions. The most commonly method was introduced by Wilhelm Wundt in [3]. As mentioned before, Wundt did not believe that it is possible to come up with a consistent list of basic emotions. That is the primary reason where he came up with certain quality dimensions of emotions, which resulted in the activation-evaluation space.

---

[3] Vowel, glide, nasal, stop, and fricative.

Activation-Evaluation Space

A simple method, that is still capable of representing a wide range of emotions, is the activation-evaluation space. It consists of two key themes. The first theme is *Valence*. People in emotional states evaluate people, events or things in a positive or a negative way. The link between valence and emotion is widely agreed upon, although authors describe it in different terms, e.g. Arnold refers to the "judgment of weal or woe" [21]. The second theme is the *Activation Level*. Starting with Darwin, research has revealed that emotions involve dispositions to act in certain ways. The emotional states are rated in the strength of the person's disposition to take some action rather than none.

The axes of the activation-evaluation space represent those two concepts. The ordinate represents the activation level and the abscissa the evaluation, as it can be seen in Fig. 1. The advantage of this representation is, that one is able to translate in and out of this space easily, i.e. words can be translated into this space and coordinates can be understood and then be translated into words. Various techniques lead to that conclusion, including factor analysis or direct scaling.

However, this representation is quite simple and yet still capable of representing a big variety of emotional states. The rest of the chapter is organized as follows: In Sect. 2, general methods to acquire emotional speech data are described along with the utilized data set. Section 3 presents the different feature types used to classify emotions. Sections 4 and 5 describe the experimental setup, while in Sect. 6, the results are presented and analyzed. Finally in Sect. 7, the chapter is concluded and open issues are discussed.

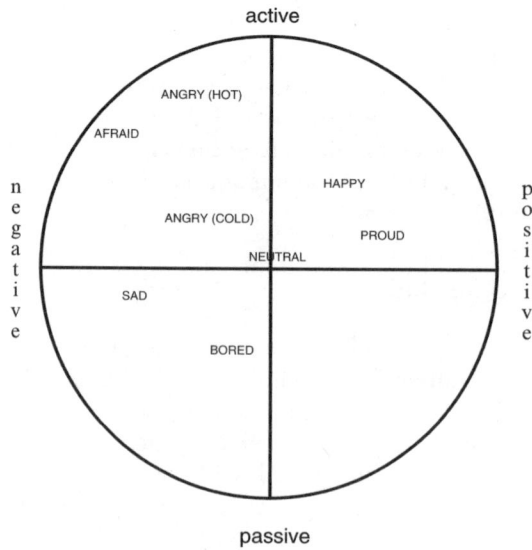

Fig. 1. The activation-evaluation space with some example emotions

2 Acquisition of Emotional Data

An important task in emotion recognition is the collection of an emotional data corpus. Depending on the tasks to be addressed, different groups designed various corpora to suit their purposes. There are three possible methods to record emotional data, which are described and discussed in Sect. 2.1. Apart from these general methods a novel approach for data collection, describing a scenario, which has been developed by the competence center Perception and Interactive Technologies (PIT) at the University of Ulm, is presented [22]. In Sect. 2.2 the recording methods are discussed and some open psychological issues are explained. An example corpus with acted emotional material will be described in detail in Sect. 2.3. Additionally a benchmark – a human performance test – has been conducted and is explained in detail, along with the results of the test.

2.1 Recording Methods

Acted

The simplest way to record emotional data is to hire professional or non-professional actors and tell them that they should play the emotions. The disadvantage of acted emotions is obvious, because of the fact, that they are not actually felt by the person acting, there is a bias in the expression. Other issues are, for example, it is very hard for actors to suppress or evoke underlying emotions that could affect the recordings. So the compromise is often made, that only the *full-blown* emotions are recorded, because of their simplicity. To minimize the bias it is necessary to give the actors a chance to put themselves into the specified emotion, for example, by acting a little play or reminding situations of the past, this method of self-induction is known as the Stanislavski method [23].

Acted emotional speech is the most common version of emotional data found in the literature [15,16,24,25], since it is the easiest way to collect data. It is important to note, that acted emotional speech may not be the best data to train automatic emotion recognizers, because it may not represent the appropriate emotions for the intended application, for example, it may never happen in an application, that people are – in the sense of *full-blown* emotions – angry or happy. Most of the time subtle emotions are experienced as it will be discussed in Sect. 2.2.

Further details regarding acted emotional data acquisition are described in Sect. 2.3. The presented corpus is used in automatic emotion recognition presented in Sects. 4 and 5.

Real

A second possibility of recording emotional data is to record *real* emotional data. Real means that the emotional data is not biased such as acted emotional data is. The people are not aware of being recorded and act naturally. This is the case in the database used by Lee et al. [10], who used a database recorded by SpeechWorks. The data was taken from a real call-center application, which is already in use. The disadvantage of this data corpus is the quality of the recordings since it was recorded over the phone. From the realistic point of view, this on the other hand may be an advantage, because it is directly connected to a possible application.

Another way of recording real emotional data is to wire people almost 24/7 to catch true emotional situations in everyday life [26]. This method however is not directly related to a certain topic or task, but is definitely a way of recording truly felt emotions.

The complexity of obtaining real emotional data is one of the main disadvantages, besides the need of filtering the data collected using the wiring method. This data needs to be segmented into emotional and non-emotional data. Furthermore, it is necessary to label the data using one of the techniques mentioned in Sect. 1.3. All these processing steps are mandatory and consume lots of time. Hence, there is no bias in real emotional data, it is probably the most valuable data for automatic emotion recognition.

Wizard-of-Oz

The third method is somehow a compromise between the two mentioned before. It is based on induced emotions within a Wizard-of-Oz (WOz) scenario. The advantages of this method are that, it can be supervised at any time by the wizard, the data can be recorded with the same high quality recording equipment as in the acted method, and the emotions are not acted. The recorded emotions lie in the discretion of the wizard and need not be directly connected to the targeted application. This is one of the reasons for appropriate planning and training to record relevant emotional data in such a scenario. The following paragraph describes a WOz arrangement in detail.

PIT scenario

In the project, Perception and Interaction in Multi-User Environments the task is to select a restaurant. Two users, sitting in front of a computer and are discussing which restaurant should be chosen. The wizard, which is located in another room, is listening passively. The arrangement of the scenario is shown in Fig. 2.

If the wizard is asked a direct question or receives an order from the main user A1, the wizard will interact with the two dialog partners via the speakers S1 and S2 or the monitor A3, which is believed to be the dialog partner. The wizard is listening live to the users A1 and A2 via the wireless microphones M1 and M2. He uses a database of restaurants and other localities to produce answers to the demands of the users. These answers are transformed using VoiceXML from text to speech. Microphone M3 is recording the whole scenario. Camera C1 records the face of user A1 for online gaze detection. The

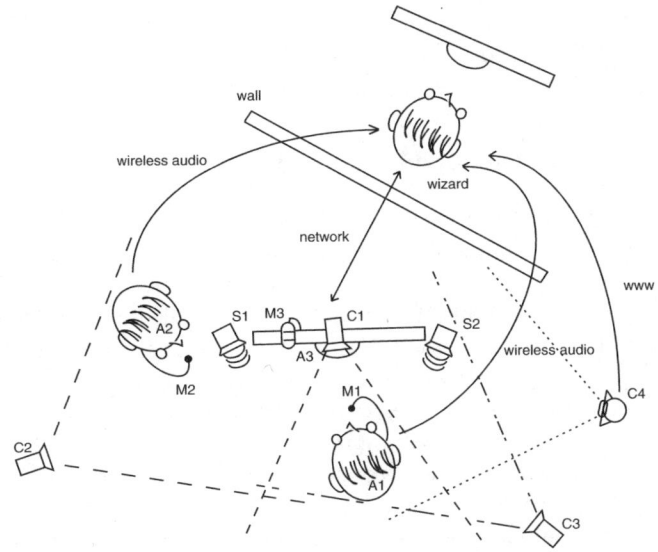

Fig. 2. The arrangement of the Wizard-of-Oz scenario [22]

- **Scenario:** Two colleagues A and B want to eat something after a long day at work
- **Person A:**
  - Dialog partner of computer
  - Loves Greek food
  - Had a bad day (is tetchy)
  - Is short in money
- **Person B:**
  - Is disgusted by garlic
  - Likes plain German food

Fig. 3. Example task of Wizard-of-Oz scenario

cameras C2 and C3 are used to record the environment and the line of sight of the users. As an additional surveillance tool for the wizard, a webcam C4 has been installed. If A1 requests a menu of a selected restaurant the wizard is able to display it on the screen. With C4 the wizard may check if the menu is displayed correctly.

The main task of the wizard is now to induce emotions into the users. In order to do that the wizard has several possibilities, which afford some training. The wizard, for example, could give false answers repetitively to induce some amount of anger or boredom, he could interrupt the dialog partners while speaking, or to induce happiness or contentment by giving right answers fast. Apart from the wizard the users receive some tasks that should be accomplished in the dialog before the recording starts. The tasks are embedded in some sort of role play, which is helpful for users that do not know each other to start a conversation. An example role play can be seen in Fig. 3.

Since the efforts are quite high in order to obtain induced emotional data the corpus is still growing. Apart from the corpus new labeling techniques similar to "FEELTRACE" are in development.

2.2 Comparison of the Acquisition Methods

Designing or planning an emotional corpus should consider about the underlying motivation, for example, if the motivation is to keep data acquisition as simple as possible, acted emotions is the best choice. However, if realistic behavior is the desired goal, then this method would not be sufficient. The first step would be to implement the application as far as possible without considering emotions at all and then record data as performed by SpeechWorks. But this would limit the data to the scenarios encountered only in the implemented application. Changing the application would probably lead to totally different emotions, e.g. consider a call-center application of a bank and of a travel agency. Furthermore, as mentioned in Sect. 2.1, recording real emotional data is unbiased and therefore includes all the facets of *true emotions*, which leads to some complicating factors mentioned in the following paragraphs.

First, there are so called "display rules". For example, as described by Ekman and Friesen there are certain culturally rooted rules, that "manage the appearance of particular emotions in particular situations" [27]. In many cultures, uninhibited expression of anger or grief is not accepted by the society and result rather in an attempted smile than in a neutral expression.

The next factor that complicates the situation is deception, which is very close to the display rules. In social life, misrepresenting emotions is often the case and therefore detecting deception has always been a main application for the psychological tradition. Because humans are often deceived by misrepresented emotions, recognizing deception is not expected from automatic emotion recognizers.

The third factor is systematic ambiguity. It is clear that lowered eyebrows may indicate concentration as well as anger. For speech signals it is not that obvious, but there are strong similarities between the prosodic characteristics associated with depression [28] and those associated with poor reading [29]. The systematic ambiguity is a serious issue in detecting emotions, which leads to the ideas of combining more sources, like facial and speech expressions in order to minimize the errors.

These are the difficulties one has to deal with while using real emotions, but they hardly occur in acted emotions. That makes it clear that recording real emotions is far more complicated than acquiring acted emotions. As mentioned before, a WOz scenario is somehow a compromise, since the emotions are not acted and the above mentioned factors could be discovered by a watchful wizard during the conversations. Nonetheless, it is not possible to define the best method, but the above considerations may be helpful in deciding, what method should be chosen in a given scenario or application.

2.3 Recording Emotional Data

The following paragraphs describe the corpus that has been tested in Sects. 4 and 5. It is a corpus that has been kept simple by recording acted emotions. The emotions were labeled discretely by using terms for *full-blown* emotions, such as anger, sadness or boredom. Furthermore, an experiment with several human subjects has been conducted in order to test the quality of the corpus and to provide a benchmark.

Scenario

The first step in creating an emotional database was to design 20 sentences, which may be read in the different emotions, e.g. "Und wohin wollen wir gehen?".[4] In order to *feel* the emotion while talking the test persons were advised to put themselves into the emotional state before recording. It is easier for the actors to simulate the emotions while recalling experienced emotional situations. To support the emotional state further, plays introduced with screenplays were performed in order to put oneself into the emotion play-fully. The 20 sentences were spoken by the test persons talking into a close room microphone.[5] The emotions were recorded using a Creative Labs Audigy 2 external USB sound card at 16 kHz, in a 16-bit resolution and mono. However, this database was not recorded in an anechoic chamber as the Berlin Database of Emotional Speech [16], but in a *normal* office with background noise. The office environment was chosen in order to analyze the performance in practice rather than in an experimental environment. The actors read the 20 sentences after each other in the same emotion, to assure that the emotions won't be confused while speaking. At the moment there are around 400 sentences recorded.

In the next section an experiment is described and evaluated, which has been used to give an idea about the quality of the recorded emotions.

Human Performance Test: A Benchmark

In addition to recording emotional data it was necessary to test the quality of the recorded data, in order to put it in relation with existing material, e.g. the Berlin Emotional Speech Database. Furthermore, it is interesting how accurate humans are in recognizing emotions. For this reason a simple perception experiment has been conducted.

The test persons were advised to listen to the recorded audio data with headphones in order to minimize the noise level. They had to label the utterances with a simply designed program, which does not confuse the user and allows him to listen to the sentences as often as necessary.

---

[4] "And where do we want to go to?"
[5] AKG C1000S.

Table 1. Confusion matrix of the human performance experiment. Correct classifications: 65.1%

	Neutral	Anger	Sadness	Fear	Disgust	Boredom	Happiness
Neutral	188	5	8	1	7	22	5
Anger	20	181	0	1	6	1	27
Sadness	19	1	120	25	18	52	1
Fear	35	14	36	111	20	5	15
Disgust	16	17	27	19	123	15	19
Boredom	27	10	14	5	17	162	1
Happiness	18	18	1	4	4	0	191

First of all, the user had to login and chose whether he was an existing user or not. After that, the experiment starts immediately with a new screen where only eight buttons[6] and a progress counter are displayed. Once all the emotion files have been listened to the evaluation was conducted. In the following paragraphs the results of this test are presented.

Results

In Table 1 the results for the human performance test are shown. The performances for the single emotions range from around 47% for fear to around 81% for happiness. Since, happiness is quite easy to recognize over the loudness of the speech besides anger, it is the emotion with the best classification results followed by neutral. Sadness was confused quite frequently with fear, disgust and boredom. This probably results from the fact that these emotions are quite close to each other on the activation axis, which is the easiest to distinguish. Happiness and anger are close on the activation axis as well and are therefore confused with each other. Neutral was chosen most of the times and it was confused quite often with all of the other emotions, which leads to the belief that neutral is chosen if one is not quite sure about the heard emotion. Another interesting fact is that the emotions disgust and fear were selected the least often and their recognition rate is quite close to the one of sadness at around 50%, which is quite far from average. According to this experiment the average classification rate is at around 65%.

3 Feature Extraction

Feature extraction is one of the central problems in pattern recognition and of course in emotion recognition. As mentioned in Sect. 1.2 many different features have been tested in emotion recognition research, with differing success. In the following sections four different types of features based on energy and prosody are introduced. The four types are the most common ones extracted from speech signals for various applications, for example MFCC features are

---

[6] The seven emotions plus a button to repeat the current sentence.

widely used in speech recognition or LPC features are used for speaker recognition. The most popular features in emotion recognition are pitch related features.

3.1 Energy Related Features

Energy related features are easily extracted from the waveform of the speech signal. Possible features would be the energy itself of each frame besides mean, median, standard deviation and other statistical features. It is obvious that, especially emotions which differ in the loudness of the speech, which reflects the activation dimension, are easy to classify utilizing these simple features.

3.2 Linear Predictive Coding Features

Linear Predictive Coding (LPC) features are very important in many speech related applications. The idea behind LPC is that a speech sample can be approximated as a linear combination of the past samples. To achieve that, it is necessary to minimize the sum of the squared differences between the actual speech sample and the predicted coefficients. The result of that analysis is a set of unique predictor coefficients [30].

The LPC features are all-round applicable, for instance one could use them to reduce the space needed to store audio signals or to estimate speech parameters. Another reason why the LPC features are very popular is that they are quite fast to compute, e.g. with an algorithm proposed by Durbin and Levinson [30].

In this implementation 12 LPC features were extracted from each Hamming windowed frame (Framesize = 32 ms; offset = 10 ms; samplerate = 16 kHz) using the Durbin–Levinson algorithm.

3.3 Mel Frequency Cepstral Coefficients Features

MFCC features are short-term spectral-based features. They are very popular in speech recognition tasks, because they are able to represent the speech amplitude spectrum in a very compact form. The MFCC features are interesting for speech recognition as well as for emotion recognition, because they are related to the human auditory system. The values on the Mel scale are approximately arranged in the same way as pitch is perceived in the auditory system. A doubling of the perceived pitch results in a doubling of the value on the Mel scale. This is approximately linear for small values, for example, the Mel value for 100 Hz is around 100 Mel and for 200 Hz it is around 200 Mel. For bigger values this is not true anymore, e.g. 1.5 kHz equals around 1,100 Mel and 10 kHz is around 2,200 Mel.

The extraction of the MFCCs in this implementation follows the standard algorithms. Twelve MFCC features were extracted from each Hamming windowed frame (Framesize = 32 ms; offset = 10 ms; samplerate = 16 kHz) [30].

3.4 Pitch Related Features

Pitch period estimation (or equivalently, fundamental frequency estimation) is one of the most important issues in speech processing. Pitch carries important prosodic information in most languages and in tonal languages it also carries semantic information. Pitch detectors are used in many applications of speech such as vocoders, speaker identification and verification systems, speech recognition systems and emotion recognition from speech.

Unfortunately, defining pitch is not a straight forward exercise, since many different definitions have been proposed over the years [31]. The ANSI definition of psycho-acoustical terminology says "Pitch is that attribute of auditory sensation in terms of which sounds may be ordered on a scale extending from low to high. Pitch depends primarily on the frequency of the sound stimulus, but it also depends on the sound pressure and the waveform of the stimulus".

In most analysis a rather simple definition is adopted based on idea that many of the sounds in our environment have acoustic waveforms that repeat over time [30]. These sounds are often perceived as having a pitch that corresponds with the repetition rate of the sound. This means that pitch evoking stimuli are usually periodic, and the pitch is related to the periodicity of these stimuli. Therefore, a pitch detection mechanism must estimate the period or its inverse (the fundamental frequency) of the stimuli.

One pitch value is extracted from each frame of the signal using the Simple Inverse Filtering Tracking (SIFT) method proposed by Rabiner and Schafer [30]. For each frame of this series of pitch values statistical features are extracted, such as the maximum or minimum of the pitch. Furthermore, prosodic features such as derivatives or trends of the pitch are extracted as well.

4 Multi-Classifier System

The idea behind a multi-classifier system (MCS) is, that one single classifier may be capable of classifying one certain class, respectively emotion in this particular case, better than another classifier and the combination of both will lead to better classification results. A MCS aims therefore on more accurate classification results at "the expense of increased complexity" [32]. The usefulness however, is not fully agreed upon especially since Ho expressed his concerns in a very critical review article:

> Instead of looking for the best set of features and the best classifier, now we look for the best set of classifiers and then the best combination method. One can imagine that very soon we will be looking for the best set of combination methods and then the best way to use them all. If we do not take the chance to review the fundamental problems arising from this challenge, we are bound to be driven into

such an infinite recurrence, dragging along more and more complicated combination schemes and theories and gradually losing sight of the original problem [33].

Nonetheless, MCSs become more and more popular in pattern recognition.

In this particular recognition task a MCS with separately trained classifiers has been implemented. The classifiers are trained as seen in Fig. 4 with different feature streams and are then combined in the fusion box after the single classification tasks have been performed.

For a more detailed understanding of Fig. 4 it has to be mentioned, that the cascade of this figure is not the classification cascade, but the training respectively the initialization cascade. As mentioned before, the resulting prototypes of the K-Means may be used as initialization for the more accurate Learning Vector Quantization (LVQ1–LVQ3) classifier and the resulting LVQ prototypes after the training again may be used as initialization of the weights between the input and the hidden layer of the Radial Basis Function (RBF) network. In the architecture – as seen in Fig. 4 – only the decisions made by the different RBF networks are considered and fed into the fusion box at the end of the block diagram. The mentioned learning algorithms K-Means, LVQ and RBF networks are implemented in the automatic emotion recognizer according to the proposals by Kohonen [34] and Schwenker et al. [35], using the Gaussian density function as transfer function in the neurons of the hidden layer of the RBF network.

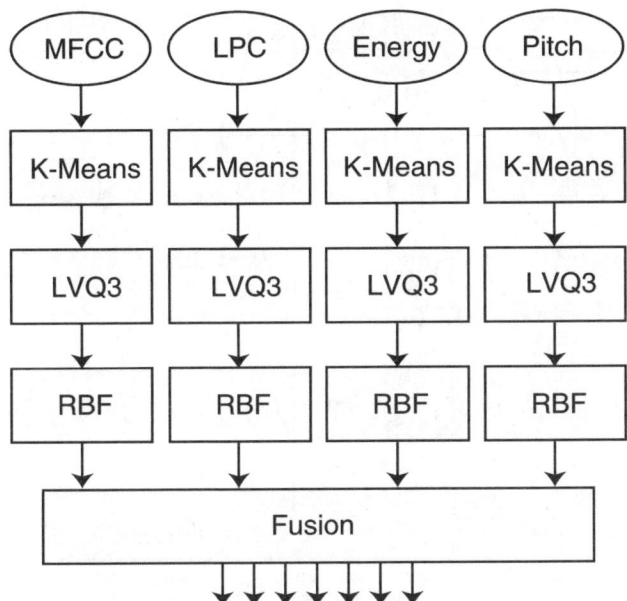

Fig. 4. Topology of the multi-classifier system

The fusion of the classifications is done by the Decision Templates (DT) fusion method [32]. The idea behind DT fusion is that Decision Profiles, calculated from the training material, are stored for each class of the given classification problem and compared to the outputs of the classifiers. The profile that is most similar to the DT receives the most support and is chosen to be the winner. The similarity measure is defined as

$$S(a, b) = \frac{\sum_{i=1}^{n} \min(a_i, b_i)}{\sum_{i=1}^{n} \max(a_i, b_i)} \tag{1}$$

with $a, b \in I\!\!R^n$.

5 RBF Ensemble

The RBF ensemble architecture is not different from a standard RBF architecture in this implementation. The topology is shown in Fig. 5. It is shown that the input and hidden layer are exactly the same as in a standard setup.

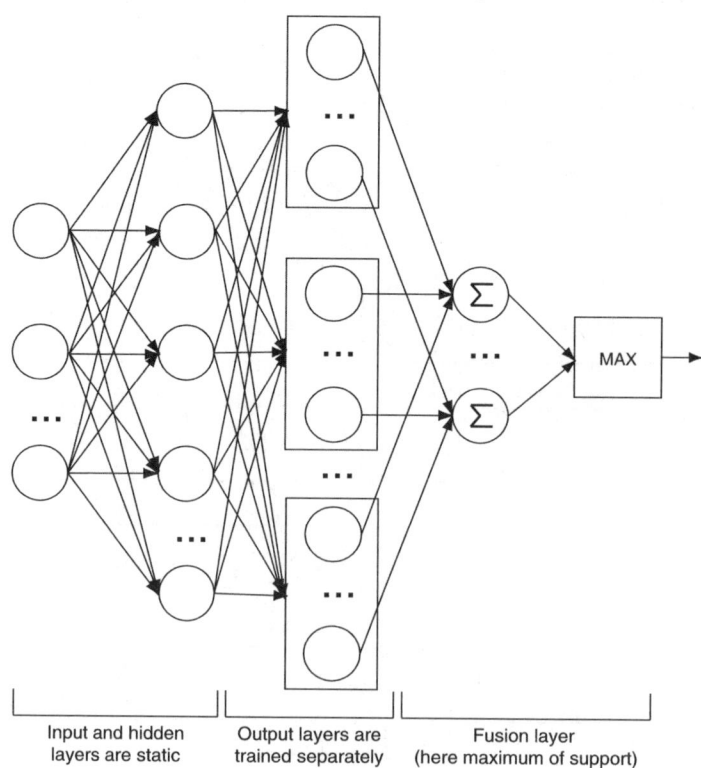

Input and hidden layers are static   Output layers are trained separately   Fusion layer (here maximum of support)

Fig. 5. RBF ensemble topology

The weights in this implementation between the input and the hidden layer are fixed by preceding K-Means and LVQ training as mentioned in Sect. 4. The σ values in the hidden layer of the Gaussian bell function are set to the mean distance toward the five nearest neighbors in the prototype space, which should help to cover a good part of the input space.

The more interesting part is the latter of the feed forward network. The weights w between the hidden and output layers are trained using the pseudo inverse learning rule

$$w = (H^t H)^{-1} H^t T, \tag{2}$$

where $H := H(X)$ is a matrix composed by the hidden layer outputs of all the chosen input vectors X and T is a matrix of the corresponding teacher signals. In this implementation the input vectors x are taken from the extracted MFCC vectors ($x \in \mathbb{R}^{12}$). Not all of the extracted vectors are taken to calculate one set of w values, but small subsets of 700 vectors are chosen randomly and 15 different sets of w values are calculated. For the classification the outputs of all output layers are summed up and the class with the maximum support is chosen. Choosing the maximum of the sums is not the only possible fusion method. It may be exchanged by DT or any other fusion method.

6 Evaluation

In this section, the experiments using the classification algorithms introduced in Sects. 4 and 5 are evaluated. The complete audio data was divided using a tenfold cross validation method into test and training sets. In a second test, the audio data was reduced by omitting two emotions (sadness and fear), that were considered to be less important for the targeted application. The following Sect. 6.1 summarizes the obtained results using the MCS introduced in Sect. 4. Furthermore, the results achieved using single feature streams and single classifiers are presented. Section 6.2 lists the confusion matrices and classification performances using the RBF ensemble.

6.1 Multi-Classifier System Results

Single Classifier Results

The results for the single classifiers, for example, K-Means or LVQ3, is shown for the PIT database in Table 2. Additionally, the classification rates for the decimated PIT (DPIT) database, i.e. omitting the emotions sadness and fear, are listed in Table 3.

The results in Tables 2 and 3 show that using energy based, LPC or Pitch features did not lead to promising results. Only MFCC features yielded better results using single classifier systems. The best result was achieved classifying MFCC features with RBF networks. These networks reached an average

Table 2. Classification results (in %) for the whole PIT database of simple classifiers according to the used features

	K-Means	LVQ3	RBF
Energy	17.3	17.0	12.0
LPC	17.0	16.0	21.0
Pitch	22.7	23.0	23.0
MFCC	31.0	39.7	51.0

Table 3. Classification results (in %) for the DPIT database of simple classifiers according to the used features

	K-Means	LVQ3	RBF
Energy	22.7	26.4	24.6
LPC	22.7	25.0	17.3
Pitch	25.1	26.4	28.6
MFCC	39.1	50.9	58.2

recognition rate of around 51% for the PIT database and 58% for the DPIT database. MFCC features possibly reached the best results, because they imitate the human auditory system, which achieved an average recognition rate of 65% in the human performance test.

Classifier Fusion: Decision Template

The MCS reached a performance of only 26.7% for the PIT database and 32.2% for the DPIT database. The factors that cause these results are shown in Tables 2 and 3, where only MFCC features reached classification rates above 50%. Since the discrepancy within the features is high, the RBF ensembles introduced in Sect. 5 are trained using MFCC features only.

6.2 RBF Ensemble Results

The confusion matrix for the PIT database, presented in Sect. 2.3, is shown in Table 4. The diagonal of the confusion matrix is clearly dominant. The averaged classification result is only about five percent worse than in the human performance test. The average result of the single RBF networks in Sect. 6.1 could be improved by around 8%. In Table 5 the results for the DPIT database are listed. In this case the diagonal of the matrix is even more dominant and the averaged classification result is at around 75%. Unfortunately, the human performance test has only been conducted using all of the seven available emotions, which leads to a lack of possible comparisons, while using only five emotions for the automatic emotion recognition experiments. The

Table 4. Confusion matrix of the automatic classification experiment using the whole dataset. Correct classifications: 59.7%

	Neutral	Anger	Sadness	Fear	Disgust	Boredom	Happiness
Neutral	20	1	2	5	14	12	0
Anger	1	35	0	0	4	1	4
Sadness	6	0	13	8	2	7	0
Fear	0	0	0	39	7	3	5
Disgust	1	2	1	4	25	2	1
Boredom	4	0	2	4	1	19	0
Happiness	0	11	1	1	4	0	28

Table 5. Confusion matrix of the experiment using the DPIT database (only five of the available seven emotions). Correct classifications: 75.0%

	Neutral	Anger	Disgust	Boredom	Happiness
Neutral	35	0	0	2	2
Anger	0	33	7	5	1
Disgust	0	7	29	1	8
Boredom	5	2	1	42	4
Happiness	6	0	0	4	26

result of the single RBF networks could be improved by around 17%, which looks impressive.

In the human performance test, neutral was chosen and confused the most, which indicates that it is selected if a test person is uncertain. In Table 4, it is observed that anger and happiness are confused as in the human performance test quite often. Since, anger and happiness are quite close to each other according to the activation evaluation space this is as expected. Furthermore, other common confusions such as neutral and disgust or sadness and boredom are also found in the results of the human performance test. These similarities in the confusion matrices are as expected, since the MFCC features imitate the human auditory system.

7 Conclusions

The objective of this chapter was to give an overview of emotion recognition, as well as a detailed description of an implementation of an automatic emotion recognizer. Strategies to record emotional speech have been introduced and problems according to the task of data acquisition and labeling of emotional data have been discussed.

In order to solve the task of automatic emotion recognition a RBF ensemble using MFCC features, was introduced and yielded the best results. This architecture is quite simple to implement and very fast to train. The fact that each of the members of the ensemble concerns only a small part of the training data leads to a computationally efficient, yet accurate classifier. The RBF ensemble could classify around 60% of the utterances correctly and therefore reached a performance, that is quite close to the human performance, which was at around 65% in a human emotion recognition experiment. The classification improved to 75% by leaving two emotions out, which were considered less important for the target application.

According to the fact that understanding emotions is important for "healthy" human to human communication, one may consider automatic emotion recognition as important for artificial systems that are able to communicate with humans as well. The topic of understanding emotions produced by humans may comprise the crucial factors for a broad acceptance of automatic systems that are capable of replacing human operators.

Acknowledgments

This work is supported by the competence center Perception and Interactive Technologies (PIT) in the scope of the project: "Der Computer als Dialogpartner: Perception and Interaction in Multi-User Environments". We would like to thank P.-M. Strauss for the collaboration in designing and recording the emotional data corpus.

References

1. E. Douglas-Cowie, R. Cowie, N. Tsapatsoulis, G. Votsis, S. Kollias, W. Fellenz, and J. Taylor "Emotion recognition in human-computer interaction," *IEEE Signal Processing Magazine*, vol. 18, no. 1, pp. 32–80, 2001.
2. C. Darwin, *The Expression of Emotions in Man and Animals*, Reprinted by Univiersity of Chicago Press, Chicago, 1965.
3. R. W. Reiber and D. K. Robinson, *Wilhelm Wundt in History: The Making of a Scientific Psychology*, Kluwer, Dordrecht, 2001.
4. R. Harre and R. Finlay-Jones, *Emotion talk across times*, pp. 220–233, Blackwell, Oxford, 1986.
5. H. Morsbach and W. J. Tyler, *A Japanese Emotion: Amae*, pp. 289–307, Blackwell, Oxford, 1986.
6. J. R. Averill, *Acquisition of Emotion in adulthood*, p. 100, Blackwell, Oxford, 1986.
7. V. Petrushin, "Emotion in speech: Recognition and application to call centers," in *Proceedings of Artificial Neural Networks Engineering*, November 1999, pp. 7–10.
8. S. Yacoub, S. Simske, X. Lin, and J. Burns, "Recognition of emotions in interactive voice response systems," in *Proceedings of Eurospeech*, 2003.
9. F. Dellaert, T. Polzin, and A. Waibel, "Recognizing emotion in speech," in *Proceedings of the ICSLP*, 1996, pp. 1970–1973.
10. C. Lee, S. Narayanan, and R. Pieraccini, "Classifying emotions in human machine spoken dialogs," in *Proceedings of International Conference on Multimedia and Expo (ICME)*, 2002, vol. 1, pp. 737–740.

11. F. Yu, E. Chang, X. Yingqing, and H.-Y. Shum, "Emotion detection from speech to enrich multimedia content," in *Proceedings of the Second IEEE Pacific Rim Conference on Multimedia*, London, UK, 2001, pp. 550–557, Springer.

12. C. M. Lee, S. Yildirim, M. Bulut, A. Kazemzadeh, C. Busso, Z. Deng, S. Lee, and S. Narayanan, "Emotion recognition based on phoneme classes," in *Proceedings of ICSLP 2004*, 2004.

13. K. R. Scherer, R. Banse, H. G. Wallbott, and T. Goldbeck, "Vocal cues in emotion encoding and decoding," *Motivation and Emotion*, vol. 15, no. 2, pp. 123–148, 1991.

14. E. Douglas-Cowie, R. Cowie, and C. Cox, "Beyond emotion archetypes: Databases for emotion modeling using neural networks," *Neural Networks*, vol. 18, no. 4, pp. 371–388, 2005.

15. A. Noam, A. Bat-Chen, and G. Ronit, "Perceiving prominence and emotion in speech – a cross lingual study," in *Proceeding of SP-2004*, 2004, pp. 375–378.

16. F. Burkhardt, A. Paeschke, M. Rolfes, W. F. Sendlmeier, and B. Weiss, "A database of german emotional speech," in *Proceedings of Interspeech*, 2005.

17. E. Douglas-Cowie, R. Cowie, and M. Schroeder, "A new emotion database: Considerations, sources and scope," in *Proceedings of the ISCA Workshop on Speech and Emotion*, 2000, pp. 39–44.

18. R. Cowie, "Describing the emotional states expressed in speech," in *Proceedings of the ISCA Workshop on Speech and Emotion*, 2000, pp. 11–18.

19. E. Douglas-Cowie, R. Cowie, and M. Schroeder, "The description of naturally occurring emotional speech," in *15th International Conference of Phonetic Sciences*, 2003, pp. 2877–2880.

20. I. H. Witten and E. Frank, *Data Mining: Practical machine learning tools and techniques*, 2nd edition, Morgan Kaufmann, San Francisco, 2005.

21. M. B. Arnold, *Emotion and Personality: Vol. 2 Physiological Aspects*, Columbia University Press, New York, 1960.

22. P.-M. Strauss, H. Hoffmann, W. Minker, H. Neumann, G. Palm, S. Scherer, F. Schwenker, H. Traue, W. Walter, and U. Weidenbacher, "Wizard-of-oz data collection for perception and interaction in multi-user environments," in *International Conference on Language Resources and Evaluation (LREC)*, 2006.

23. C. Stanislavski, *An Actor Prepares*, Routledge, New York, 1989.

24. S. T. Jovicic, Z. Kasic, M. Dordevic, and M. Rajkovic, "Serbian emotional speech database: design, processing and evaluation," in *Proceedings of SPECOM-2004*, 2004, pp. 77–81.

25. T. Seppnen, J. Toivanen, and E. Vyrynen, "Mediateam speech corpus: a first large finnish emotional speech database," in *Proceeding of 15th International Congress of Phonetic Sciences*, 2003, vol. 3, pp. 2469–2472.

26. N. Campbell, "The recording of emotional speech; jst/crest database research," in *Proceedings of International Conference on Language Resources and Evaluation (LREC)*, 2002, vol. 6, pp. 2026–2032.

27. P. Ekman and W. Friesen, *Unmasking the Face*, Prentice-Hall, Englewood Cliffs, 1975.

28. A. Nilsonne, "Speech characteristics as indicators of depressive illness," *Acta Psychiatrica Scandinavica*, vol. 77, pp. 253–263, 1988.

29. R. Cowie, A. Wichmann, E. Douglas-Cowie, P. Hartley, and C. Smith, "The prosodic correlates of expressive reading," in *14th International Congress of Phonetic Sciences*, 1999, pp. 2327–2330.

30. L. R. Rabiner and R. W. Schafer, *Digital Processing of Speech Signals*, Prentice-Hall Signal Processing Series, Englewood Cliffs, NJ, 1978.
31. C. J. Plack, A. J. Oxenham, R. R. Fay, and A. N. Popper, Eds., *Pitch – Neural Coding and Perception, Series: Springer Handbook of Auditory Research*, vol. 24, Springer, Berlin Heidelberg New York, 2005.
32. L. Kuncheva, *Combining Pattern Classifiers: Methods and Algorithms*, Wiley, New York, 2004.
33. T. K. Ho, *Multiple Classifier Combination: Lessons and Next Steps*, chapter 7, World Scientific, Singapore, 2002.
34. T. Kohonen, *Self-Organizing Maps*, Springer, Berlin Heidelberg New York, 1995.
35. F. Schwenker, H. A. Kestler, and G. Palm, "Three learning phases for radial basis function networks," *Neural Networks*, vol. 14, pp. 439–458, 2001.

Modeling Supra-Segmental Features
of Syllables Using Neural Networks

K. Sreenivasa Rao

Department of Electronics and Communication Engineering, Indian Institute of
Technology Guwahati, Guwahati 781039, Assam, India, ksrao@iitg.ernet.in

Summary. In this chapter we discuss modeling of supra-segmental features (intonation and duration) of syllables, and suggest some applications of these models.
These supra-segmental features are also termed as prosodic features, and hence the
corresponding models are known as prosody models. Neural networks are used to
capture the implicit duration and intonation knowledge in the sequence of syllables
of an utterance. A four layer feedforward neural network trained with backpropagation algorithm is used for modeling the duration and intonation knowledge of
syllables separately. Labeled broadcast news data in the languages Hindi, Telugu
and Tamil is used to develop neural network models in order to predict the duration
and F0 of syllables in these languages. The input to the neural network consists
of a feature vector representing the positional, contextual and phonological constraints. For improving the accuracy of prediction, further processing is done on the
predicted values. We also propose a two-stage duration model for improving the
accuracy of prediction. The performance of the prosody models is evaluated using
objective measures such as average prediction error, standard deviation and correlation coefficient. The prosody models are examined for applications such as speaker
recognition, language identification and text-to-speech synthesis.

1 Introduction

During production of speech, human beings seem to impose automatically
the required duration, intonation and coarticulation patterns on the sequence
of sound units. Since these patterns are associated with larger units (than
phonemes) such as syllables, words, phrases and sentences, these are often
considered as suprasegmental features also known as prosodic features. These
patterns reflect the linguistic and production constraints in the prosody (duration and intonation) and coarticulation in speech. The knowledge of these
constraints is implicit in the speech signal in the sense that it is difficult
to articulate the rules governing this knowledge. But lack of this knowledge
can easily be perceived while attempting to communicate the desired message through speech. Acoustic analysis and synthesis experiments have shown

K. Sreenivasa Rao: *Modeling Supra-Segmental Features of Syllables Using Neural Networks*,
Studies in Computational Intelligence (SCI) **83**, 71–95 (2008)
www.springerlink.com   © Springer-Verlag Berlin Heidelberg 2008

that duration and intonation patterns are the two most important prosodic features responsible for the quality of synthesized speech [1]. The prosody constraints are not only characteristic of the speech message and the language, but they also characterize a speaker uniquely. Even for speech recognition, human beings seem to rely on the prosody cues to disambiguate errors in the perceived sounds. Thus acquisition and incorporation of prosody knowledge becomes important for developing speech systems. In this chapter we discuss mainly the models to capture the duration and intonation knowledge, and demonstrate the use of prosody (intonation and duration) models for text-to-speech synthesis, speaker and language identification studies.

The implicit knowledge of prosody is usually captured using modeling techniques. In speech signal the duration and the average F_0 value (pitch) of each sound unit is dictated by the linguistic context of the unit and the production constraints. In this section, we review the methods for modeling duration and intonation (sequence of F_0 values) patterns.

1.1 Review of Duration Models

Modeling duration may be rule-based or statistical. Rule-based methods involve analysis of segment durations manually to determine the duration constraints on the sequence of sound units. The derived rules get better in terms of accuracy, as the amount of speech data used in the analysis is increased. But with large amount of data the process of deriving the rules manually becomes tedious and time consuming. Hence rule-based methods are limited to small amount of data. One of the early attempts for developing rule-based duration models was in 1970s [2]. More recently, a rule-based duration model was developed for a text-to-speech system for the Indian language Hindi [3]. The model was developed using the broadcast news speech data in Hindi. In general, the rule-based methods are difficult to study, due to complex interaction among the linguistic features at various levels [4]. Therefore the rule inference process is restricted to controlled experiments, where only a limited number of contextual factors are involved.

Statistical methods are attractive when large phonetically labeled databases are available. The statistical methods can be based on parametric or nonparametric regression models [5]. Examples of parametric regression models are sums-of-products (SOP) model, generalized linear model and multiplicative model [6]. In the SOP model the duration of a segment is represented as a sum of factors that affect the duration, and their interactions (product terms) [7]. The generalized linear model is a variant of the SOP model. The advantage of the SOP model is that the model can be developed using small amount of data. But in the SOP model the number of different SOP grows exponentially with the number of factors. Thus it is difficult to find an SOP model that best describes the data. In addition, the SOP model requires significant amount of preprocessing the data to correct the interaction among factors and data imbalance [8, 9].

Nonlinear regression models are either classification and regression tree (CART) models or neural network models [6,10]. In the CART model a binary branching tree is constructed by feeding the attributes of the feature vectors from top node and passing through the arcs representing the constraints [10]. The feature vector of a segment represents the positional, contextual and phonological information. The segment duration is predicted by passing the feature vector through the tree so as to minimize the variance at each terminal node. The tree construction algorithm usually guarantees that the tree fits the training data well. But there is no guarantee that the new and unseen data will be predicted properly. The prediction performance of the CART model depends on the coverage of the training data.

Neural network models are known for their ability to capture the functional relation between input–output pattern pairs [11,12]. Several models based on neural network principles are described in the literature for predicting the durations of syllables in continuous speech [13–21]. Campbell used a feedforward neural network (FFNN) trained with feature vectors, each representing six features of a syllable [14]. The six features are: number of phonemes in a syllable, the nature of syllable, position in the tone group, type of foot, stress, and word class. Barbosa and Bailly used a neural network model to capture the perception of rhythm in speech [16]. The model predicts the duration of a unit, known as inter-perceptual center group (IPCG). The IPCG is delimited by the onset of a nuclear vowel and the onset of the following vowel. The model is trained with the following information: boundary marker at phrase level, sentence mode, accent marker, number of consonants in IPCG, number of consonants in coda and nature of the vowel.

1.2 Review of Intonation Models

The temporal changes of the fundamental frequency (F_0) value, i.e., intonation, depends on the information contained at various levels: (a) Segmental coarticulation at the phonemic level. (b) Emphasis of the words and phrases in a sentence. (c) Syntax, semantics, prominence and presence of new information in an utterance [22]. In addition to these, variations in fundamental frequency is also influenced by age, gender, mood and the emotional state of the speaker.

Intonation models can be grouped into rule-based methods and data-based methods. Rule-based methods model the intonation patterns using a set of phonological rules [23–26]. These rules are inferred by observing the intonation patterns for a large set of utterances with the help of linguists and phoneticians. The relationship between the linguistic features of input text and the intonation (F_0 contour) pattern of an utterance is examined to derive the rules. Although this is done by induction, it is generally difficult to analyze the effect of mutual interaction of linguistic features at different levels. Hence, the inferred phonological rules for intonation modeling are always imprecise and incomplete.

Some of the prominent intonation models reported in the literature are: Tone sequence model, Fujisaki model, Institute of Perception Research (IPO) model and tilt model [25, 27–30]. In the tone sequence model, the F_0 contour is generated from a sequence of phonologically distinctive tones, which are determined locally, and they do not interact with each other. Tone sequence models do not properly represent actual pitch variations. No distinction is made on the differences in tempo or acceleration of pitch moments. The Fujisaki model is hierarchically organized, and includes several components of different temporal scopes. In this model, the F_0 contour is generated by a slowly varying phrase component and a fast varying accent component. The main drawback of this model is that, it is difficult to deal with low accents and slowly raising section of the contour. In the IPO model, intonation analysis is performed in three steps: (1) Perceptually relevant movements in F_0 contour are stylized by straight lines (known as *copy contour*). (2) Features of the copy contours are expressed in terms of duration and range of F_0 moments. (3) A grammar of possible and permissible combination of F_0 movements is derived. The major drawback observed in the IPO model is that the contours derived from the straight line approximations are quite different from the original. This is due to the difficulty in modeling curves with straight lines. The tilt model provides a robust analysis and synthesis of intonation contours. In the tilt model, intonation patterns are represented by a sequence of events. The event may be an accent, a boundary, silence or a connection between events. The events are described by the following parameters: Starting F_0, rise amplitude, fall amplitude, duration, peak position and tilt. The tilt model gives a good representation of natural F_0 contour and a better description of the type of accent. But the model does not provide the linguistic interpretation for different pitch accents.

Data-based methods generally depend on the quality and quantity of available data. Among several data-based methods for modeling intonation, CART models and neural network models are more popular [20, 31–38]. Scordis and Gowdy used neural networks in a parallel and distributed manner to predict the average F_0 value for each phoneme and the temporal variations of F_0 within a phoneme [35]. The network consists of two levels: Macroscopic level and microscopic level. At the macroscopic level, a FFNN is used to predict the average F_0 value for each phoneme. The input to the FFNN consists of the set of phonemic symbols, which represents the contextual information. At the microphonemic level, a recurrent neural network (RNN) is used to predict the temporal variations of F_0 within a phoneme. Marti Vainio and Toomas Altosaar used a three layer FFNN to predict F_0 values for a sequence of phonemes in Finnish language [33]. The features used for developing the models are: Phoneme class, length of a phoneme, identity of a phoneme, identities of previous and following syllables (context), length of a word and position of a word. Buhmann et al. used a RNN for developing multi-lingual intonation models [37]. The features used in this work are the universal (language independent) linguistic features such as part-of-speech and type of punctuation, along with prosodic features such as word boundary, prominence of the word and duration of the phoneme.

In this work, we use a four layer FFNN to predict the durations and F_0 contour for a sequence of syllables separately. The linguistic and production constraints associated with a sequence of syllables are represented with positional, contextual and phonological features. These features are used to train the neural network to capture the implicit intonation and duration knowledge. Details of these features are discussed in Sect. 4.

There are two objectives in this study. The first one is to determine whether neural network models can capture the implicit knowledge of the prosody patterns of syllables in a language. One way to infer this is to examine the error for the training data. If the error is reducing for successive training cycles, then one can infer that the network is indeed capture the implicit relations in the input–output pairs. We propose to examine the ability of the neural network models to capture the prosody knowledge for speech spoken by various speakers in different Indian languages. We consider three Indian languages (Hindi, Telugu and Tamil) using syllable as the basic sound unit. The second objective is to demonstrate the use of prosody models for the applications such as, text-to-speech synthesis, speaker recognition and language identification.

The prediction performance of the neural network model depends on the nature of training data used. Distributions of the durations of syllables (Fig. 1) indicate that majority of the durations are concentrated around mean of the distribution. The distribution of the syllable durations for the three languages used in this work is shown in Fig. 1. This kind of training data force the model to bias towards mean of the distribution. To avoid this problem, some postprocessing and preprocessing methods are proposed. Postprocessing methods modify the predicted values further using some durational constraints. Preprocessing methods involve use of multiple models, one for each limited range of duration. This requires a two-stage duration model, where first stage is used to segregate the input into groups according to number of models, and second stage is used for prediction.

The chapter is organized as follows: Sect. 2 discusses the factors that affect the duration and average F_0 of the syllables. The database used for the proposed prosody models is described in Sect. 6.1. Section 4 discusses the features used as input to the neural network for capturing the knowledge of the prosody for the sequence of syllables. Section 5 gives the details of the neural network model, and discusses its performance in predicting the durations and F_0 values of syllables. A two-stage duration model is proposed in Sect. 6 to reduce the error in the predicted durations of the syllables. In Sect. 7 we discuss three applications namely, speaker recognition, language identification and text-to-speech synthesis, using prosody models. A summary of this work is given in Sect. 8 of the chapter along with a discussion on some issues that need to be addressed.

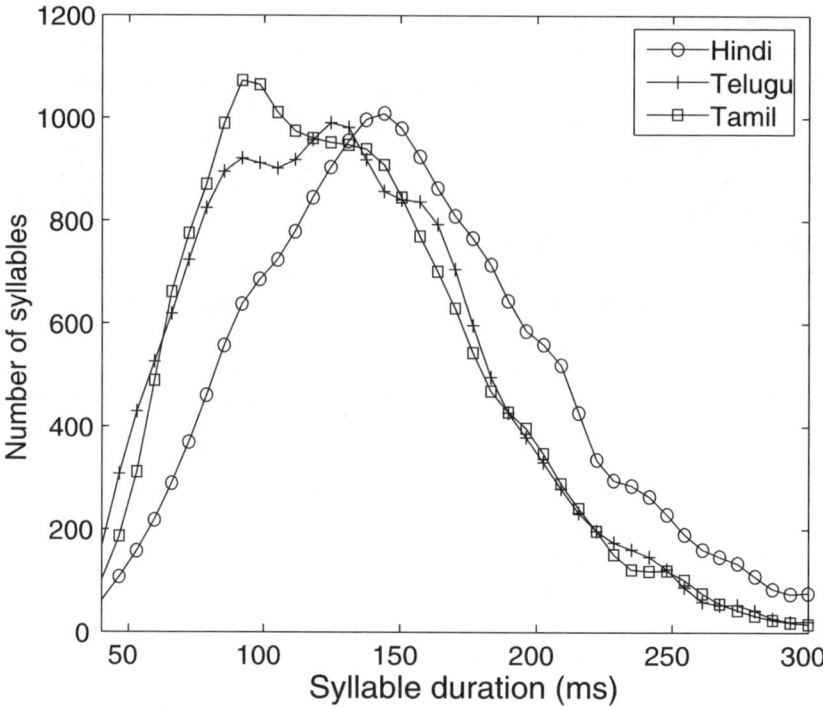

Fig. 1. Distributions of the durations of the syllables in the languages Hindi, Telugu and Tamil

2 Factors Affecting the Duration and Intonation Patterns

The factors affecting the durations and F_0 values of the basic sound units in continuous speech can be broadly categorized into phonological, positional and contextual [1]. The vowel is considered as the nucleus of a syllable, and consonants may be present on either side of the vowel. The duration and the average F_0 value of a syllable is influenced by the position of the vowel, the category of the vowel and the type of the consonants associated with the vowel. The positions that affect the durations and F_0 values of the syllables are: Word initial position, word final position, phrase boundary and sentence ending position. The contextual factors include the preceding and following syllables. The manner and place of articulation of the syllables in the preceding and following positions also affect the suprasegmental features of the present syllable. In addition, the gender of the speaker, psychological state of the speaker (happy, anger, fear, etc.), age, relative novelty in the words and words with relatively large number of syllables, also affect the duration and mean F_0 value.

3 Speech Database

The database for this study consists of 19 Hindi, 20 Telugu and 33 Tamil broadcast news bulletins [39]. In each language these news bulletins were read by male and female speakers. Total durations of speech in Hindi, Telugu and Tamil are 3.5, 4.5 and 5 h, respectively. The speech signal was sampled at 16 kHz and represented as 16 bit numbers. The speech utterances are manually transcribed into text using common transliteration code (ITRANS) for Indian languages [40]. The speech utterances are segmented and labeled manually into syllable-like units. Each bulletin is organized in the form of syllables, words and orthographic text representations of the utterances. Each syllable and word file contains the text transcriptions and timing information in number of samples. The fundamental frequencies of the syllables are computed using the autocorrelation of the Hilbert envelope of the linear prediction residual [41]. The average pitch (F_0) for male speakers and female speakers in the database was found to be 129 and 231 Hz, respectively.

4 Features for Developing Prosody Models

In this study we use 25 features (which form a feature vector) for representing the linguistic context and production constraints of each syllable. These features represent positional, contextual and phonological information of each syllable. Features representing the positional information are further classified based on the position of a word in the phrase and the position of the syllable in a word and phrase.

Syllable position in the phrase. Phrase is delimited by orthographic punctuation. The syllable position in a phrase is characterized by three features. The first one represents the distance of the syllable from the starting position of the phrase. It is measured in number of syllables, i.e., the number of syllables ahead of the present syllable in the phrase. The second feature indicates the distance of the syllable from the terminating position of the phrase. The third feature represents the total number of syllables in the phrase.

Syllable position in the word. In Indian languages words are identified by spacing between them. The syllable position in a word is characterized by three features similar to the phrase. The first two features are the positions of the syllable with respect to the word boundaries. The third feature is the number of syllables in a word.

Position of the word. The F_0 and duration of a syllable may depend on the position of the word in an utterance. Therefore the word position is used for developing duration and intonation models. The word position in an utterance is represented by three features. They are the positions of the word with respect to the phrase boundaries, and the number of words in the phrase.

Syllable identity. A syllable is a combination of segments of consonants (C) and vowels (V). In this study, syllables with more than four segments (Cs or Vs) are ignored, since the number of such syllables present in the database is very less (less than 1%). Each segment of a syllable is encoded separately, so that each syllable identity is represented by four features.

Context of the syllable. The duration and mean F_0 of a syllable may be influenced by its adjacent syllables. Hence for modeling the duration and F_0 of a syllable, the contextual information is represented by the previous and following syllables. Each of these syllables is represented by a four-dimensional feature vector, representing the identity of the syllable.

Syllable nucleus. Another important feature is the vowel position in a syllable, and the number of segments before and after the vowel in a syllable. This feature is represented with a three-dimensional feature vector specifying the consonant–vowel structure present in the syllable.

Pitch. The F_0 value of a syllable may be influenced by the pitch value of the preceding syllable. Therefore this information is used in the feature vector of the syllable for modeling the F_0 of a syllable.

Gender identity. The database contains speech from both male and female speakers. This gender information is represented by a single feature.

The list of features and the number of input nodes in a neural network needed to represent the features are given in Table 1. Among these, features in the rows 1–6 are (24 features) common in both duration and intonation models. The 25th feature in the duration model is the gender of the speaker, and in the intonation model it is F_0 of the previous syllable.

Table 1. List of factors affecting the duration and F_0 of a syllable, features representing the factors and the number of nodes needed for neural network to represent the features

Factors	Features	# Nodes
Syllable position in the phrase	Position of syllable from beginning of the phrase Position of syllable from end of the phrase Number of syllables in the phrase	3
Syllable position in the word	Position of syllable from beginning of the word Position of syllable from end of the word Number of syllables in the word	3
Word position in the phrase	Position of word from beginning of the phrase Position of word from end of the phrase Number of words in the phrase	3
Syllable identity	Segments of the syllable (consonants and vowels)	4
Context of the syllable	Previous syllable Following syllable	4 4
Syllable nucleus	Position of the nucleus Number of segments before the nucleus Number of segments after the nucleus	3
Pitch	F_0 of the previous syllable	1
Gender identity	Gender of the speaker	1

5 Prosody Modeling with Feedforward Neural Networks

A four layer FFNN is used for modeling the duration and intonation patterns of syllables. The general structure of the FFNN is shown in Fig. 2. The first layer is the input layer with linear units. The second and third layers are hidden layers. The second layer (first hidden layer) of the network has more units than the input layer, and it can be interpreted as capturing some local features in the input space. The third layer (second hidden layer) has fewer units than the first layer, and can be interpreted as capturing some global features [11,12]. The fourth layer is the output layer having one unit representing the F_0 (for modeling intonation) or duration (for modeling duration) of a syllable. The activation function for the units at the input layer is linear, and for the units at the hidden layers, it is nonlinear. Generalization by the network is influenced by three factors: The size of the training set, the architecture of the neural network, and the complexity of the problem. We have no control over the first and last factors. Several network structures are explored in this study. The (empirically arrived) final structure of the network is $25L\ 50N\ 12N$ $1N$, where L denotes a linear unit, and N denotes a nonlinear unit. The integer value indicates the number of units used in that layer. The nonlinear units use $tanh(s)$ as the activation function, where s is the activation value of that unit. All the input and output features are normalized to the range $[-1, +1]$ before presenting to the neural network. The backpropagation learning algorithm is used for adjusting the weights of the network to minimize the mean squared

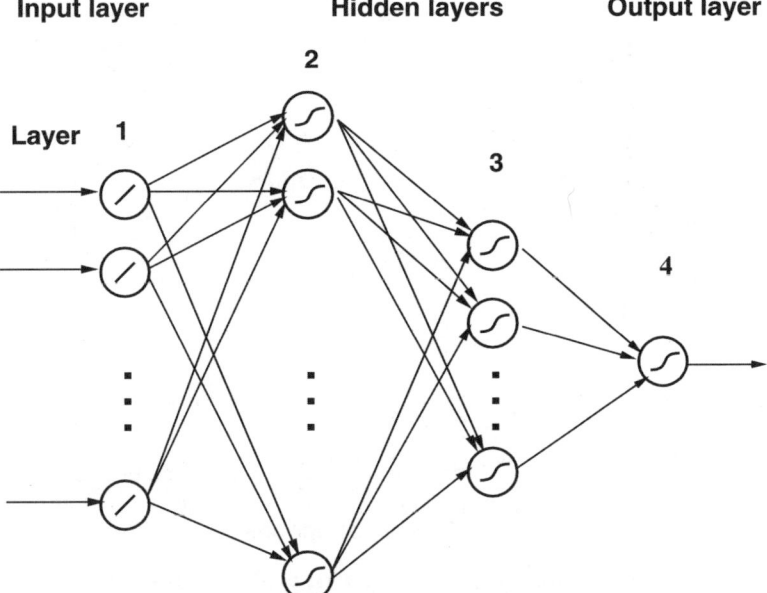

Fig. 2. Four layer feedforward neural network

error at the output for each duration or F_0 value of a syllable [12]. In this section, first we discuss the modeling of the durations of syllables, and then the modeling of intonation will be briefly discussed.

A separate model is developed for each of the three languages. For Hindi 35,000 syllables are used for training the network, and 11,222 syllables are used for testing. For Telugu 64,000 syllables are used for training, and 17,630 are used for testing. For Tamil 75,000 syllables are used for training, and 21,493 are used for testing. For each syllable a 25 dimension input vector is formed, representing the positional, contextual and phonological features. The duration of each syllable is obtained from the timing information available in the database. Syllable durations seem to follow a logarithmic distribution, and hence the logarithm of the duration is used as the target value [14]. The number of epochs needed for training depends on the behaviour of the training error. It was found that 500 epochs are adequate for this study. The learning ability of the network from training data can be observed from training error. Training errors for neural network models for each of the three languages are shown in Fig. 3. The decreasing trend in the training error indicate that the network is capturing the implicit relation between the input and output.

The duration model is evaluated using syllables in the test set. For each syllable in the test set, the duration is predicted using the FFNN by giving the feature vector of each syllable as input to the neural network. The deviation of the predicted duration from the actual duration is obtained. The prediction performance of the models for the three languages is shown in Fig. 4. Each plot

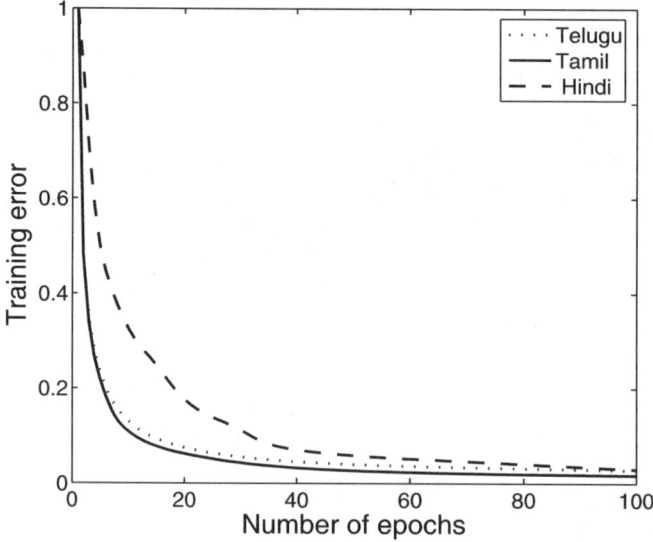

Fig. 3. Training errors for the neural network models in three Indian languages (Hindi, Telugu and Tamil)

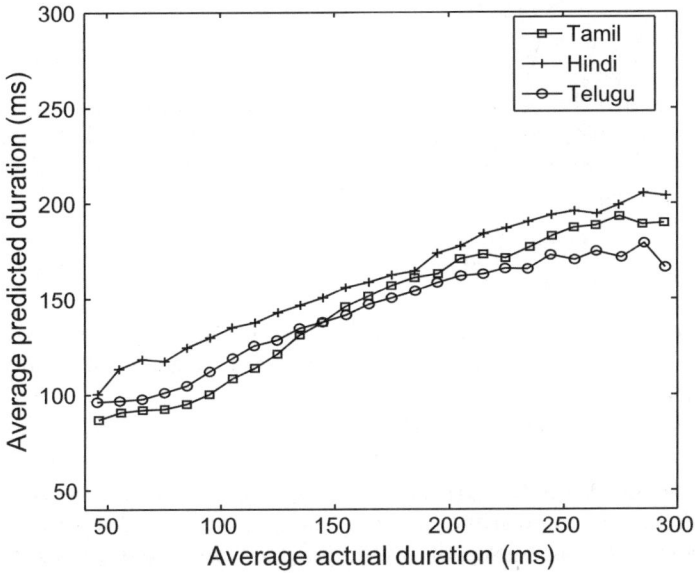

Fig. 4. Prediction performance of the neural network models for the languages Hindi, Telugu and Tamil

Table 2. Performance of the FFNN models for predicting the durations of syllables for different languages

Language	% Predicted syllables within deviation			Objective measures		
	10%	25%	50%	μ (ms)	σ (ms)	γ
Hindi	29	68	84	32	26	0.75
Telugu	29	66	86	29	23	0.78
Tamil	34	75	96	26	22	0.82

represents the average predicted duration vs. the average duration of a syllable. The percentages of syllables predicted within different deviations from their actual durations is given in Table 2. In order to evaluate the prediction accuracy, the average prediction error (μ), the standard deviation (σ), and the correlation coefficient ($\gamma_{X,Y}$) are computed using actual and predicted duration values. These results are given in Table 2.

The first column indicates different languages used in the analysis. Columns 2–4 indicate the percentage of syllables predicted within different deviations from their actual durations. Columns 5–7 indicate the objective measures (μ, σ and $\gamma_{X,Y}$). The definitions of average prediction error (μ), standard deviation (σ) and linear correlation coefficient ($\gamma_{X,Y}$) are given below:

$$\mu = \frac{\sum_i |x_i - y_i|}{N},$$

$$\sigma = \sqrt{\frac{\sum_i d_i^2}{N}}, \quad d_i = e_i - \mu, \quad e_i = x_i - y_i,$$

where x_i, y_i are the actual and predicted durations, respectively, and e_i is the error between the actual and predicted durations. The deviation in error is d_i, and N is the number of observed syllable durations. The correlation coefficient is given by

$$\gamma_{X,Y} = \frac{V_{X,Y}}{\sigma_X \cdot \sigma_Y}, \quad \text{where} \quad V_{X,Y} = \frac{\sum_i |(x_i - \bar{x})| \cdot |(y_i - \bar{y})|}{N}.$$

The quantities σ_X, σ_Y are the standard deviations for the actual and predicted durations, respectively, and $V_{X,Y}$ is the correlation between the actual and predicted durations.

The accuracy of prediction (Fig. 4) of the duration models is not uniform for the entire duration range. Better prediction is observed around the mean of the distribution of the original training data. Syllables with long durations tend to be underestimated, and short durations to be overestimated. For improving the accuracy of prediction, the predicted values are further modified by imposing piecewise linear transformation [42–44]. The prediction performance is observed to be improved slightly by using the transformation.

For modeling the intonation patterns, a separate model is developed for each speaker in the database. The extracted input vectors representing positional, contextual and phonological features are presented as input, and the corresponding F_0 values of the syllables are presented as desired outputs to the FFNN model. The network is trained for 500 epochs. The prediction performance of the intonation models is illustrated in Table 3 for one male (M) and one female (F) speaker for each language.

Figure 5 shows the predicted and actual (original) pitch contours for the utterance *"pradhAn mantri ne kahA ki niyantran rekhA se lekar"* in Hindi spoken by a male speaker. It shows that the predicted pitch contour is close to the original contour. This indicates that the neural network predicts the F_0 values reasonably well for the sequence of syllables in the given text.

Table 3. Performance of the FFNN models for predicting the F_0 values for different languages

Language	Gender	% Predicted syllables within deviation			Objective measures		
	# syllables	10%	15%	25%	μ(Hz)	σ(Hz)	γ
Hindi	F(660)	67	82	96	20	18	0.78
	M(1143)	74	92	98	12	9	0.79
Telugu	F(1276)	72	91	99	16	13	0.78
	M(984)	64	82	96	10	9	0.79
Tamil	F(741)	77	91	99	18	14	0.85
	M(1267)	77	90	97	13	12	0.80

The results for female (F) and male (M) speakers are given separately.

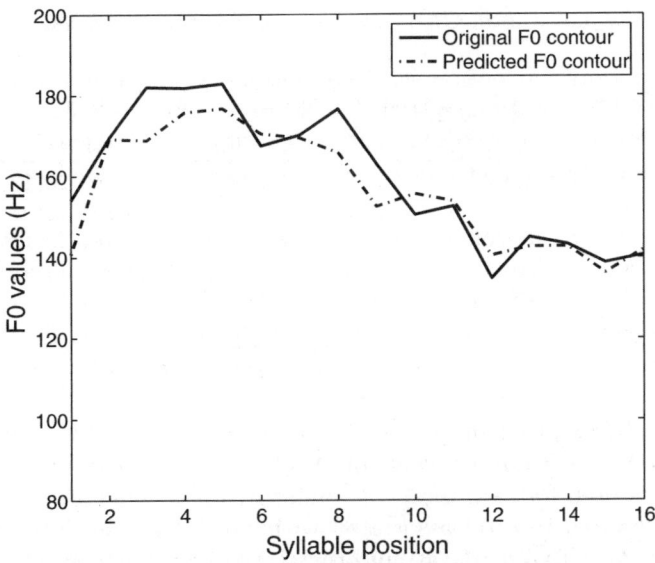

Fig. 5. Comparison of predicted F_0 contour with original contour for the utterance "*pradhAn mantri ne kahA ki niyantran rekhA se lekar*" in Hindi spoken by a male speaker

Table 4. Performance of the CART models for predicting the durations of syllables for different languages

Language	% Predicted syllables within deviation			Objective measures		
	10%	25%	50%	μ(ms)	σ(ms)	γ
Hindi	31(29)	67(68)	92(84)	32(32)	26(26)	.76(.75)
Telugu	30(29)	64(66)	88(86)	29(29)	24(23)	.78(.78)
Tamil	33(34)	71(75)	93(96)	25(26)	21(22)	.81(.82)

The numbers within brackets indicate the performance of FFNN models.

The prediction performance of the neural network models can be compared with the results obtained by CART models. The performance of the CART models for predicting the duration and intonation patterns is given in Tables 4 and 5, respectively. The performance of FFNN models (shown within brackets in Tables 4 and 5) is comparable to CART models.

6 Duration Modeling Using Two-Stage Approach

Since, a single FFNN model is used for predicting the durations of syllables for the entire range of 40–300 ms, the accuracy of prediction (Fig. 4) is biased towards the distribution of the training data. This leads to poor prediction for long and short duration syllables which lie at the tail portions of the

Table 5. Performance of the CART models for predicting the F_0 values for different languages

Language	Gender	% Predicted syllables with deviation			Objective measures		
		10%	15%	25%	μ(Hz)	σ(Hz)	γ
Hindi	F	58(67)	73(82)	90(96)	21(20)	20(18)	0.76(0.78)
	M	68(74)	90(92)	98(98)	14(12)	11(9)	0.77(0.79)
Telugu	F	68(72)	87(91)	99(99)	18(16)	13(13)	0.74(0.78)
	M	64(64)	83(82)	96(96)	11(10)	11(9)	0.74(0.79)
Tamil	F	69(77)	90(91)	99(99)	21(18)	15(14)	0.80(0.85)
	M	72(77)	88(90)	95(97)	15(13)	13(12)	0.77(0.80)

The numbers within brackets indicate the performance of FFNN models.

distributions. This problem can be alleviated to some extent by using multiple models, one for each limited range of duration [36, 45]. This ensures that the training data used for each model is uniformly distributed and well balanced within the limited interval associated with the model. Here the number of models (number of intervals) is not crucial. Optimum number of models and range of each interval can be arrived at experimentally. But this requires a preprocessing of data, which categorizes syllables into different groups based on duration.

For implementing this concept, a two-stage duration model is proposed. The first stage consists of a syllable classifier which groups the syllables based on their duration. The second stage is a function approximator for modeling the syllable duration, which consists of specific models for the given duration interval. The block diagram of the proposed two-stage duration model is shown in Fig. 6. The performance of the proposed model depends on the performance of the syllable classifier (1st stage), since the error at the 1st stage will route the syllable features to unintended model for prediction. The duration intervals arrived at empirically for the language Hindi are 40–120, 120–170 and 170–300 ms, whereas for the languages Telugu and Tamil, the intervals are 40–100, 100–150 and 150–300 ms.

Support vector machine (SVM) models are used for syllable classification [11,46]. The block diagram of the syllable classifier using SVM models is shown in stage 1 of Fig. 6. The SVMs are designed for two-class pattern classification. Multi-class (n-class) pattern classification problems can be solved using a combination of binary (2-class) SVMs. One-against-the-rest approach is used for decomposition of the n-class pattern classification problem into several (n) two-class classification problems. The classification system consists of three SVMs. The set of training examples $\{\{(\mathbf{x}_i, k)\}_{i=1}^{N_k}\}_{k=1}^{n}$ consists of N_k number of examples belonging to kth class, where the class label $k \in \{1, 2, \ldots, n\}$. All the training examples are used to construct the SVM for a class. The SVM for class k is constructed using the set of training examples and their desired outputs, $\{\{(\mathbf{x}_i, y_i)\}_{i=1}^{N_k}\}_{k=1}^{n}$. The desired output y_i for a training example $\mathbf{x}_i$ is defined as follows:

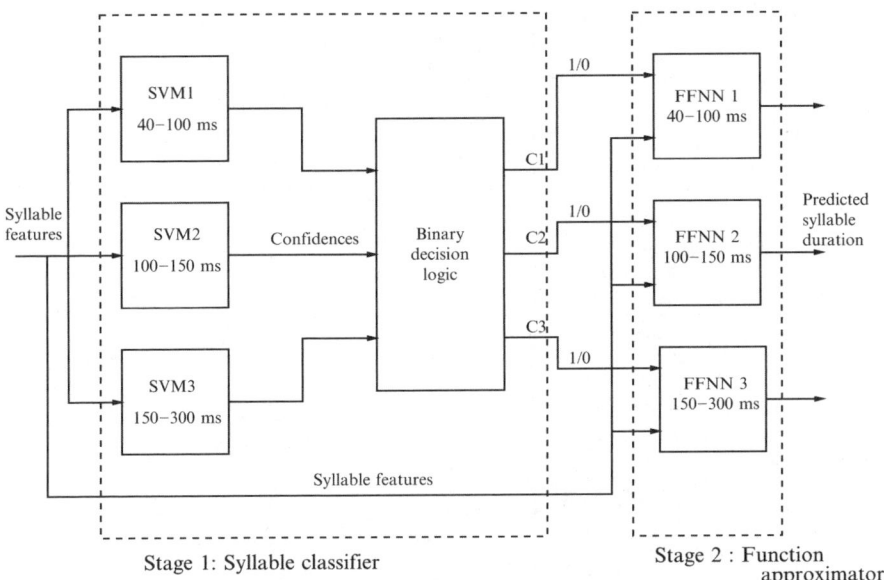

Stage 1: Syllable classifier

Stage 2 : Function
approximator

Fig. 6. Two-stage duration model

$$y_i = \begin{cases} +1 & : & if \quad \mathbf{x}_i \in k\text{th } class \\ -1 & : & otherwise. \end{cases}$$

The examples with $y_i = +1$ are called positive examples, and those with $y_i = -1$ are called negative examples. An optimal hyperplane is constructed to separate positive examples from negative examples. The separating hyperplane (margin) is chosen in such a way as to maximize its distance from the closest training examples of different classes [11, 46]. Figure 7 illustrates the geometric construction of a hyperplane for two dimensional input space. The support vectors are those data points that lie closest to the decision surface, and therefore the most difficult to classify. They have a direct bearing on the optimum location of the decision surface. For a given test pattern $\mathbf{x}$, the evidence $D_k(\mathbf{x})$ is obtained from each of the SVMs. In the decision logic, the class label k associated with the SVM which gives maximum evidence, is hypothesized as the class (C) of the test pattern. That is,

$$C(\mathbf{x}) = \arg\max_k \quad (D_k(\mathbf{x})).$$

The decision logic provides three outputs (corresponding to the number of duration intervals), which are connected to the corresponding models in the following stage (function approximator). For each syllable the decision logic activates only the duration model corresponding to the maximum confidence. The selected duration model now predicts the duration of the syllable in that

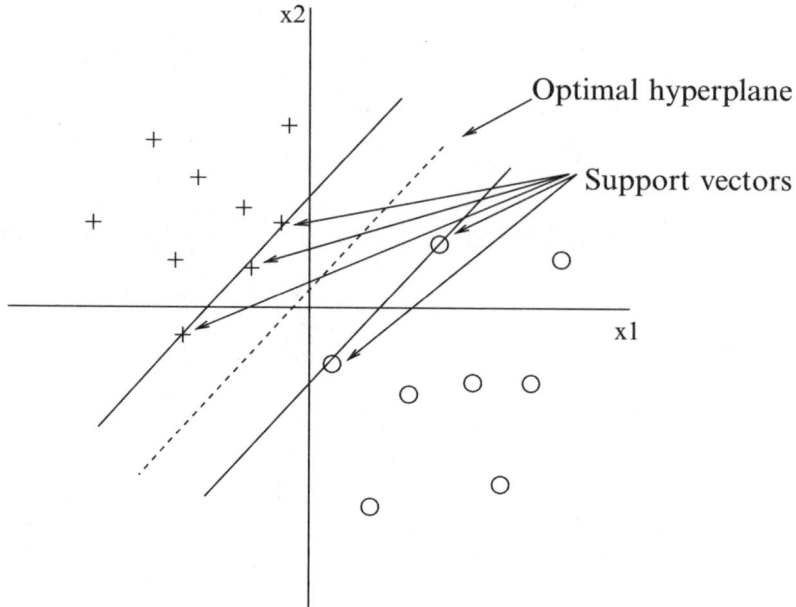

Fig. 7. Illustration of the idea of support vectors and an optimum hyper plane for linearly separable patterns

Table 6. Classification performance using SVM models for the languages Hindi, Telugu and Tamil

Language	% Syllables correctly classified
Hindi	81.92
Telugu	80.17
Tamil	83.26

limited interval. The performance of classification using SVM models is shown in Table 6.

For modeling the syllable duration, syllable features are presented to all the models of the syllable classifier. The decision logic in the syllable classifier routes the syllable features to one of the duration models present in the second stage (function approximator) for predicting the duration of the syllable in a specific limited duration interval. The performance of the proposed two-stage duration model is given in Table 7. The numbers within brackets indicate the performance of the single FFNN model.

Results show that the prediction accuracy is improved with two-stage model compared to single model for the entire duration range. For comparing the performance of two-stage model with single FFNN model, Tamil broadcast news data is chosen. The prediction performance of single FFNN model and two-stage model are shown in Fig. 8. Performance curves in the figure

Table 7. Performance of the two-stage model

Language	% Predicted syllables within deviation			Objective measures		
	10%	25%	50%	μ(ms)	σ(ms)	γ
Hindi	36(29)	79(68)	96(84)	25(32)	20(26)	0.82(0.75)
Telugu	38(29)	81(66)	95(86)	23(29)	23(23)	0.82(0.78)
Tamil	44(34)	85(75)	97(96)	20(26)	20(22)	0.85(0.82)

The numbers within brackets indicate the performance of the single FFNN model.

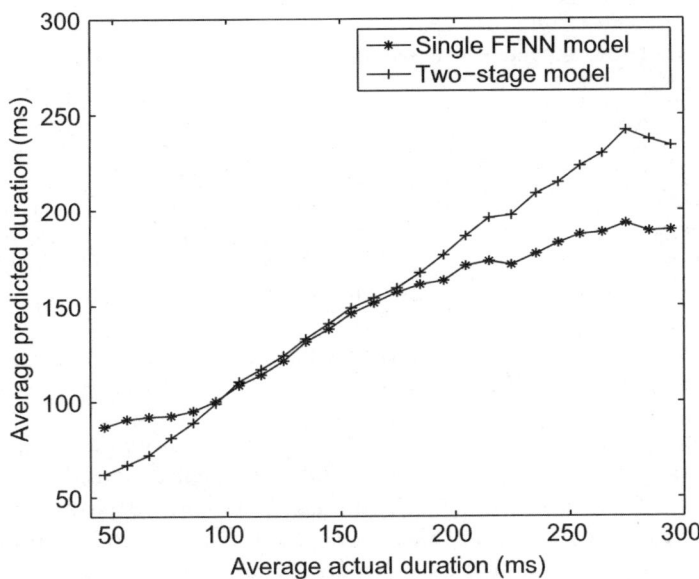

Fig. 8. Prediction performance of single FFNN model and two-stage model

show that the syllables having duration around the mean of the distribution are estimated better in both the models, whereas the short and long duration syllables are poorly predicted in the case of single FFNN model. The prediction accuracy of these extreme (long and short) syllables is improved in the two-stage model, because of specific models for each of the duration interval.

7 Applications of the Prosody Models

The prosody models (duration models and intonation models) can be used for different applications. In particular, prosody models are exploited for recognizing some specific categories of speech sounds in speech recognition [47]. Prosody models can be used in other applications such as speaker and language identification, and text-to-speech synthesis. In this section we discuss speaker recognition, language identification and text-to-speech synthesis applications using prosody models.

Table 8. Speaker discrimination using intonation models

	% Syllables within 15% deviation				Mean error (Hz)				Std. dev. (Hz)				Corr. coeff.			
	S_1	S_2	S_3	S_4	S_1	S_2	S_3	S_4	S_1	S_2	S_3	S_4	S_1	S_2	S_3	S_4
S_1	92	65	83	81	16	28	24	21	14	19	18	19	0.79	0.73	0.74	0.75
S_2	75	92	66	69	27	18	31	26	23	14	24	23	0.77	0.83	0.73	0.74
S_3	68	71	85	76	25	26	19	24	20	27	15	19	0.72	0.75	0.82	0.76
S_4	72	80	79	88	25	23	24	18	21	19	18	15	0.76	0.77	0.78	0.82

Table 9. Speaker discrimination using duration models

	% Syllables within 25% deviation				Mean error (ms)				Std. dev. (ms)				Corr. coeff.			
	S_1	S_2	S_3	S_4	S_1	S_2	S_3	S_4	S_1	S_2	S_3	S_4	S_1	S_2	S_3	S_4
S_1	58	51	53	55	33	38	37	36	27	30	29	29	0.76	0.69	0.69	0.71
S_2	54	56	52	51	35	32	36	37	29	24	26	27	0.71	0.73	0.70	0.69
S_3	53	52	60	52	36	35	30	34	30	26	23	26	0.72	0.71	0.75	0.71
S_4	53	54	56	59	36	35	34	30	28	27	27	24	0.73	0.70	0.68	0.76

7.1 Speaker Recognition

For demonstrating the speaker recognition using prosody models, speech data of four different female speakers for Hindi language is considered. Prosody models include both duration and intonation models. Separate models are developed to capture the speaker-specific intonation and duration knowledge from the given labeled speech data. The speaker models are denoted by S_1, S_2, S_3 and S_4. The models were tested with an independent set of the data for determining the speaker discrimination. The results of the speaker discrimination using the intonation models are given in Table 8.

The results show that the percentage of syllables predicted within 15% deviation from the actual F_0 value is highest for the intended speaker compared to others. Likewise the average prediction error (μ), standard deviation (σ) values are low, and the correlation coefficient (γ) values are high for the intended speaker. A similar study is performed using duration models separately for each of the speakers. The results of speaker discrimination using duration models are given in Table 9. Similar performance is observed as that of the speaker-specific intonation models.

7.2 Language Identification

For demonstrating the language identification using intonation models, speech data of male speakers in Hindi (Hi), Telugu (Te), and Tamil (Ta) was chosen. Separate models were developed to capture the language-specific intonation

Table 10. Language discrimination using intonation models

	% Syllables within 15% deviation			Mean error (Hz)			Std. dev. (Hz)			Corr. coeff.		
	Hi	Te	Ta	Hi	Te	Ta	Hi	Te	Ta	Hi	Te	Ta
Hi	90	59	63	11	22	20	9	14	14	0.79	0.75	0.68
Te	42	85	61	20	10	16	13	10	12	0.74	0.78	0.73
Ta	56	65	81	19	17	12	14	13	10	0.74	0.76	0.80

Table 11. Language discrimination using duration models

	% Syllables within 25% deviation			Mean error (ms)			Std. dev. (ms)			Corr. coeff.		
	Hi	Te	Ta	Hi	Te	Ta	Hi	Te	Ta	Hi	Te	Ta
Hi	68	59	57	32	37	38	24	28	30	0.76	0.71	0.69
Te	51	66	58	37	29	32	27	23	25	0.74	0.79	0.76
Ta	55	61	74	35	33	26	27	26	22	0.74	0.75	0.88

knowledge from the given labeled speech data. The results for language identification are given in Table 10. The results show that the intonation models
also captures the language-specific knowledge.

For demonstrating the language identification using duration models, separate models are developed for each of the languages. The results show that
the performance of the models is superior, when the test data is from the
same language. The results for language identification using duration models
are given in Table 11.

We have used four speaker-specific duration and intonation models for
demonstrating the ability of the speaker discrimination by the prosody models. The results show that both duration and intonation models provide
speaker recognition. The intonation models show better discrimination among
the speakers compared to duration models. This may be due to the fact
that intonation patterns for the sequence of syllables are more speaker-
specific compared to duration patterns associated with the syllables. In this
study language identification using prosody models was demonstrated with
three languages. Both duration and intonation models show a clear language
discrimination among the languages by capturing the language-specific information along with the prosody knowledge. The performance of the intonation
models for language identification is better compared to the performance of
the duration models. Even though in this study we demonstrated the speaker
and language identification using prosody models with small number of speakers and languages, a similar performance may be expected with large number
of speakers and languages.

7.3 Text-to-Speech Synthesis

For text-to-speech (TTS) synthesis, prosody models (duration and intonation models) provide the specific duration and intonation information associated with the sequence of sound units present in the given text. In this study we employ waveform concatenation for synthesizing the speech. The basis of concatenative synthesis is to join short segments of speech, usually taken from a pre-recoreded database, and then impose the associated prosody by appropriate signal processing methods. We considered syllable as a basic unit for synthesis. The database used for concatenative synthesis consists of syllables, that are extracted from a carrier words spoken in isolation. We have selected nonsense words as carrier words, because they offer minimum bias towards any of the factors that affect on the basic characteristics of the syllable. These syllables are considered as neutral syllables. Each carrier word consists of three characters. For example, for the syllable mA we can form a carrier word as $pamAt$. That is the characters with stop consonants are adjacent to the given basic unit. Using some guidelines the carrier words can be formed for different categories of syllables. Some of the guidelines are prepared while developing the TTS system for Hindi at IIT Madras [48].

In text-to-speech synthesis, we need to synthesize the speech for the given text. Firstly, the given text is analyzed using a text analysis module and derives the positional, contextual and phonological features (linguistic and production constraints) for each of the syllables present in the given text. These features are presented to the duration and intonation models, which will generate the appropriate duration and intonation information corresponding to the syllables. The duration and intonation models are developed by presenting the features representing the linguistic and production constraints of the syllables as input, and their associated duration or pitch (F_0) as output. The details of developing the models are discussed in Sect. 5. The derived duration and intonation knowledge is incorporated in the sequence of concatenated syllables (neutral syllables).

At synthesis stage, firstly, the pre-recorded (neutral) syllables are concatenated according to the sequence present in the text. The derived duration and intonation knowledge, corresponding to the sequence of syllables is incorporated in the sequence of concatenated syllables using prosody modification methods. The prosody modification method used here, modifies the prosody parameters (duration and pitch) by manipulating the excitation source information (linear prediction residual) using the instants of significant excitation of the vocal tract system during the production of speech [49]. Instants of significant excitation are computed from the linear prediction (LP) residual of the speech signals by using the average group-delay of minimum phase signals [50, 51]. As we are modifying the prosody parameters (duration and intonation) of the syllables in their residual domain without altering the vocal tract characteristics (spectral shape), the phase, spectral and audible distortions are not present in the synthesized speech.

Table 12. Ranking used for judging the quality of the speech signal

Rating	Speech quality
1.	Unsatisfactory
2.	Poor
3.	Fair
4.	Good
5.	Excellent

Table 13. Mean opinion scores for the quality of synthesized speech in the languages Hindi, Telugu and Tamil

Language	Mean opinion score	
	Intelligibility	Naturalness
Hindi	3.87	2.93
Telugu	4.15	3.24
Tamil	4.19	3.18

The quality (intelligibility and naturalness) of synthesized speech is evaluated by perceptual analysis. Perceptual evaluation is performed by conducting subjective tests with 25 research scholars in the age group of 25–35. The subjects have sufficient speech knowledge for proper assessment of the speech signals. Five sentences are synthesized from text for each of the languages Hindi, Telugu and Tamil. Each of the subjects were given a pilot test about perception of speech signals with respect to intelligibility and naturalness. Once they are comfortable with judging, they were allowed to take the tests. The tests were conducted in the laboratory environment by playing the speech signals through headphones. In the test, the subjects were asked to judge the intelligibility and naturalness of the speech. Subjects have to assess the quality on a 5-point scale for each of the sentences. The 5-point scale for representing the quality of speech is given in Table 12 [52]. The mean opinion scores for assessing the intelligibility and naturalness of the synthesized speech in each of the languages Hindi, Telugu and Tamil are given in Table 13. The scores indicate that the intelligibility of the synthesized speech is fairly acceptable for all the languages, where as the naturalness seems to be poor. Naturalness mainly attributed to individual perception. Naturalness can be improved to some extent by incorporating the coarticulation and stress information along with duration and intonation.

8 Summary

FFNN models were proposed for predicting the duration and intonation patterns (prosody) of the syllables. Duration and intonation patterns of the syllables are constrained by their linguistic context and production constraints. These were represented with positional, contextual and phonological features, and used for developing the models. Suitable neural network structures were arrived at empirically. The models were evaluated by computing the average prediction error (μ), the standard deviation (σ) and the correlation coefficient (γ) between the predicted and actual values. The accuracy in prediction is also analyzed in terms of percentage of syllables predicted within different deviations with respect to their actual values. The performance of the neural network models was compared with the performance of the CART models. A two-stage duration model was proposed to alleviate the problem of poor prediction due to single FFNN model. The performance of the two-stage model was improved by appropriate syllable classification model and the selection criterion of duration intervals. The intonation and duration models also capture the speaker and language information besides information about the message in speech. This was verified with the models of different speakers and different languages. The speaker-specific and language-specific knowledge may be useful for speaker and language identification tasks. The developed prosody models were shown to be useful in synthesizing the speech from the text input. In this case the prosodic parameters of the basic sound units were modified according to the derived prosody from the models. The performance of the prosody models can be further improved by including the accent and prominence of the syllable in the feature vector. Weighting the constituents of the input feature vectors based on linguistic and phonetic importance may further improve the performance. The accuracy of labeling, diversity of data in the database, and fine tuning of neural network parameters, all of these factors may also play a role in improving the prediction of the intonation patterns of the syllables.

References

1. Huang, X., Acero, A., Hon, H.W.: In: Spoken Language Proceesing. Prentice-Hall, New York, NJ, USA (2001)
2. Klatt, D.H.: Linguistic uses of segmental duration in English: Acoustic and perceptual evidence. Journal of Acoustic Society of America **59** (1976) 1209–1221
3. Yegnanarayana, B., Murthy, H.A., Sundar, R., Ramachandran, V.R., Kumar, A.S.M., Alwar, N., Rajendran, S.: Development of text-to-speech system for Indian languages. In: Proceedings of the International Conference on Knowledge Based Computer Systems, Pune, India (1990) 467–476
4. Chen, S.H., Lai, W.H., Wang, Y.R.: A new duration modeling approach for Mandarin speech. IEEE Transactions on Speech and Audio Processing **11** (2003) 308–320
5. Mixdorff, H., Jokisch, O.: Building an integrated prosodic model of German. In: Proceeding of the European Conference on Speech Communication and Technology. Volume 2, Aalborg, Denmark (2001) 947–950
6. Mixdorff, H.: An integrated approach to modeling German prosody. PhD Thesis, Technical University, Dresden, Germany (2002)

7. Santen, J.P.H.V.: Assignment of segment duration in text-to-speech synthesis. Computer Speech and Language **8** (1994) 95–128
8. Goubanova, O., Taylor, P.: Using Bayesian belief networks for modeling duration in text-to-speech systems. In: Proceedings of the International Conference on Spoken Language Processing. Volume 2, Beijing, China (2000) 427–431
9. Sayli, O.: Duration analysis and modeling for Turkish text-to-speech synthesis. Master's Thesis, Department of Electrical and Electronics Engineering, Bogaziei University (2002)
10. Riley, M.: Tree-based modeling of segmental durations. Talking Machines: Theories, Models and Designs (1992) 265–273
11. Haykin, S. In: Neural Networks: A Comprehensive Foundation. Pearson Education Asia, Inc., New Delhi, India (1999)
12. Yegnanarayana, B. In: Artificial Neural Networks. Prentice-Hall, New Delhi, India (1999)
13. Campbell, W.N.: Analog i/o nets for syllable timing. Speech Communication **9** (1990) 57–61
14. Campbell, W.N.: Syllable based segment duration. In: Bailly, G., Benoit, C., Sawallis, T.R., eds.: Talking Machines: Theories, Models and Designs. Elsevier (1992) 211–224
15. Campbell, W.N.: Predicting segmental durations for accommodation within a syllable-level timing framework. In: Proceedings of the European Conference on Speech Communication and Technology. Volume 2, Berlin, Germany (1993) 1081–1084
16. Barbosa, P.A., Bailly, G.: Characterization of rhythmic patterns for text-to-speech synthesis. Speech Communication **15** (1994) 127–137
17. Barbosa, P.A., Bailly, G.: Generating segmental duration by P-centers. In: Proceedings of the Fourth Workshop on Rhythm Perception and Production, Bourges, France (1992) 163–168
18. Cordoba, R., Vallejo, J.A., Montero, J.M., Gutierrezarriola, J., Lopez, M.A., Pardo, J.M.: Automatic modeling of duration in a Spanish text-to-speech system using neural networks. In: Proceedings of the European Conference on Speech Communication and Technology, Budapest, Hungary (1999)
19. Hifny, Y., Rashwan, M.: Duration modeling of Arabic text-to-speech synthesis. In: Proceedings of the International Conference on Spoken Language Processing, Denver, Colorado, USA (2002) 1773–1776
20. Sonntag, G.P., Portele, T., Heuft, B.: Prosody generation with a neural network: Weighing the importance of input parameters. In: Proceedings of the IEEE International Conference on Acoustics, Speech, Signal Processing, Munich, Germany (1997) 931–934
21. Teixeira, J.P., Freitas, D.: Segmental durations predicted with a neural network. In: Proceedings of the European Conference on Speech Communication and Technology, Geneva, Switzerland (2003) 169–172
22. Klatt, D.H.: Review of text-to-speech conversion for English. Journal of Acoustic Society of America **82**(3) (1987) 737–793
23. Olive, J.P.: Fundamental frequency rules for the synthesis of simple declarative English sentences. Journal of Acoustic Society of America (1975) 476–482
24. Fujisaki, H., Hirose, K., Takahashi, N.: Acoustic characteristics and the underlying rules of the intonation of the common Japanese used by radio and TV anouncers. In: Proceedings of the IEEE International Conference on Acoustics, Speech, Signal Processing (1986) 2039–2042

25. Taylor, P.A.: Analysis and synthesis of intonation using the Tilt model. Journal of Acoustic Society of America **107** (2000) 1697–1714
26. Madhukumar, A.S., Rajendran, S., Sekhar, C.C., Yegnanarayana, B.: Synthesizing intonation for speech in Hindi. In: Proceedings of the Second European Conference on Speech Communication and Technology. Volume 3, Geneva, Italy (1991) 1153–1156
27. Pierrehumbert, J.B.: The Phonology and Phonetics of English Intonation. PhD Thesis, MIT, MA, USA (1980)
28. Fujisaki, H.: Dynamic characteristics of voice fundamental frequency in speech and singing. In: MacNeilage, P.F., ed.: The Production of Speech. Springer-Verlag, New York, USA (1983) 39–55
29. Fujisaki, H.: A note on the physiological and physical basis for the phrase and accent components in the voice fundamental frequency contour. In: Fujimura, O., ed.: Vocal Physiology: Voice Production, Mechanisms and Functions. Raven Press, New York, USA (1988) 347–355
30. t'Hart, J., Collier, R., Cohen, A.: A Perceptual Study of Intonation. Cambridge University Press, Cambridge
31. Cosi, P., Tesser, F., Gretter, R.: Festival speaks Italian. In: Proceedings of EUROSPEECH 2001, Aalborg, Denmark (2001) 509–512
32. Tesser, F., Cosi, P., Drioli, C., Tisato, G.: Prosodic data driven modeling of a narrative style in Festival TTS. In: Fifth ESCA Speech Synthesis Workshop, Pittsburgh, USA (2004) 185–190
33. Vainio, M., Altosaar, T.: Modeling the microprosody of pitch and loudness for speech synthesis with neural networks. In: Proceedings of the International Conference on Spoken Language Processing, Sidney, Australia (1998)
34. Vegnaduzzo, M.: Modeling intonation for the Italian festival TTS using linear regression. Master's Thesis, Department of Linguistics, University of Edinburgh (2003)
35. Scordilis, M.S., Gowdy, J.N.: Neural network based generation of fundamental frequency contours. In: Proceedings of the IEEE International Conference on Acoustics, Speech, Signal Processing. Volume 1, Glasgow, Scotland (1989) 219–222
36. Vainio, M.: Artificial neural network based prosody models for Finnish text-to-speech synthesis. PhD Thesis, Department of Phonetics, University of Helsinki, Finland (2001)
37. Buhmann, J., Vereecken, H., Fackrell, J., Martens, J.P., Coile, B.V.: Data driven intonation modeling of 6 languages. In: Proceedings of the International Conference on Spoken Language Processing. Volume 3, Beijing, China (2000) 179–183
38. Hwang, S.H., Chen, S.H.: Neural-network-based F0 text-to-speech synthesizer for Mandarin. IEEE Proceedings on Image Signal Processing **141** (1994) 384–390
39. Khan, A.N., Gangashetty, S.V., Yegnanarayana, B.: Syllabic properties of three Indian languages: Implications for speech recognition and language identification. In: International Conference on Natural Language Processing, Mysore, India (2003) 125–134
40. Chopde, A.: (Itrans Indian language transliteration package version 5.2 source) http://www.aczone.con/itrans/.

41. Prasanna, S.R.M., Yegnanarayana, B.: Extraction of pitch in adverse conditions. In: Proceedings of the IEEE International Conference on Acoustics, Speech, Signal Processing, Montreal, Canada (2004)
42. Bellegarda, J.R., Silverman, K.E.A., Lenzo, K., Anderson, V.: Statistical prosodic modeling: From corpus design to parameter estimation. IEEE Transactions on Speech and Audio Processing 9 (2001) 52–66
43. Bellegarda, J.R., Silverman, K.E.A.: Improved duration modeling of English phonemes using a root sinusoidal transformation. In: Proceedings of the International Conference on Spoken Language Processing (1998) 21–24
44. Silverman, K.E.A., Bellegarda, J.R.: Using a sigmoid transformation for improved modeling of phoneme duration. In: Proceedings of the IEEE International Conference on Acoustics, Speech, Signal Processing, Phoenix, AZ, USA (1999) 385–388
45. Siebenhaar, B., Zellner-Keller, B., Keller, E.: Phonetic and timing considerations in a Swiss high German TTS system. In: Keller, E., Bailly, G., Monaghan, A., Terken, J., Huckvale, M., eds.: Improvements in Speech Synthesis. Wiley, Chichester (2001)
46. Burges, C.J.C.: A tutorial on support vector machines for pattern recognition. Data Mining and Knowledge Discovery 2 (1998) 121–167
47. Shriberg, Elizabeth, Stolcke, Andreas: Prosody modeling for automatic speech understanding: An overview of recent research at SRI. In: Prosody in Speech Recognition and Understanding, ISCA Tutorial and Research Workshop (ITRW), Molly Pitcher Inn, Red Bank, NJ, USA (2001)
48. Srikanth, S., Kumar, S.R.R., Sundar, R., Yegnanarayana, B. In: A text-to-speech conversion system for Indian languages based on waveform concatenation model. Technical report no. 11, Project VOIS, Department of Computer Science and Engineering, Indian Institute of Technology Madras (1989)
49. Rao, K.S., Yegnanarayana, B.: Prosodic manipulation using instants of significant excitation. In: Proceedings of the IEEE International Conference on Multimedia and Expo, Baltimore, Maryland, USA (2003) 389–392
50. Smits, R., Yegnanarayana, B.: Determination of instants of significant excitation in speech using group delay function. IEEE Transactions on Speech and Audio Processing 3 (1995) 325–333
51. Murthy, P.S., Yegnanarayana, B.: Robustness of group-delay-based method for extraction of significant excitation from speech signals. IEEE Transactions on Speech and Audio Processing 7 (1999) 609–619
52. Deller, J.R., Proakis, J.G., Hansen, J.H.L. In: Discrete-Time Processing of Speech Signals. Macmillan, New York, USA (1993)

Objective Speech Quality Evaluation Using an Adaptive Neuro-Fuzzy Network

Guo Chen[1] and Vijay Parsa[2]

[1] Department of Electrical & Computer Engineering, University of Western Ontario, London, Ontario, Canada, `guo.chen@nca.uwo.ca`
[2] Department of Electrical & Computer Engineering and National Centre for Audiology, University of Western Ontario, London, Ontario, Canada, `parsa@nca.uwo.ca`

Summary. A speech quality measure is a valuable assessment tool for the development of speech coding and enhancing techniques. Commonly, two approaches, subjective and objective, are used for measuring the speech quality. Subjective measures are based on the perceptual ratings by a group of listeners while objective metrics assess speech quality using the extracted physical parameters. Objective metrics that correlate well with subjective ratings are attractive as they are less expensive to administer and give more consistent results. In this work, we investigated a novel non-intrusive speech quality metric based on adaptive neuro-fuzzy network techniques. In the proposed method, a first-order Sugeno type fuzzy inference system (FIS) is applied for objectively estimating the speech quality. The features required for the proposed method are extracted from the perceptual spectral density distribution of the input speech by using the co-occurrence matrix analysis technique. The performance of the proposed method was demonstrated through comparisons with the state-of-the-art non-intrusive quality evaluation standard, the ITU-T P.563.

1 Speech Quality Evaluation

Speech quality evaluation is a valuable assessment tool for the development of speech coding and enhancing techniques. During the course of designing a speech compression system, it is desirable to have a speech quality measure for indicating the amount of distortion introduced by the compression algorithm, for determining the parameter settings, and for optimizing the system structure. For example, in mobile communications and voice over Internet protocol (VoIP), the speech signal is compressed into a compact representation before transmission and is reconstructed at the other end. The speech coding/decoding process invariably introduces distortion and a speech quality measure allows for the relative comparison of various speech coding techniques.

G. Chen and V. Parsa: *Objective Speech Quality Evaluation Using an Adaptive Neuro-Fuzzy Network*, Studies in Computational Intelligence (SCI) **83**, 97–116 (2008)
`www.springerlink.com` © Springer-Verlag Berlin Heidelberg 2008

Table 1. Mean opinion score (MOS) subjective listening test

Rating	Speech quality
5	Excellent
4	Good
3	Fair
2	Poor
1	Unsatisfactory

Commonly, two approaches, subjective and objective, are used for measuring the speech quality. Subjective measures are based on the perceptual ratings by a group of listeners, who subjectively rank the quality of speech along a predetermined scale. The most widely used subjective test is the absolute category rating (ACR) method as stated in the International Telecommunication Unit (ITU) Recommendation P.800 (1996) [1], which results in a mean opinion score (MOS). In the ACR test, listeners rank the speech quality by using a five-point scale, in which the quality is represented by five grades as shown in Table 1. Typically, the ratings are collected from a pool of listeners and the arithmetic mean of their ratings forms the MOS ratings. While subjective opinions of speech quality are preferred as the most trustworthy criterion for speech quality, they are also time-consuming, expensive and not exactly reproducible. In contrast, objective measures, which assess speech quality by using the extracted physical parameters and computational models, are less expensive to administer, save time, and give more consistent results. Also, the results conducted at different times and with different testing facilities can be directly compared. Thus, good objective measures are highly desirable in practical applications.

Objective speech quality measures can also be divided into two groups: intrusive and non-intrusive. The intrusive measures assess speech quality based on the use of known, controlled test signals which are processed through the system (or "condition") under test. The intrusive measures are also termed input–output-based or double-ended measures. In the intrusive measures, both the original and processed signal are available for the computational models. The difference between the original and processed signal will be mapped into a speech quality score. Note that it is always assumed that the original input signal is itself of perfect or near-to perfect quality. In addition, it is crucial that the input and output signals of the system under test be synchronized before performing the speech quality computations. Otherwise, the calculation of the difference between these two signals does not provide meaningful results. In practice, a perfect synchronization between the input and output signals is always difficult to achieve, e.g., due to the fading and error bursts that are common in the wireless communications. In addition to this, there are many applications where the original speech is not steadily available, as in mobile communications, satellite communications and voice over IP. In these situations non-intrusive speech quality measures, which assess speech quality based only on the output signal of the system under test, are attractive. Non-intrusive evaluation, which is also termed single-ended or output-based evaluation, is a challenging problem in the field of speech quality measurement in that the measurement of speech quality has to be performed with only the output speech signal of the system under test, without using the original signal as a reference. In this paper, we focus on non-intrusive speech quality measurement combining the neural network and fuzzy logic techniques.

In the past three decades, objective quality measurement has received considerable attention [2–5]. A majority of these objective methods are based on input/output comparisons, i.e. intrusive methods, [6–16]. However, as mentioned before, in some applications, we need the other type of speech quality evaluation, i.e. non-intrusive speech quality evaluation, which assesses speech quality based only on the output signal of the system under test [17–37]. Regardless of the type of quality measurement, the speech quality evaluation consists of two blocks. The first is a feature extraction block which aims to find a set of observation features for effectively representing different kinds of speech quality and the second is a cognitive modeling block which aims to construct a mapping function to combine the extracted features into the speech quality scores (such as the MOS ratings). A survey of the previous literature shows that considerable research effort has been spent on developing perceptual models to extract the pertinent features for representing the speech quality. In contrast, scant research has been performed on cognitive models. In particular, neural networks and fuzzy logic paradigms, which have been successfully used in other areas, have not been widely applied to speech quality evaluation.

In this work, we present a new non-intrusive speech quality evaluation method which is based on neuro-fuzzy network techniques. In the proposed method, a first-order Sugeno type fuzzy inference system (FIS) is applied to objectively estimating the speech quality and this FIS is implemented by an adaptive neuro-fuzzy inference system (ANFIS). The features required for the proposed method are extracted from the perceptual spectral density distribution of the input speech by using the co-occurrence matrix analysis technique, which is a two dimensional histogram analysis method used for indicating the joint probability of gray-level occurrence at a certain displacement along the speech spectrogram. The extracted features are then fed into a first-order Sugeno type FIS where each input feature was fuzzified by Gaussian membership functions. Correspondingly, the IF-THEN rules were constructed for the speech quality estimation. The performance of the proposed method was demonstrated through comparisons with the state-of-the-art non-intrusive quality evaluation standard, the ITU-T P.563, using seven ITU subjective quality databases, which contain a total of 1328 MOS ratings.

The remainder of this Chapter is organized as follows. A brief description of the ANFIS is given in the next section. In Sect. 3, we outline the proposed new non-intrusive quality evaluation method. Section 4 presents the feature extraction based on measuring distributive characteristics of speech perceptual spectral density. In Sect. 5, we describe the first-order Sugeno type FIS implemented by the ANFIS. We present the experimental results in Sect. 6. Finally, conclusions are drawn in Sect. 7.

2 A Brief Introduction of ANFIS

The adaptive neuro-fuzzy inference system (ANFIS), which was proposed by Jang [38–40], can be thought of as a class of adaptive neural networks that are functionally equivalent to fuzzy inference systems. The ANFIS integrates the interpretability of a fuzzy inference system with the adaptability of a neural network. This characteristic results in the advantage of easy implementation and good learning ability for the speech quality evaluation method. In this work, a Sugeno fuzzy inference system is implemented by using the ANFIS. The Sugeno fuzzy model is a fuzzy inference system proposed by Takagi, Sugeno, and Kang [41, 42] in an effort to develop a systematic approach to generating fuzzy rules from a given input–output data set. A typical fuzzy rule in the Sugeno fuzzy system has the form:

$$\text{If } x \text{ is } A \text{ and } y \text{ is } B \text{ then } z = f(x, y), \tag{1}$$

where A and B are fuzzy sets in the premise, and $z = f(x, y)$ is a consequent. Usually $f(x, y)$ is a polynomial in the input variables x and y, but it can be any function as long as it can appropriately describe the output of the system within the fuzzy region specified by the premises of the rule. When $f(x, y)$ is a first-order polynomial, the resulting fuzzy inference system is called a first-order Sugeno fuzzy model. The present work uses the first order Sugeno fuzzy model as the fuzzy inference engine.

3 The Outline of the Proposed Method

Figure 1 shows the schematic diagram of the proposed method. The input features were first extracted by measuring the distributive characteristics of speech perceptual spectral density (a detailed description of feature extraction is given in the following Section). The extracted features, x_1, x_2 and x_3, were then fed to the first-order Sugeno type FIS where each input feature was fuzzified by K Gaussian membership functions. Correspondingly, $I(= K^3)$ if-then rules were constructed for the proposed method. In the proposed method, the algebraic product is used as a T-norm operator for the fuzzy AND operator and the overall output is calculated using the weighted average. The fuzzy if-then rules are of the form:

$$\text{Rule } i: \text{ IF } x_1 \text{ is } A_k \text{ AND } x_2 \text{ is } B_k \text{ AND } x_3 \text{ is } C_k$$
$$\text{THEN } f_i = p_i x_1 + q_i x_2 + r_i x_3 + s_i, \tag{2}$$

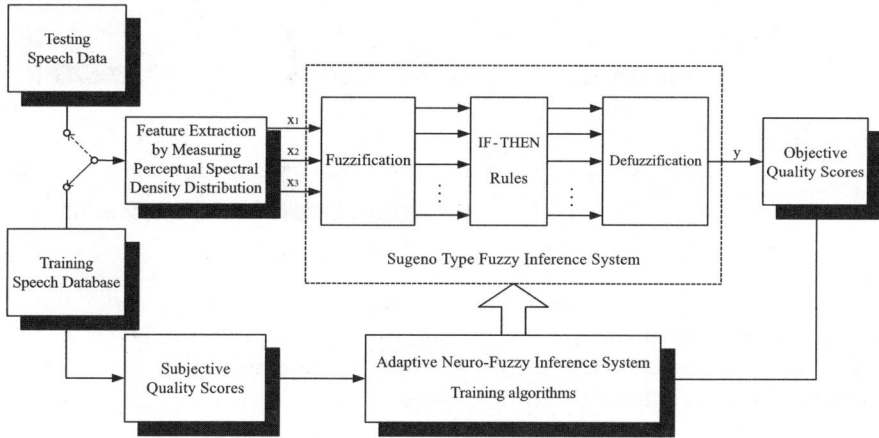

Fig. 1. The schematic diagram of the proposed method

where A_k, B_k and C_k are the fuzzy sets of the input feature x_1, x_2 and x_3, respectively. The parameters, p_i, q_i and r_i are the consequent parameters associated with x_1, x_2 and x_3, respectively, and s_i is the bias vector associated with the ith rule.

During the training phase, a hybrid learning algorithm was used to update all the parameters. This hybrid learning rule is based on the decomposition of the parameter set and through interleaving two separate phases. In the first phase, the consequent parameters were adjusted using a least squares algorithm assuming the premise parameters are fixed. In the second phase, the premise parameters were adjusted using the gradient descent assuming the consequent parameters are fixed. In the testing phase, the input features extracted from the speech under test were mapped to a speech quality score according to the reasoning rules of the trained Sugeno type FIS. The detailed operation of the proposed method is described in the following sections.

4 Feature Extraction by Measuring Spectral Density Distribution in Time-Perception Domain

The block diagram of the feature extraction is shown in Fig. 2. In the proposed method, the input speech signals were divided into frames and each frame was transformed to the frequency domain using a Hamming window and a short term FFT. The transformation to the Bark representation was done by grouping the spectral coefficients into critical Bark bands. Next, the power spectrum at each pixel on the time-Bark plane was quantized to a fixed integer level, i.e., a gray level. Following that, a co-occurrence matrix analysis was carried out to describe the characteristics of the spectral density distribution in the plane. Based on the co-occurrence matrix analysis, the feature

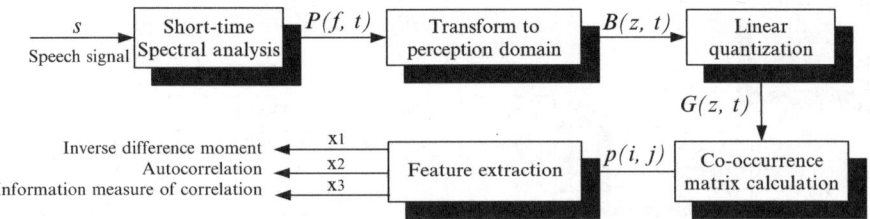

Fig. 2. The block diagram of feature extraction

parameters indicating the spectral density distribution were extracted. Speech quality was then estimated using these feature parameters. The process of the feature extraction is formulated as follows.

4.1 Short-Time Spectral Analysis

The input speech signal with in 8 kHz sampling frequency s was segmented into frames of 32 ms with an overlap of 50%, denoted by $s(k)$, $k = 1, 2, \ldots, K$. Each frame was transformed to the frequency domain using the Hamming window and a short term FFT. The real and imaginary components of the short-term speech spectrum were squared and added to obtain the short-term power spectrum, $E(f, k)$, where f is frequency scale. Here, more specifically, let the k-th frame of speech signal be denoted by samples $s(l, k)$ with l running from 1 to L ($L = 256$). The signal $s(l, k)$ was windowed by a Hamming window, i.e.

$$s_w(l, k) = h(l)x(l, k), \tag{3}$$

where $h(l)$ is the windowing function as

$$h(l) = 0.54 - 0.46cos(2\pi l/L), \quad 1 \le l \le L. \tag{4}$$

The windowed data was converted to the frequency domain using the DFT as below:

$$E(f, k) = \frac{1}{256} \sum_{l=1}^{256} s_w(l, k)e^{-j2\pi kl/256}. \tag{5}$$

The real and imaginary components were squared and added to get the short-term spectral power density $E(f, k)$, i.e.

$$E(f, k) = (\text{Re } S_w(t, k))^2 + (\text{Im } S_w(t, k))^2, \tag{6}$$

where f is in frequency Hertz scale.

4.2 Transform to Perception Domain

The short-term power spectrum $E(f, k)$ was partitioned into critical bands and the energies of each critical band were added up. The relationship between the linear frequency and Bark scales is given by

$$z = 7asinh(f/650). \tag{7}$$

In order to improve the resolution of time-perception distribution, an interval of 0.25 Bark scale was selected in our study. This leads to 70 critical bands covering the 20 Hz - 4 kHz bandwidth. The energy in each critical band was summed as follows.

$$B(z, k) = \sum_{v=bl_z}^{bh_z} E(v, k), \tag{8}$$

where bl_z and bh_z are the lower and upper boundaries of critical band z, respectively.

4.3 Quantization

In order to calculate the co-occurrence matrices of the power spectra distributed over the time-perception plane, a linear quantization was used to convert the values of power spectra into N_g gray levels. The N_g parameter was empirically chosen to be 32. Correspondingly, $B(z, k)$ becomes $G(m, n)$ which can be regarded as a 32-level gray image, i.e.,

$$G(z, k)\frac{31}{max[B(z, k)] - min[B(z, k)]}B(z, k) - min[B(z, k)] + 1. \tag{9}$$

4.4 The Calculation of Co-Occurrence Matrices

The computation of the co-occurrence matrices follows the procedure described in [43–45] and an example of this computational procedure is given in Fig. 3. Note that the gray level appearing at each pixel on G is from 1 to N_g. The distribution of G can be specified by the matrix of relative frequencies, P_{ij}, with two neighboring pixels separated by distance D occur on G, one with gray level

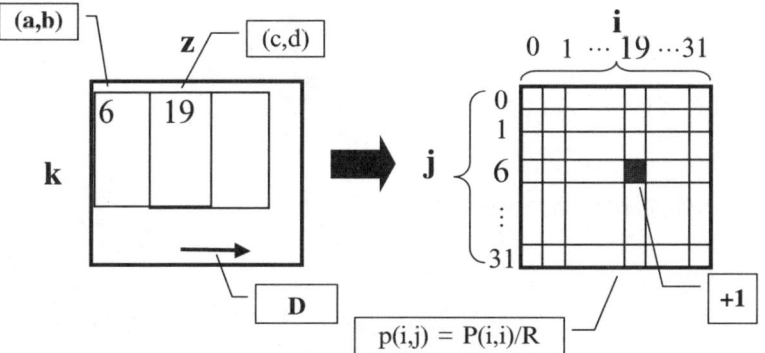

Fig. 3. The calculation of the co-occurrence matrix

i and the other with gray level j. Such matrices of gray-level co-occurrence frequencies are a function of the angular relationship and distance between the neighboring pixels. Considering the characteristics of speech signal, together with our experiments, the displacement was chosen to be three while the orientations were selected to be horizon and vertical. The un-normalized frequencies are defined as below.

$$P_h(i,j) = \#\{[(a,b),(c,d)] \in (z \times k) \times (z \times k)$$
$$: |a - c| = D, b - d = 0, G(a,b) = i, G(c,d) = j\} \qquad (10)$$

$$P_v(i,j) = \#\{[(a,b),(c,d)] \in (z \times k) \times (z \times k)$$
$$: a - c = 0, |b - d| = D, G(a,b) = i, G(c,d) = j\} \qquad (11)$$

where $\#$ denotes the number of elements in the set. The normalized co-occurrence matrix is

$$p(i,j) = P(i,j)/R, \qquad (12)$$

where R is the total number of pairs in the co-occurrence matrix. Figure 4 depicts the time-Bark scale waterfall spectral plot and the corresponding co-occurrence matrix for the clean speech signal while Fig. 5 shows the corresponding plots for the degraded speech signal. From these two figures, it can be observed that the element values of the co-occurrence matrix of clean speech concentrate along the diagonal and conform to a continual and symmetrical distribution. But for distorted speech, the element values of the co-occurrence matrix exhibit more divergence and are distributed asymmetrically and discontinuously.

4.5 Extracting the Feature Parameters

Based on our previous work [33], three feature parameters were selected in this work. The following equations define these parameters.

$$Inverse\ Difference\ Moment:\ x_1 = \sum_i \sum_j \frac{p(i,j)}{1+(i-j)^2}. \qquad (13)$$

$$Autocorrelation:\ x_2 = \sum_i \sum_j (ij)p(i,j). \qquad (14)$$

$$Information\ Measure\ of\ Correlation:\ x_3 \frac{HXY - HXY1}{max(HX, HY)}, \qquad (15)$$

where

$$HX = -\sum_i p_1(i)log(p_1(i)),\ HY = -\sum_j p_2(j)log(p_2(j)),$$

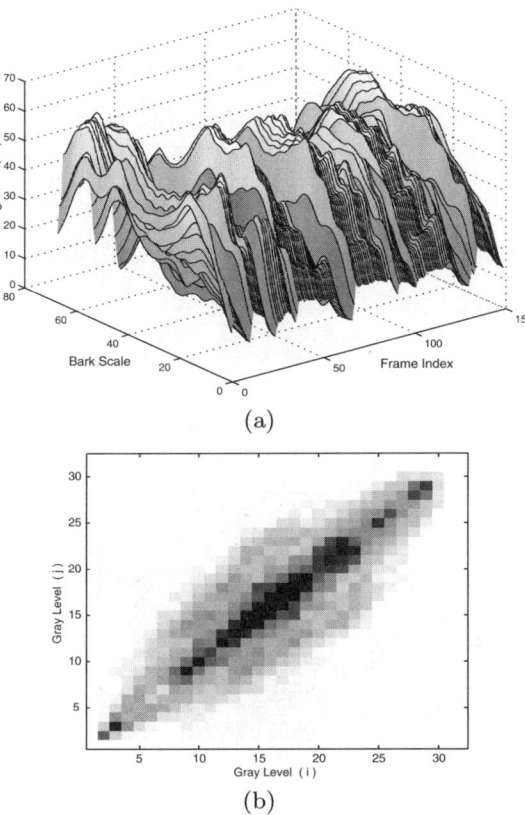

(a)

(b)

Fig. 4. An example of time-bark scale spectral plots for clean speech. (**a**) Time-Bark scale spectral plot of clean speech; (**b**) Co-occurrence matrix of clean speech

$$p_1(i) = \sum_{j=1}^{N_g} p(i,j),\ p_2(j) \sum_{i=1}^{N_g} p(i,j),$$

$$HXY = -\sum_i \sum_j p(i,j) log(p(i,j)),$$

$$HXY1 = -\sum_i \sum_j p(i,j) log(p_1(i)p_2(i)).$$

5 Implementation of the Proposed Method

Figure 6 displays the ANFIS architecture used in the proposed method when two membership functions are used. A detailed description of the ANFIS technique can be found in [38, 40]. The ANFIS technique enhances the basic fuzzy system with a self-learning capability for achieving the optimal nonlinear

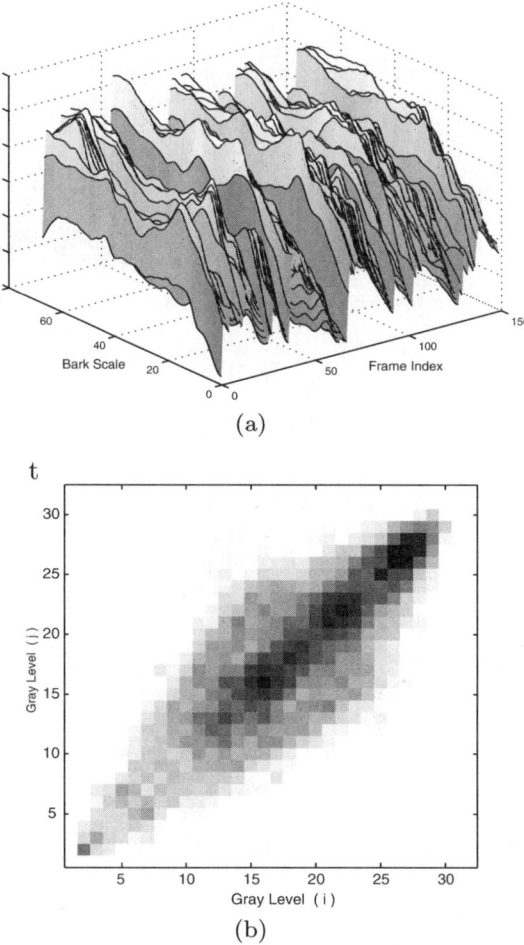

Fig. 5. An example of time-bark scale spectral plots for degraded speech (MNRU condition: SNR = 5 dB). (**a**) Time-Bark scale spectral plot of degraded speech; (**b**) Co-occurrence matrix of degraded speech

mapping. The proposed method consists of five layers, where the output of the ith node in the jth layer is denoted by O_i^j.

The outputs of the first layer are the membership degrees associated with the inputs x_1, x_2 and x_3. That is, the outputs associated with x_1, x_2 and x_3 are

$$O_i^1 = \mu_{A_k}(x_1), \quad i = 1, 2;$$
$$O_i^1 = \mu_{B_k}(x_2), \quad i = 3, 4;$$
$$O_i^1 = \mu_{C_k}(x_3), \quad i = 5, 6; \tag{16}$$

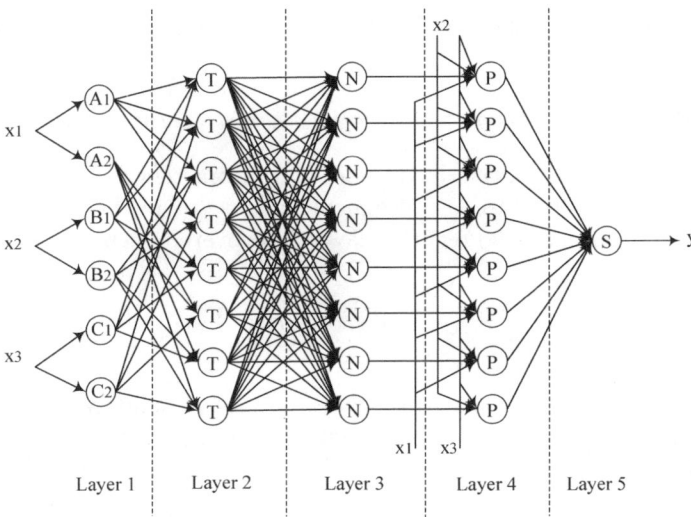

Fig. 6. The architecture of the proposed method

where x_1, x_2 and x_3 are the inputs to the nodes of this layer. A_k, B_k and C_k are the linguistic values associated with the nodes ($k = 1, 2$). The Gaussian membership functions were used in the proposed method. They are defined as

$$\mu_{A_k}(x_1) = \exp\left[-\frac{(x_1 - m_{A_k})^2}{(2\sigma_{A_k}^2)}\right],$$

$$\mu_{B_k}(x_2) = \exp\left[-\frac{(x_2 - m_{B_k})^2}{(2\sigma_{B_k}^2)}\right],$$

$$\mu_{C_k}(x_3) = \exp\left[-\frac{(x_3 - m_{C_k})^2}{(2\sigma_{C_k}^2)}\right], \tag{17}$$

where m_{A_k}, m_{B_k} and m_{C_k} are the centers of the Gaussian functions, σ_{A_k}, σ_{B_k} and σ_{C_k} are the widths of the Gaussian functions ($k = 1, 2$). These parameters are referred to as the premise parameters of the FIS.

The second layer is the composition layer which performs the fuzzy AND operation. This layer has one output per rule. The output for the ith rule is given by

$$O_i^2 = W_i = \mu_{A_k}(x_1) \times \mu_{B_k}(x_2) \times \mu_{C_k}(x_3), \quad k = 1, 2; \quad i = 1, 2, \ldots, I, \tag{18}$$

where W_i is the firing strength of the ith rule. I is the number of total rules.

The third layer is the normalization layer which normalizes the firing strength of the ith rule by the sum of the firings of all rules, i.e.,

$$O_i^3 = \overline{W_i} = \frac{W_i}{\sum_{j=1}^{I} W_j}, \quad i = 1, 2, \ldots, I. \tag{19}$$

The fourth layer is the consequent layer. It calculates the implication for every rule, i.e.,

$$O_i^4 = \overline{W}_i f_i = \overline{W}_i(p_i x_1 + q_i x_2 + r_i x_3 + s_i), \quad i = 1, 2, \ldots, I, \qquad (20)$$

where $\overline{W}_i$ is the output of the third layer. The parameters f_i, i.e., $\{p_i, q_i, r_i, s_i\}$, in this layer are referred to as consequent parameters of the FIS.

The final layer is the aggregation layer. The overall output is calculated as

$$O_1^5 = y = \sum_{i=1}^{I} O_i^4 = \sum_{i=1}^{I} \overline{W}_i(p_i x_1 + q_i x_2 + r_i x_3 + s_i), \quad i = 1, 2, \ldots, I. \quad (21)$$

The premise parameters $(m_{A_k}, m_{B_k}, m_{C_k}, \sigma_{A_k}, \sigma_{B_k},$ and $\sigma_{C_k})$ and the consequent parameters (p_i, q_i, r_i, s_i) associated with the ith rule constitute a set of parameters which will be set up in the training phase. Functionally, the output y can be written as

$$y = F(x_1, x_2, x_3, \Theta) = \overline{W}_i(p_i x_1 + q_i x_2 + r_i x_3 + s_i), \qquad (22)$$

where Θ is a vector of the premise and consequent parameters.

In the training phase, the hybrid learning algorithm is used to adapt all the parameters. In the forward pass, the inputs x_1, x_2 and x_3 go forward until layer 4, and the consequent parameters are identified by the least squares estimate, while the premise parameters are fixed. In the backward pass, the error rates, defined as the derivative of the squared error with respect to each node's output, propagate backward and the premise parameters are updated by the gradient descent, while the consequent parameters remain fixed.

6 Performance Evaluation

The proposed method was compared to the current state-of-the-art non-intrusive quality evaluation standard, the ITU-T P.563 Recommendation [35] using the seven ITU coded speech databases. The correlation coefficient (ρ) between subjective and objective values and the standard deviation of error (τ), which are widely used for evaluating the performance of an objective speech quality measure [2], were employed to quantify the speech quality evaluation performance. Generally, a subjective MOS test contains different testing conditions (or testing systems) and it is typical to use the correlation coefficient and the standard deviation of error per-condition (i.e. after the condition-averaged operation) to demonstrate the resultant performance. These two parameters can be calculated by

$$\rho = \frac{\sum_{i=1}^{N}(t_i - \bar{t})(d_i - \bar{d})}{\sqrt{\sum_{i=1}^{N}(t_i - \bar{t})^2 \sum_{i=1}^{N}(d_i - \bar{d})^2}}, \qquad (23)$$

Table 2. The speech databases used in our experiments

Database	Language	Number of files	Number of conditions	Length (seconds)	Minimum MOS	Maximum MOS
Exp1A	French	176	44	176*8	1.000	4.583
Exp1D	Japanese	176	44	176*8	1.000	4.208
Exp1O	English	176	44	176*8	1.208	4.542
Exp3A	French	200	50	200*8	1.292	4.833
Exp3D	Japanese	200	50	200*8	1.042	4.417
Exp3O	English	200	50	200*8	1.167	4.542
Exp3C	Italian	200	50	200*8	1.083	4.833
Total		1328	332	10624		

$$\tau = \sqrt{\frac{\sum_{i=1}^{N}(t_i - d_i)^2}{(N - 1)}} \tag{24}$$

where t_i is the MOS ratings for condition i (averaged over all sentences and listeners), $\bar{t}$ is the average over all t_i's, d_i is the estimated quality by an objective measure for condition i, $\bar{d}$ is the average of all d_i's, and N is the number of conditions.

The speech databases used in our experiments are listed in Table 2. The experimental data consist of 1328 subjective MOS ratings which come from seven subjective MOS databases obtained in the subjective listening tests described in the Experiment One (three databases) and Experiment Three (four databases) of the ITU-T P-Series Supplement 23 [46]. Each of these databases contains a number of speech sentence pairs spoken by four talkers (two female and two male) and each sentence pair stands for one condition under test (or one system under test). The three databases of Experiment One in the ITU-T Supplement 23 contain speech signals coded by various codecs (G.726, G.728, G.729, GSM-FR, IS-54 and JDC-HR), singly or in tandem configurations under clean channel condition. The four databases of Experiment Three in the ITU-T Supplement 23 contain single and multiple-encoded G.729 speech under various channel error conditions (BER 0–10%; burst and random frame erasure 0–5%) and input noise conditions at 20 dB signal-to-noise ration (SNR) including clean, street, vehicle, and hoth noises. In Experiment One, each database contains 44 sentence pairs while in Experiment Three each database contains 50 sentence pairs.

We combined these seven databases into a global database, giving a total of 1,328 MOS scores. Three-quarters of the global database was used for training the model, while the remaining was used as a validation data set. The experimental results are shown in Tables 3 and 4. Table 3 shows the correlations and standard errors of a series of models as a function of the number of membership functions. For each model, an optimum structure of the proposed method was chosen to have parameters associated with the minimum validation data error, which was used to prevent the model from over-fitting the training data

Table 3. Results of the models with different number of membership functions without condition-averaged operation

Number of MF	Training Data Set		Validation Data Set		$|\Delta\rho|$
	ρ	ϵ	ρ	ϵ	
9	0.8168	0.4466	0.4933	1.1087	0.3235
8	0.8080	0.4562	0.6651	0.6516	0.1429
7	0.7938	0.4708	0.6892	0.6243	0.1046
6	0.7723	0.4918	0.6942	0.5959	0.0781
5	0.7531	0.5093	0.6869	0.5956	0.0662
4	0.7386	0.5219	0.7164	0.5557	0.0222
3	0.7202	0.5371	0.7090	0.5610	0.0112
2	0.6617	0.5805	0.7123	0.5590	0.0506

Table 4. Performance of the NISQE–ANFIS compared with ITU-T P.563 Recommendation with condition-averaged operation

Speech Database	ρ		ϵ		ρ
	Proposed method	P.563	Proposed method	P.563	Improved %
P.Sup23 Exp1A (French)	0.8972	0.8833	0.3309	0.3964	101.57
P.Sup23 Exp1D (Japanese)	0.9254	0.8100	0.2395	0.3756	114.25
P.Sup23 Exp1O (English)	0.8394	0.8979	0.4170	0.3935	93.48
P.Sup23 Exp3A (French)	0.8752	0.8730	0.3265	0.3520	100.25
P.Sup23 Exp3D (Japanese)	0.8561	0.9241	0.3462	0.2994	92.64
P.Sup23 Exp3O (English)	0.8005	0.9067	0.4304	0.3925	88.29
P.Sup23 Exp3C (Italian)	0.8153	0.8337	0.4830	0.5489	97.79
Total Data (7 Databases)	0.8812	0.8422	0.3647	0.4493	104.63

set. The results from both the training and validation data sets are shown in Table 4. Based on these results, the model with three membership functions was chosen for the following experiments, since it performed equivalently on both the training and validation data sets.

In order to demonstrate the performance of the proposed method, a comparison experiment with the ITU-T P.563 Recommendation was done using the complete MOS data (including seven subjective quality databases) and each database individually. The condition-averaged results of the P.563 and proposed method are shown in Table 4. Figure 7 displays the plots of the estimated condition-averaged MOSs by the P.563 and proposed method vs. the actual subjective MOSs for the Experiment One of the ITU-T P. Supplement 23 [46] (including three databases) while Fig. 8 displays the plots of the predicted condition-averaged MOSs by the P.563 and the proposed method vs. the actual subjective MOSs for the Experiment Three of the ITU-T P. Supplement 23 [46] (including four databases). The predicted condition-averaged MOSs by the P.563 and proposed method for the entire data are shown in

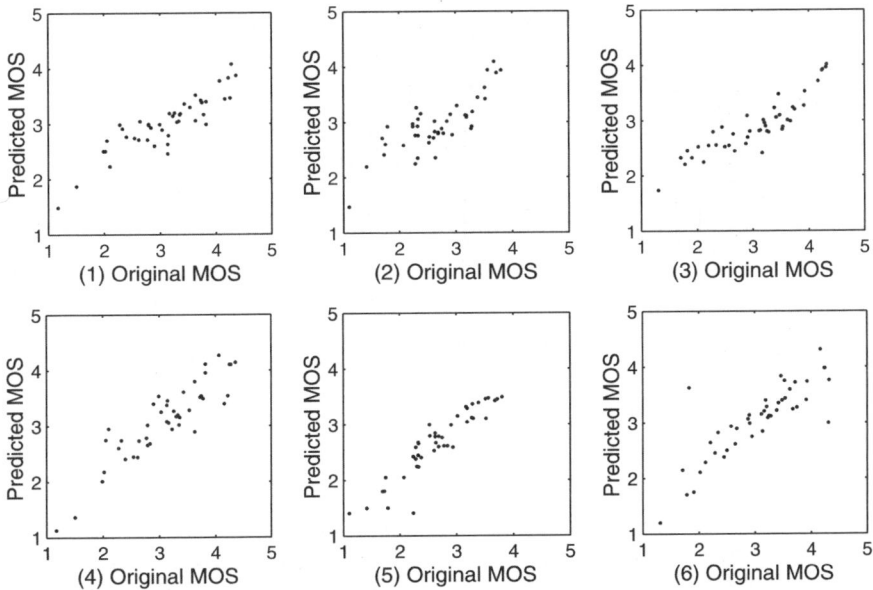

Fig. 7. The estimated condition-averaged MOSs vs. the actual MOSs for Experiment One (three databases). **(1)** The results of the proposed method for the Exp1A; **(2)** The results of the proposed method for the Exp1D; **(3)** The results of the proposed method for the Exp1O; **(4)** The results of the P.563 for the Exp1A; **(5)** The results by the P.563 for the Exp1D; **(6)** The results by the P.563 for the Exp1O (The associated correlations are shown in Table 4.)

Fig. 9. It can be seen from the results that the correlation coefficient between the actual MOS data and the estimated MOS data by our method was 0.8812 with a standard error of 0.3647 across the entire database. This compares favorably with the standard P.563, which provided a correlation coefficient and standard error of 0.8422 and 0.4493, respectively.

7 Conclusions

In this work, we presented a non-intrusive speech quality evaluation method. The proposed method applied a first-order Sugeno type FIS to evaluate speech quality using only the output signal of a system under test. The features used by the proposed method were extracted from the perceptual spectral density distribution of the input speech. The premise and consequent parameters of the FIS constructed by the ANFIS technique were learned by back-propagation and least squares algorithms. The proposed method was compared with the ITU-T P.563 Recommendation using seven subjective

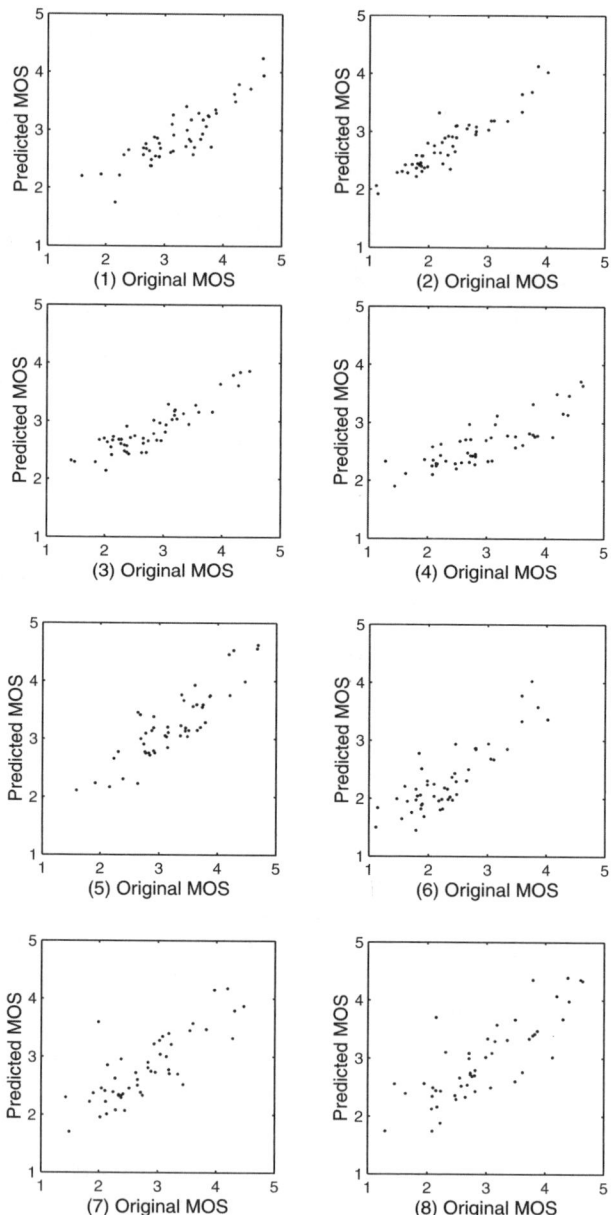

Fig. 8. The predicted condition-averaged MOSs vs. the actual MOSs for Experiment three (four databases). (**1**) The results of the proposed method for the Exp3A; (**2**) The results of the proposed method for the Exp3D; (**3**) The results of the proposed method for the Exp3O; (**4**) The results of the proposed method for the Exp3C; (**5**) The results of the P.563 for the Exp3A; (**6**) The results of the P.563 for the Exp3D; (**7**) The results of the P.563 for the Exp3O; (**8**) The results of the P.563 for the Exp3C (The associated correlations are shown in Table 4.)

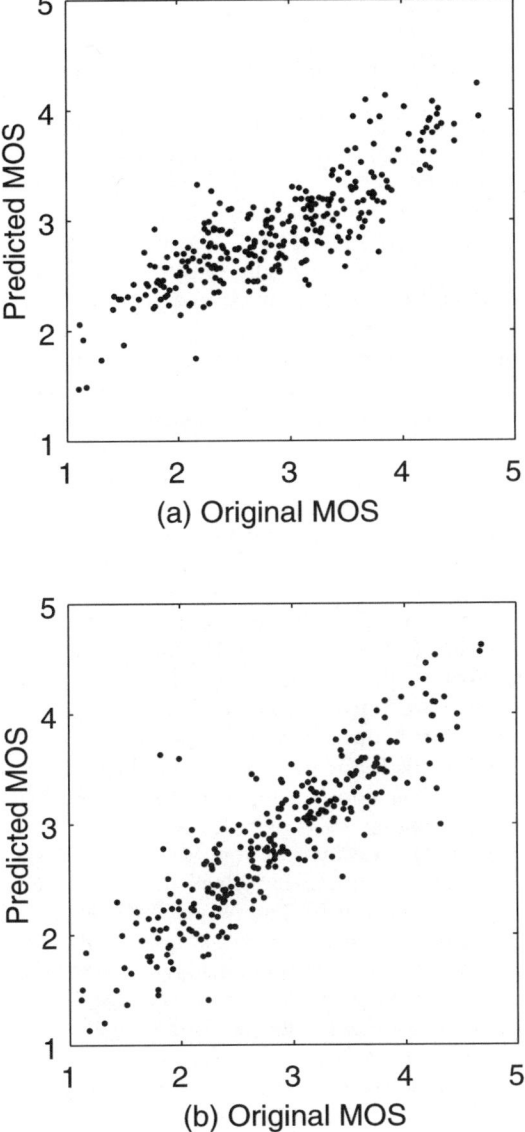

Fig. 9. The predicted condition-averaged MOSs. (a) The results of the P.563 ($\rho =$ 0.8422, $\epsilon = 0.4493$); (b) The results of the proposed method ($\rho = 0.8812$, $\epsilon = 0.3647$). Note that predicted scores by the proposed method better align with the diagonal

quality MOS databases of the ITU-T P-series Supplement 23. Experimental results showed that the correlation of the proposed method with subjective quality scores reached 0.8812. This compares favorably with the standardized P.563, which provided a correlation of 0.8422.

References

1. ITU (1996) Methods for subjective determination of transmission quality. *ITU-T P.800*.
2. Quackenbush SR, Barnwell-III TP, and Clements MA (1988) *Objective Measures of Speech Qaulity*, Prentice-Hall, Englewood Cliffs, NJ.
3. Dimolitsas S (1989) Objective speech distortion measures and their relevance to speech quality assessments. *IEE Proceedings - Communications, Speech and Vision*, vol. 136, no. 5, pp. 317–324.
4. Rix A (2004) Perceptual speech quality assessment - a review. In *Proceedings of IEEE International Conference on Acoustics, Speech and Signal Processing*, Montreal, Canada, vol. 3, pp. 1056–1059.
5. Rix A, Beerends JG, Kim DS, Kroon P, and Ghitza O (2006) Objective assessment of speech and audio quality—technology and applications. *IEEE Transactions on Audio, Speech and Language Processing*, vol. 14, no. 6, pp. 1890–1901.
6. Wang S, Sekey A, and Gersho A (1992) An objective measure for predicting subjective quality of speech coders. *IEEE Journal on selected areas in communications*, vol. 10, no. 5, pp. 819–829.
7. Beerends JG and Stemerdink JA (1994) A perceptual speech-quality measure based on a psychoacoustic sound representation. *Journal of the Audio Engineering Society*, vol. 42, no. 3, pp. 115–123.
8. Yang W, Benbouchta M, and Yantorno R (1998) Performance of the modified bark spectral distortion as an objective speech quality measure. In *Proceedings of IEEE International Conference on Acoustics, Speech and Signal Processing*, Washington, USA, vol. 1, pp. 541–544.
9. Voran S (1999) Objective estimation of perceived speech quality - part i. development of the measuring normalizing block technique. *IEEE Transactions on speech and audio processing*, vol. 7, no. 4, pp. 371–382.
10. Voran S (1999) Objective estimation of perceived speech quality - part ii. evaluation of the measuring normalizing block technique. *IEEE Transactions on speech and audio processing*, vol. 7, no. 4, pp. 383–390.
11. ITU (2001) Perceptual evaluation of speech quality. *ITU-T P.862*.
12. Zha W and Chan WY (2004) A data mining approach to objective speech quality measurement. In *Proceedings of IEEE International Conference on Acoustics, Speech and Signal Processing*, Montreal, Canada, vol. 3, pp. 461–464.
13. Kates JM and Arehart KH (2005) A model of speech intelligibility and quality in hearing aids. In *IEEE Workshop on Applications of Signal Processing to Audio and Acoustics*, New York, USA, pp. 53–56.
14. Karmakar A, Kumar A, and Patney RK (2006) A multiresolution model of auditory excitation pattern and its application to objective evaluation of perceived speech quality. *IEEE Transactions on Audio, Speech and Language Processing*, vol. 14, no. 6, pp. 1912–1923.
15. Chen G, Koh S, and Soon I (2003) Enhanced itakura measure incorporating masking properties of human auditory system. *Signal Processing*, vol. 83, pp. 1445–1456.
16. Chen G, Parsa V, and Scollie S (2006) An erb loudness pattern based objective speech quality measure. In *Proceedings of Iternational Conference on Spoken Language Processing*, Pittsburg, USA, pp. 2174–2177.

17. Liang J and Kubichek R (1994) Output-based objective speech quality. In *Proceedings of IEEE 44th Vehicular Technology Conference*, Stockholm, Sweden, vol. 3, pp. 1719–1723.
18. Jin C and Kubichek R (1996) Vector quantization techniques for output-based objective speech quality. In *Proceedings of IEEE International Conference on Acoustics, Speech and Signal Processing*, Atlanta, GA, vol. 1, pp. 491–494.
19. Picovici D and Mahdi AE (2003) Output-based objective speech quality measure using self-organizing map. In *Proceedings of IEEE International Conference on Acoustics, Speech and Signal Processing*, Hongkong, China, vol. 1, pp. 476–479.
20. Picovici D and Mahdi AE (2004) New output-based perceptual measure for predicting subjective quality of speech. In *Proceedings of IEEE International Conference on Acoustics, Speech and Signal Processing*, Montreal, Canada, vol. 5, pp. 633–636.
21. Falk T, Xu Q, and Chan WY (2005) Non-intrusive gmm-based speech quality measurement. In *Proceedings of IEEE International Conference on Acoustics, Speech and Signal Processing*, Philadelphia, USA.
22. Falk T and Chan WY (2006) Nonintrusive speech quality estimation using gaussian mixture models. *IEEE Signal Processing Letters*, vol. 13, no. 2, pp. 108–111.
23. Falk T and Chan WY (2006) Single-ended speech quality measurement using machine learning methods. *IEEE Transactions on Audio, Speech and Language Processing*, vol. 14, no. 6, pp. 1935–1947.
24. Falk T and Chan WY (2006) Enhanced non-intrusive speech quality measurement using degradation models. In *Proceedings of IEEE International Conference on Acoustics, Speech and Signal Processing*, Toulouse, France, vol. 1, pp. 837–840.
25. Nielsen LB (1993) Objective scaling of sound quality for normal-hearing and hearing-impaired listerners. Tech. Rep. No. 54, The acoustics laboratory, Technical University of Denmark, Denmark.
26. Gray P, Hollier MP, and Massara RE (2000) Non-intrusive speech quality assessment using vocal-tract models. *IEE Proceedings - Vision, Image and Signal Processing*, vol. 147, no. 6, pp. 493–501.
27. Kim DS and Tarraf A (2004), Perceptual model for non-intrusive speech quality assessment. In *Proceedings of IEEE International Conference on Acoustics, Speech and Signal Processing*, Montreal, Canada, vol. 3, pp. 1060–1063.
28. Kim DS (2004) A cue for objective speech quality estimation in temporal envelope representations. *IEEE Signal Processing Letters*, vol. 1, no. 10, pp. 849–852.
29. Kim DS (2005) Anique: An auditory model for single-ended speech quality estimation. *IEEE Transactions on Speech and Audio Processing*, vol. 13, no. 4, pp. 1–11.
30. Kim DS and Tarraf A (2006) Enhanced perceptual model for non-intrusive speech quality assessment. In *Proceedings of IEEE International Conference on Acoustics, Speech and Signal Processing*, Toulouse, France, vol. 1, pp. 829–832.
31. Chen G and Parsa V (2004) Output-based speech quality evaluation by measuring perceptual spectral density distribution. *IEE Electronics Letter*, vol. 40, no. 12, pp. 783–784.
32. Chen G and Parsa V (2004) Neuro-fuzzy estimator of speech quality. In *Proceedings of International Conference on signal processing and communications (SPCOM)*, Bangalore, India, pp. 587–591.

33. Chen G and Parsa V (2005) Non-intrusive speech quality evaluation using an adaptive neuro-fuzzy inference system. *IEEE Signal Processing Letters*, vol. 12, no. 5, pp. 403–406.

34. Chen G and Parsa V (2005) Bayesian model based non-intrusive speech quality evaluation. In *Proceedings of IEEE International Conference on Acoustics, Speech and Signal Processing*, Philadelphia, USA, vol. 1, pp. 385–388.

35. ITU (2004) Single ended method for objective speech quality assessment in narrow-band telephony applicaitons. *ITU-T P.563*.

36. Ding L, Radwan A, El-Hennawey MS, and Goubran RA (2006) Measurement of the effects of temporal clipping on speech quality. *IEEE Transactions on Instrumentation and Measurement*, vol. 55, no. 4, pp. 1197–1203.

37. Grancharov V, Zhao DY, Lindblom J, and Kleijn WB (2006) Low-complexity, nonintrusive speech quality assessment. *IEEE Transactions on Audio, Speech and Language Processing*, vol. 14, no. 6, pp. 1948–1956.

38. Jang JS (1993) Anfis: adaptive-network-based fuzzy inference systems. *IEEE Transactions on System, Man, and Cybernetics*, vol. 23, no. 3, pp. 665–685.

39. Jang JS and Sun CT (1995) Neuro-fuzzy modeling and control. *The Proceedings of the IEEE*, vol. 83, no. 3, pp. 378–406.

40. Jang JS, Sun CT, and Mizutani E (1997) *Neuro-Fuzzy and Soft Computing: A Computational Approach to Learning and Machine Intelligence*, Prentice-Hall, Englewood Cliffs, NJ.

41. Sugeno M and Kang GT (1988) Structure identificaiton of fuzzy model. *Fuzzy Sets and Systems*, vol. 28, pp. 15–33.

42. Takagi T and Sugeno M (1985) Fuzzy identification of systems and its application to modelling and control. *IEEE Transactions on Systems, Man, and Cybernetics*, vol. 15, pp. 116–132.

43. Haralick RM, Shanmugan K, and Dinstein IH (1973) Textural features for image classification. *IEEE Transactions on System, Man, and Cybernetics*, vol. SMC-3, pp. 610–621.

44. Haralick RM (1979) Statistical and structural approaches to texture. *Proceedings of IEEE*, vol. 67, pp. 786–804.

45. Terzopoulos D (1985) Co-occurrence analysis of speech waveforms. *IEEE Transactions on acoustics, speech and signal processing*, vol. ASSP-33, no. 1, pp. 5–30.

46. ITU (1998) ITU-T coded-speech database. *ITU-T P-series Supplement 23*.

A Novel Approach to Language Identification Using Modified Polynomial Networks

Hemant A. Patil[1] and T.K. Basu[2]

[1] Dhirubhai Ambani Institute of Information and Communication Technology, DA-IICT, Gandhinagar, Gujarat, India, hemant-patil1977@yahoo.com
[2] Department of Electrical Engineering, Indian Institute of Technology, IIT Kharagpur, West Bengal, India, tkb@ee.iitkgp.ernet.in

Summary. In this paper, a new method of classifier design, viz., Modified Polynomial Networks (MPN) is developed for the Language Identification (LID) problem. The novelty of the proposed method consists of building up language models for each language using the normalized mean of the training feature vectors for all the speakers in particular language class with discriminatively trained polynomial network having based on Mean-Square Error (MSE) learning criterion. This averaging process in transformed feature domain (using polynomial basis) represents in some sense the common acoustical characteristics of a particular language. This approach of classifier design is also interpreted as designing a neural network by viewing it as a curve-fitting (approximation) problem in a high-dimensional space with the help of Radial-Basis Functions (RBF) (polynomials in the present problem). The experiments are shown for LID problem in four Indian languages, viz., Marathi, Hindi, Urdu and Oriya using Linear Prediction Coefficients (LPC), Linear Prediction Cepstral Coefficients (LPCC) and Mel Frequency Cepstral Coefficients (MFCC) as input spectral feature vectors to the second order modified polynomial networks. Confusion matrices are also shown for different languages.

1 Introduction

Language Identification (LID) refers to the task of identifying an unknown language from the test utterances. LID applications fall into two main categories: pre-processing for machine understanding systems and preprocessing for human listeners. As suggested by Hazen and Zue, consider the hotel lobby or international airport of the future, in which one might find a multi-lingual voice-controlled travel information retrieval system [20]. If the system has no mode of input other than speech, then it must be capable of determining the language of the speech commands either while it is recognizing the commands or before recognizing the commands. To determine the language during recognition would require running many speech recognizers in parallel, one for each language. As one might wish to support tens or even hundreds of

H.A. Patil and T.K. Basu: *A Novel Approach to Language Identification Using Modified Polynomial Networks*, Studies in Computational Intelligence (SCI) **83**, 117–140 (2008)
www.springerlink.com

input languages, the cost of the required real-time hardware might prove pro-
hibitive. Alternatively, an LID system could be run in advance of the speech
recognizer. In this case, the LID system would quickly output a list containing
the most likely languages of the speech commands, after which the few, most
appropriate, language-dependent speech recognition models could be loaded
and run on the available hardware. A final LID determination would only be
made once speech recognition was complete. Alternatively, LID might be used
to route an incoming telephone call to a human switchboard operator fluent
in the corresponding language. Such scenarios are already occurring today:
for example, AT and T offers the Language Line interpreter service to, among
others, police departments handling emergency calls. When a caller to Language Line does not speak any English, a human operator must attempt to
route the call to an appropriate interpreter. Much of the process is trial and
error (for example, recordings of greetings in various languages may be used)
and can require connections to several human interpreters before the appropriate person is found. As reported by Muthusamy et al. [25], when callers
to Language Line do not speak any English, the delay in finding a suitable
interpreter can be of the order of minutes, which could prove devastating in
an emergency situation. Thus, an LID system that could quickly determine
the most likely languages of the incoming speech might cut the time required
to find an appropriate interpreter by one or two orders of magnitude [26, 46].
In addition to this, in the multi-lingual countries like India, automatic LID
systems have an important significance because multi-lingual inter-operability
is an important issue for many applications of modern speech technology. The
need for development of multi-lingual speech recognizers and spoken dialogue
systems are very important in Indian scenario. An LID system can be connected as an excellent front-end device for multi-lingual speech recognizers or
language translation systems [24]. Human beings and machines use different
perceptual cues (such as phonology, morphology, syntax and prosody) to distinguish one language from the other. Based on this, to solve LID problem,
following approaches are used [24]:

- Spectral similarity approaches
- Prosody-based approaches
- Phoneme-based approaches
- Word level based approaches
- Continuous speech recognition approaches

It has been observed that human beings often can identify the language
of an utterance even when they have no strong linguistic knowledge of that
language. This suggests that they are able to learn and recognize language-
specific patterns directly from the signal [22, 24]. In the absence of higher level
knowledge of a language, a listener presumably relies on lower level constraints
such as acoustic-phonetic, syntactic and prosody. In this paper, spectral similarity approach for language identification is used which concentrates on the
differences in spectral content among languages. This is for exploiting the

fact that speech spoken in different languages contains different phonemes and phones. The training and testing spectra could be used directly as feature vectors or they could be used instead to compute cepstral feature vectors [24]. Till recently, Gaussian Mixture Model (GMM), single language phone recognition followed by language modelling (PRLM), parallel PRLM (PPRLM), GMM tokenization and Gaussian Mixture B-gram Model (GMBM) [41], Support Vector Machine (SVM) [9] and Autoassociative Neural Network (AANN) [24] based techniques were applied to the LID problem. Although the GMM-based approach has been successfully employed for speaker recognition, its language identification performance has consistently lagged that of phone-based approaches but it has advantage of being most efficient in terms of time complexity [39, 41]. In contrast, parallel PRLM yielded highest LID rate but was the most complex system for identifying languages. The recent approach proposed by Mary and Yegnanarayana, uses autoassociative neural network (AANN) which performs an identity mapping of the input feature space [19,44] and is a special class of feed-forward neural network architecture having some interesting properties which can be exploited for some pattern recognition tasks [45]. It can be used to capture the non-linear distribution of feature vectors which characterizes each language. In their work, separate AANN models (having structure as 12L 38N 4N 38N 12L, where L denotes linear units and N denotes non-linear units with the activation function of the non-linear unit as hyperbolic tangent function) are used to capture distribution of 12-dimensional spectral feature vectors for each language. The AANN network is trained using error propagation learning algorithm for 60 epochs [44]. The learning algorithm adjusts weights of the network to minimize the Mean Square Error (MSE) for each feature vectors. Once the AANN is trained, it is used as language model [24]. In this paper, the problem of LID is viewed from the standpoint of speaker recognition and a new method for classifier design is suggested for LID problem by modifying the structure of the polynomial networks for preparing language models for different Indian languages, viz., Marathi, Hindi, Urdu and Oriya. We refer this method of classifier design as Modified Polynomial Networks (MPN). The organization of the paper is as follows. Section 2 discusses the motivation and details of training algorithm for polynomial networks along with the proposed modification in the network structure, viz., MPN for LID problem followed by discussion on training algorithm and statistical interpretation of scoring. Section 3 discusses the interpretation of proposed method as designing a neural network with the help of RBF. Section 4 discusses the details of data collection, acquisition and corpus design procedure and different speech features used in this study. Finally, Sect. 5 describes the results on LID for four Indian languages while Sect. 6 concludes the paper with short discussion.

2 Polynomial Networks

As discussed in Sect. 1, in this paper the problem of LID is viewed from the stand point of speaker recognition. The idea is to prepare a language model by adopting and or modifying the techniques of speaker modelling used in speaker recognition task. Hence, the motivation and techniques of polynomial networks for speaker recognition is first discussed here [6,8], and then its structure is modified for the proposed LID problem. Modern speaker recognition task requires high accuracy at low complexity. These features are essential for embedded devices, where memory is at a premium. On server based systems, this property increases the capacity to handle multiple recognitions task simultaneously. Of late, speaker recognition has become an attractive biometric for implementation in computer networks. Speaker recognition has the advantage that it requires little custom hardware and is non-intrusive [8]. Many classifiers have been proposed for speaker recognition. The two most popular techniques are statistical and connectionist-based methods. For the former, Gaussian Mixture Models (GMM) [33] and Hidden Markov Model (HMM) systems with cohort normalization [21, 34] or background model normalization [35] have been the techniques of choice. For connectionist systems, neural tree network and multi-layer perceptrons have been commonly used. GMM is a parametric approach to speaker modelling which assumes that training feature vectors are drawn from the mixture of Gaussian density functions with a priori fixed number of mixtures. These methods are based on only first and second order statistics meaning it involves estimation of the mean and covariance of the speaker data for a parametric model. However, most practical data encountered in speech and other application areas has complex distribution in the feature space and hence cannot be adequately described by GMM, which used only first and second order statistics and mixture weights. Moreover, to have detailed description of the acoustic space and also to achieve good recognition performance, the number of Gaussian mixture components in each model is usually large, especially when diagonal covariance matrices are used [43]. The surface representing the distribution may be highly non-linear in the multi-dimensional feature space, and hence it is difficult to model the surface by simplified models (i.e., in this case Gaussian) [44,45]. In addition to this, the main drawback of this approach is that the out-of-class data is not used to maximize the performance. Moreover, for small amount of training data, the assumption of Gaussian density for feature vectors may not be satisfactory [8, 14]. The discriminative methods to speaker recognition involve the use of discriminative neural networks (NN). Rather than train individual models to represent particular speakers as in case of GMM, discriminative NN are trained to model the decision function which best discriminates speakers within a known set. Oglesby and Mason were the first to exploit the use of neural network techniques in speaker recognition literature [27]. Subsequently, several different neural networks such as most popular multi-layer perceptrons (MLP) (with the weights trained using backpropagation algorithm) [36], radial basis functions (RBF) which can be viewed as clustering followed by a perceptron with the mean of clusters used in kernel function which are typically Gaussian kernels or sigmoids [27, 28, 32], time delay NN which capture the temporal correlations of the input data (i.e., supervised counterpart of HMM) by using layered feed-forward neural network [3], modified neural tree networks (MNTN) [16], NN with encoding scheme based on interspeaker information [42] and probabilistic RAM (pRAM) neural network for a hardware solution to reduce the processing time for a practical system [10]. The most important advantage of NN is that they require a smaller number of model/weights parameters than independent speaker models and have produced good speaker recognition performance, comparable to that of VQ systems. The major aspect of many of the conventional NN techniques is that the complete network must be retrained when a new speaker is added to the system involving a large training time. Moreover, such discriminative methods usually involve large matrices, which lead to large intractable problems [8, 26, 33]. On the other hand approach based on polynomial networks has several advantages over traditional methods such as [6, 8]:

- Polynomial networks have been known in the literature for many years and were introduced for speaker verification and identification recently by Campbell et al. [8]. Because of the Weierstrass–Stone approximation theorem, polynomial networks are universal approximators to the optimal Bayes classifier which is based upon minimizing the average MSE learning criterion by using out of class data to optimize the performance.
- The approach is extremely computationally efficient for LID task meaning that the training and recognition algorithms are based on simple multiply-add architectures which fit well with modern DSP implementations.
- The network is based on discriminative training which eliminates the need for computing background or cohort model (either for speaker or language).
- The method generates small class models which then require relatively small memory.
- The training framework is easily adapted to support new class addition, adaptation, and iterative methods. Above salient features of polynomial network motivated us to employ this network for LID problem. In the next subsections, the details of training method of polynomial networks for speaker recognition, proposed modification in their structure for LID task, training algorithm and statistical interpretation of scoring in LID problem is discussed.

2.1 Training Method

Training techniques for polynomial networks fall into several categories. First being estimates of parameters for the polynomial expansion based on inclass data [15, 40]. These methods approximate the class-specific probabilities [38].

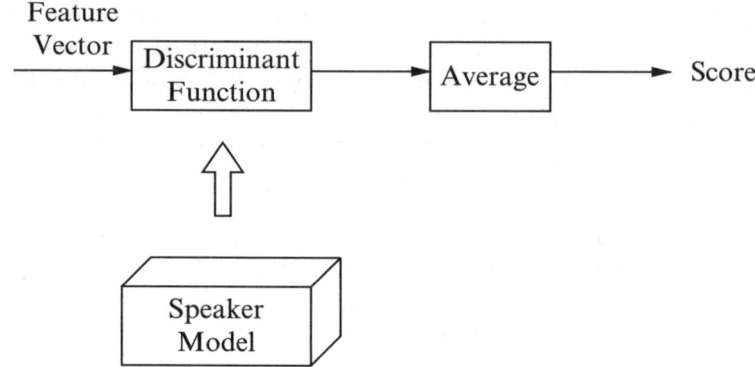

Fig. 1. Audio feature extraction procedure (adapted from [8])

Since out-of-class data is not used for training a specific model, accuracy is limited. A second category of training involves discriminative training with MSE criterion which approximates a posteriori probabilities and is used in this work [6,8,38]. The basic structure of the polynomial network for speaker identification is shown in Fig. 1. The testing feature vectors are processed by the polynomial discriminant function. Every speaker has speaker specific vector $\mathbf{w_j}$, to be determined during machine learning and the output of a discriminant function is averaged over time resulting in a score for every $\mathbf{w_j}$ [8]. The score for jth speaker is then given by

$$s_j = \frac{1}{M} \sum_{i=1}^{M} \mathbf{w}_j^T p(bf x_i), \tag{1}$$

where $x_i = i$th input test feature vector, $\mathbf{w_j} = j$th speaker model and $p(\mathbf{x}_i) =$ vector of polynomial basis terms of the ith input test feature vector.

Training polynomial network is accomplished by obtaining the optimum speaker model for each speaker using discriminatively trained network with MSE criterion, i.e., for a speaker's feature vector, an output of one is desired, whereas for an impostor data an output of zero is desired.

For the two class problem, let $\mathbf{w_{spk}}$ be the optimum speaker model, w the class label, and $y(w)$ the ideal output, i.e., $y(spk) = 1$ and $y(imp) = 0$. The resulting problem using MSE is

$$\mathbf{w}_{spk} = \arg \min_{w} E\left\{ \left(\mathbf{w}^T p(\mathbf{x}) - y(\omega)\right)^2 \right\}, \tag{2}$$

where $E[.]$ means expectation over $\mathbf{x}$ and w. This can be approximated using the training feature set as

$$\mathbf{w}_{spk} = \arg \min_{\mathbf{w}} \left[\sum_{i=1}^{N_{spk}} \left|\mathbf{w}^T p(\mathbf{x}_i) - 1\right|^2 + \sum_{i=1}^{N_{imp}} \left|\mathbf{w}^T p(y_i)\right|^2 \right], \tag{3}$$

where $\mathbf{x}_1, \ldots, \mathbf{x}_{N_{spk}}$ are speaker's training data and $y_{1,\ldots,y_{N_{imp}}}$ is the impostor's training data. This training algorithm can be expressed in matrix form. Let $\mathbf{M}_{spk} = \left[p(\mathbf{x}_1)\, p(\mathbf{x}_2)\, \ldots\, p(\mathbf{x}_{N_{spk}}) \right]^T$ and a similar matrix for $\mathbf{M}_{imp}$. Also let $\mathbf{M} = \left[\mathbf{M}_{spk}\ \mathbf{M}_{imp} \right]^T$ and thus the training problem in (1a) is reduced to the well known linear approximation problem in normalized space as

$$\mathbf{w}_{spk} = \arg\min_{w} \|\mathbf{M}\mathbf{w} - \mathbf{o}\|_2 , \tag{4}$$

where $\mathbf{o}$ consists of N_{spk} ones followed by zeros. We define $\mathbf{R}_{spk} = \mathbf{M}_{spk}^T \mathbf{M}_{spk}$ and define $\mathbf{R}_{imp}$ similarly; and then the problem can be solved using the method of normal equations,

$$\mathbf{M}^T\mathbf{M}\mathbf{w} = \mathbf{M}^T\mathbf{o} \Rightarrow (\mathbf{M}_{\mathbf{spk}}^T\mathbf{M}_{\mathbf{spk}} + \mathbf{M}_{\mathbf{imp}}^T\mathbf{M}_{\mathbf{imp}})\mathbf{w} = \mathbf{M}_{\mathbf{spk}}^T\mathbf{1}, \tag{5}$$

where $\mathbf{1}$ is the vector of all ones. Let us define $\mathbf{R}_{spk} = \mathbf{M}_{spk}^T\mathbf{M}_{spk}$, $\mathbf{R}_{imp} = \mathbf{M}_{imp}^T\mathbf{M}_{imp}$ and $\mathbf{R} = \mathbf{R}_{spk} + \mathbf{R}_{imp}$. Thus (2) reduces to

$$(\mathbf{R}_{spk} + \mathbf{R}_{imp})\mathbf{w}_{spk} = \mathbf{M}_{spk}^T\mathbf{1} \Rightarrow \mathbf{w}_{spk} = \mathbf{R}^{-1}\left(\mathbf{M}_{spk}^T\mathbf{1} \right), \tag{6}$$

where w_{spk} is called as speaker model. Thus the training method for polynomial networks will be based on (3) and since $\mathbf{R} = \mathbf{R}_{spk} + \mathbf{R}_{imp}$, calculation of $\mathbf{R}_{\mathbf{spk}}$ and $\mathbf{R}_{\mathbf{imp}}$ is the most significant part of the machine learning.

2.2 Modified Polynomial Network (MPN) Structure for LID

In this and next subsections, different designs issues considered for using polynomial networks for LID task are described. The basic formulation of the problem for LID using polynomial network is based on MSE criterion similar to the one used in speaker identification or verification; the difference here is that the model for a particular language class (called as Language Model) has been prepared rather than for a speaker. The method of LID is as follows. An unknown language spoken in an utterance is introduced to the system. The unknown language is tested with all the language models stored in the machine. The language model for which the unknown language produces the maximum score is accepted as the genuine language being spoken. Basic notations and terminologies are similar to that of the polynomial network discussed in Sect. 2.1. From (3), the matrix formulation of the MSE criterion for the polynomial network is expressed as

$$\begin{aligned} \mathbf{R}\mathbf{w}_{\mathbf{spk}} &= \mathbf{M}_{\mathbf{spk}}^T\mathbf{1} = \left[p(\mathbf{x}_1)\, p(\mathbf{x}_2)\, \ldots\, p(\mathbf{x}_{N_{spk}}) \right]^T \left[\mathbf{1}\, \ldots\, \mathbf{1} \right]^T \\ &\Rightarrow \mathbf{R}\mathbf{w}_{\mathbf{spk}} = \left[p(\mathbf{x}_1) + p(\mathbf{x}_2) + \cdots + p(\mathbf{x}_{N_{spk}}) \right] = \mathbf{A}_{\mathbf{spk}} \end{aligned} \tag{7}$$

hence $\mathbf{w}_{\mathbf{spk}} = \mathbf{R}^{-1}\mathbf{A}_{\mathbf{spk}}$.

Thus, the vector $\mathbf{A_{spk}}$ represents the normalized mean of the polynomial expansion of training feature vector for a speaker. Now the suggested modification prepares the model for each language by using modified $\mathbf{A_{spk}}$ For proposed LID problem, we define

$$\mathbf{A_i} = \mathbf{M^T_{Lspk,i}}1 = \left[\mathbf{p(x_{1,i})}^T \ \mathbf{p(x_{2,i})}^T \ \cdots \ \mathbf{p(x_{Nspk,i})}^T\right][1 \ \cdots \ 1]^T \quad (8)$$

$$\mathbf{A_i} = \left[\mathbf{p(x_{1,i})^T} + \mathbf{p(x_{2,i})^T} + \cdots + \mathbf{p(x_{Nspk,i})^T}\right] \quad (9)$$

$$\mathbf{A_L} = \sum_{i=1}^{N_{Lspk}} \mathbf{M^T_{Lspk,i}}1 = \sum_{i=1}^{N_{Lspk}} \mathbf{A_i}$$
$$\mathbf{A_L} = \sum_{i=1}^{N_{Lspk}} \left[\mathbf{p(x_{1,i})^T} + \mathbf{p(x_{2,i})^T} + \cdots + \mathbf{p(x_{Nspk,i})^T}\right]. \quad (10)$$

Thus, the A_L vector in (4) represents the normalized mean of the entire training vectors for N_{Lspk} number of speakers in a language class. And hence, in some sense, it tracks common acoustical characteristics of a particular language spoken by N_{Lspk} number of speakers in a linguistic group, i.e., Marathi or Hindi or Urdu or Oriya. Finally, we get the model for each language as

$$\mathbf{w_L} = \mathbf{R^{-1}A_L}, \quad (11)$$

where $\mathbf{w}_L = language model$. Then the training algorithm reduces to finding matrix and Fig. 2 shows the proposed polynomial network structure for

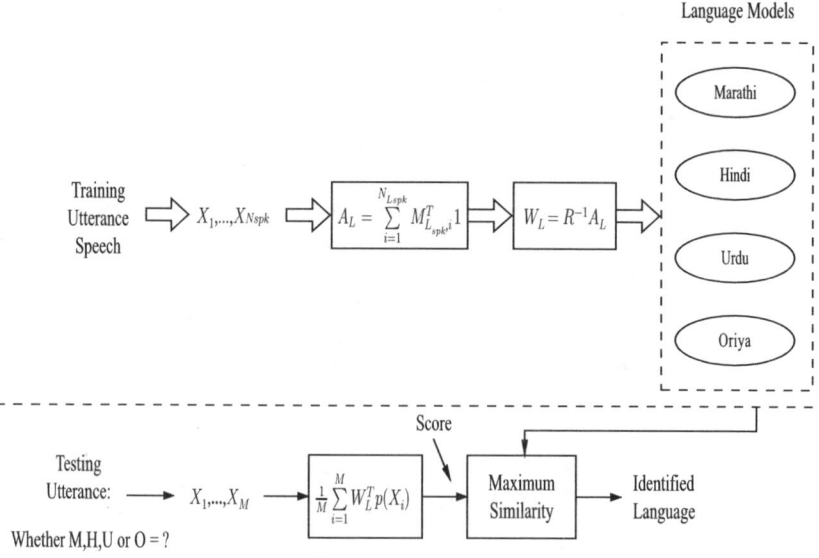

Fig. 2. Proposed MPN architecture for LID

where

$X_1, X_2, \ldots, X_{NPK}$	are feature vectors from training utterances for a particular language,
$W_L =$	language model for Marathi (M), Hindi (H), Urdu (U), or Oriya (O),
$X_1, X_2, \ldots, X_M$	are M testing feature vectors of utterances from an unknown language,
$P(X_i) = 2^{\text{nd}}$	order polynomial expansion of feature vectors of utterances from unknown language.

LID problem. Here the language model is prepared by averaging polynomial coefficients of feature vectors for all the speakers in a particular language. In this work training speech duration is varied from 30 s to 120 s (i.e., more than 2,500–10,000 training feature vectors with 50% overlap in consecutive frames) and hence the vector $\mathbf{A_L}$ represents better language-specific characteristics.

2.3 Algorithm for Polynomial Basis Expansion

The vector of polynomial basis terms, i.e., $p(\mathbf{x})$ can be computed iteratively as discussed below. Suppose one has the polynomial basis terms of order k, and wishes to calculate the terms for order $k + 1$. Assume that every term is of the form $x_{i_1} x_{i_2} \ldots x_{i_k}$, where $i_1 \leq i_2 \leq \cdots \leq i_k$. Now if one has the kth order term of the polynomial basis terms of order k with end terms having $i_k = l$ as a vector u_l then the $(k + 1)$th order terms ending with $i_{k+l} = l$ can be obtained as

$$\begin{bmatrix} x_l u_1 \ x_l u_2 \ \ldots \ x_l u_l \end{bmatrix}^T. \tag{12}$$

Using the above iterative process vector for $p(\mathbf{x})$ can be constructed as follows. Starting with the basis terms of first order, i.e., 1 and the terms of the feature vector, then iteratively calculating the $(k + 1)$th order terms from the kth terms and concatenating the different order terms, the desired $p(\mathbf{x})$ of a given order can be computed [8]. This method is explained in Table 1 with an example for 2-D feature vector and second-order polynomial approximation.

2.4 Training and Mapping Algorithm

The matrix $\mathbf{R}$ in (5) contains many redundant terms and this redundancy is very structured. Storage space and computation time can be reduced by calculating only the unique terms. These unique terms are denoted by $p_u(\mathbf{x})$. One can plot the index of terms in the vector $p_u(\mathbf{x})$ versus the index of the terms in the matrix $\mathbf{R}$. We can compute the vector $p(\mathbf{x})$ and $p_u(\mathbf{x})$. We index the element in $\mathbf{R}$ as a one dimensional array using column major form. The structure in this index map is self similar and it becomes more detailed as the polynomial degree is increased (as shown in Fig. 3a [8]).

Table 1. Algorithm for polynomial basis expansion for second order with $\mathbf{x} = [\mathbf{x_1} \ \mathbf{x_2}]^\mathbf{T}$

Step 1: Terms of order 1: $[1 \ \ x_1 \ \ x_2]^T$

Step 2: First-order terms: $[x_1 \ \ x_2]$

Step 3: $u_1 = [x_1]$, $u_2 = [x_2]$

Step 4: For $l = 1$ the $(k+1)$th terms with $i_{k+1} = l = 1$ is given by $[x_1 u_1] = [x_1 x_1]$

For $l = 2$ the $(k+1)$th terms with $i_{k+1} = l = 2$ is given by
$$\begin{bmatrix} x_2 u_1 \\ x_2 u_2 \end{bmatrix} = \begin{bmatrix} x_2 x_1 \\ x_2 x_2 \end{bmatrix}$$
Thus the second-order terms are $\begin{bmatrix} x_1^2 & x_1 x_2 & x_2^2 \end{bmatrix}$

Step 5: Concatenating all the terms we get the polynomial basis terms of order 2
as $p(\mathbf{x}) = \begin{bmatrix} 1 & x_1 & x_2 & x_1^2 & x_1 x_2 & x_2^2 \end{bmatrix}^T$

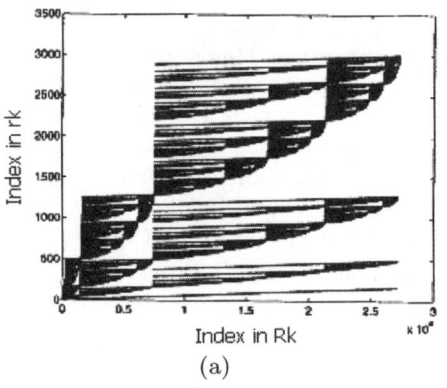

(a)

T: Isomorphism

$p(X) \longrightarrow N; r \longrightarrow R$

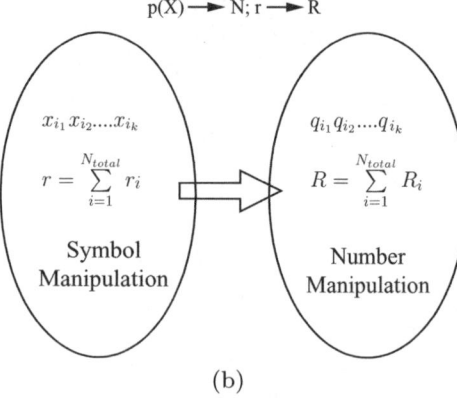

(b)

Fig. 3. (a) Index function for eight features and degree 3 (after Campbell et al. [8]). (b) Semigroup structure of monomials and mapping from to. N_{total} is the sum of total number speakers in all the language classes

To exploit the redundancy in the matrix $\mathbf{R}$, let us consider a specialized case when we map a class's vector of unique terms to a matrix structure $\mathbf{R_k}$. The matrix $\mathbf{R_k}$ is obtained from a sum of outer products

$$R_k = \sum_{i=1}^{N_k} p(\mathbf{x}_{k,i})p(\mathbf{x}_{k,i})^T, \tag{13}$$

where N_k = number of training feature vectors for kth class (which may be speaker or a language). The mapping is based upon the fact that it is not necessary to compute the sum of outer products in (6) directly. Instead one can directly compute the subset of unique entries (i.e., the vector $p_u(\mathbf{x})$), and then map this result to the final matrix. This mapping can be done for all the speakers in each language class individually but the most efficient way will be to map the single matrix r (which is the sum of vectors of unique entries for all speakers in actual language class and impostors' language classes) into $\mathbf{R}$ as shown in Fig. 3b. One straightforward way is to use an index function and computer algebra package. However, the difficulty in using an index function is that the index map must be stored. To avoid this problem, an alternative method based upon a simple property, the semigroup structure of the monomials (an algebraic structure which only imposes the property of multiplication on the elements of the set) is used. Suppose we have an input vector with n variables. The mapping is

$$x_{i_1} x_{i_2} \ldots x_{i_k} \mapsto q_{i_1} q_{i_2} \ldots q_{i_k}, \tag{14}$$

where q_{i_j} is the i_jth prime, and defines a semigroup isomorphism between the monomials and the natural numbers, i.e., the algebraic structure is preserved in range space. For example, we map $x_1^2 x_2 = x_1 x_1 x_2$ to $q_1 q_1 q_2 = 2^2 3 = 12$ (2 is the first prime). We can implement this mapping procedure efficiently using this semigroup isomorphism since it transforms symbol manipulation into number manipulation and the mapping algorithm for this is given in [8]. Combining all of the methods results in the training algorithm (or deign of MPN) for LID as shown in Table 2. The algorithm is not limited to polynomials. One key enabling property is that the basis elements form a semigroup as discussed in mapping procedure and this property allows us to compute only unique elements in the $\mathbf{R_k}$ matrix for the kth speaker. Another key property is partitioning of the problem. If a linear approximation space (e.g., Hilbert space of measurable functions which is known to be convex and hence point of extremum is guaranteed to exist with minimum approximation error [30]) is used then we can obtain the same problem as in (1a) and the problem can be broken up to ease memory usage [8].

2.5 Statistical Interpretation of Scoring for LID

Let us first formulate the problem of LID in a Bayesian framework and then its relation to the discriminant function-based approach to pattern recognition. The motivation for the probabilistic modelling of language classes is that

Table 2. Proposed training algorithm (or design of MPN) for LID

1. For $i = 1$ to $N_{Lclasses}$	[$N_{Lclasses}$ = total number of language classes]
2. Let $\mathbf{r_{i,k}}, \mathbf{A}_{L_i} = 0$ and $sum = 0$	[$\mathbf{r_{i,k}}$ corresponds to $p_u(\mathbf{x}_{i,k,j})$, $\mathbf{A_{i,k}}$ corresponds to $\mathbf{M}^T_{Lspk,i,k}\mathbf{1}$]
3. For $k = 1$ to N_{total}	[N_{total} = sum of total number of speakers for all the language classes]
4. For $j = 1$ to N_{spk}	[corresponds to the number of training feature vectors for a speaker in a language class]
5. $\mathbf{r_{i,k}} = \mathbf{r_{i,k}} + p_u(\mathbf{x_{i,k,j}})$	[$\mathbf{x_{i,k,j}}$ refers to the jth training feature vector of kth speaker belonging to i^{th} language class]
6. Next j	
7. Compute $sum = sum + r_k$	
8. Next k	
9. Next i	
10. $r \leftarrow sum$	
11. Map $\mathbf{r}$ to $\mathbf{R}$	[mapping algorithm]
12. For $i = 1$ to $N_{Lclasses}$	
13. $\mathbf{A}_{L_i} = \sum\limits_{k=1}^{N_{Lspk,i}} \mathbf{M}^{\mathbf{T}}_{\mathbf{Lspk},k}\mathbf{1}$	[proposed modification for LID problem] [$N_{Lspk,i}$=no. of speakers in language class]
14. $\mathbf{w}_{L_i} = \mathbf{R}^{-1}\mathbf{A}_{L_i}$	[$\mathbf{w}_{L_i}$ = language model for i^{th} class]
15. Next i	

the traditional method of using a minimum distance classifier (Euclidean or Mahalanobis distance) to find the distance between the means of feature vectors is not effective in the presence of channel noise and background noise [18]. On the other hand, we can model a broad class of language or speaker-specific phonemes such as vowels, nasals, fricatives and nasal-to-vowel coarticulation with mathematically tractable probabilistic models (e.g., Gaussian pdf in case of GMM [33, 41]).

If $L_1, L_2, \ldots, L_n$ is the number of languages (to be identified) in the language space L (as shown in Fig. 4), then given the test utterance, the probability that the unknown language to be identified as the ith language of a test utterance is given by the Bayes theorem [14]:

$$p(L_i/L_{test}) = \frac{p(L_{test}/L_i)P(L_i)}{\sum\limits_{i=1}^{n} p(L_{test}/L_i)P(L_i)}, \tag{15}$$

where $L_{test} = x =$ test feature vector per frame from an utterance of unknown language, $p(L_i/L_{test}) = p(L_i/\mathbf{x}) =$ a posteriori probability that the test

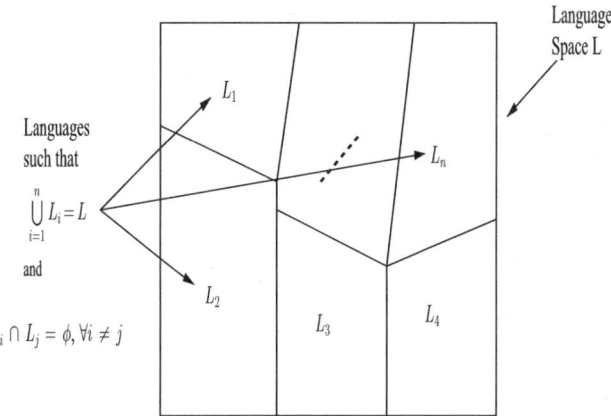

Fig. 4. Language space for n-class LID problem in Bayesian framework

feature vector per frame belongs to the ith language, $p(L_{test}/L_i) = p(\mathbf{x}/L_i) = $ a priori probability of the test feature vector per frame to be generated from ith language. This is also known as the likelihood function of the ith language, $P(L_i) = \pi_{Lspk,training} = $ class priors or training priors for different language classes, i.e., the prior probability that the test feature vector per frame came from the ith language. If we assume that the prior probabilities are equal, then the denominator term in (7) becomes an average probability, i.e., $P_{av}(\mathbf{x})$ which will be constant for all the language classes. Hence the task of identifying a language reduces to finding the language whose a posteriori probability and hence the likelihood, i.e., $p(\mathbf{x}/L_i)\,\forall i \in [1,n]$, scores the maximum for the given test feature vector per frame. The likelihood function approach requires knowledge of estimation of conditional pdfs of all the language classes. These estimation procedures are complex, and require a large number of samples to give accurate results, and hence are difficult to implement and therefore time and storage requirements for the recognition process may be excessive. To alleviate this problem, a discriminant function-based approach is used. Now let us find the relationship between the likelihood function and the discriminant function Let there be two languages and (one actual language and another as impostors' language for a given test utterance). The decision can be expressed by an algorithm for this two class problem as

$$if\, g_{L_1}(\mathbf{x}) > g_{L_2}(\mathbf{x})\mathbf{x} \in L_1 else\mathbf{x} \in L_2. \tag{16}$$

Let the discriminant function of an ith language be given by

$$g_{L_i}(\mathbf{x}) = \mathbf{w}_{L_i}^T\mathbf{x} = \langle \mathbf{w}_{L_i}, \mathbf{x}\rangle, \forall i \in [1,n], \tag{17}$$

where $\mathbf{w}_{L_i} = $ ith language model and $\mathbf{x} = $ test feature vector feature vector per frame. Thus represents the inner product of a test utterance with the model of the ith language. Hence the closer the vector $\mathbf{x}$ is to $\mathbf{w}_{L_i}$, the more

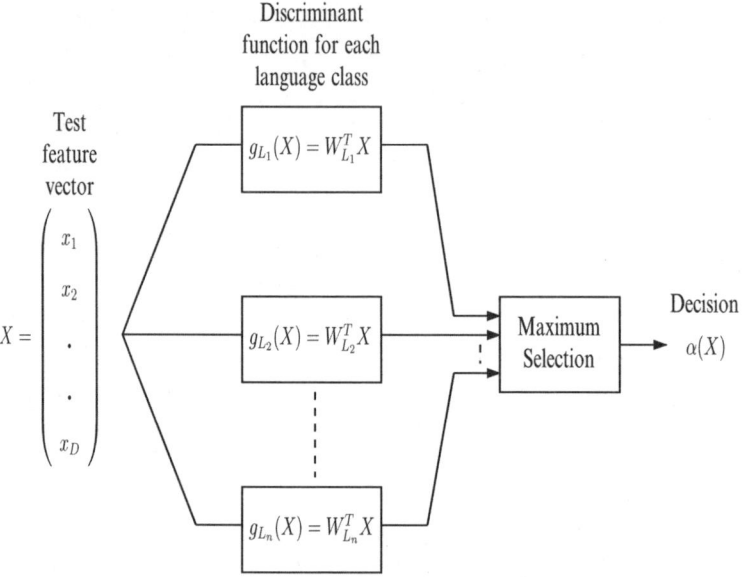

Fig. 5. Discriminant function approach for n-class LID problem

probable is the situation that it belongs to L_i, i.e., the ith language class (as shown in Fig. 5), and thus the discriminant function can be conceived as a posterior probability of unknown language in a Bayesian framework

$$\Rightarrow g_{L_i}(\mathbf{x}) = \mathbf{w}_{L_i}^T \mathbf{x} \cong p(L_i/\mathbf{x}) = p(\mathbf{x}/L_i). \tag{18}$$

Now for a sequence of M test feature vectors (each of D-dimensional), i.e., $\mathbf{X} = [\mathbf{x_1}, \mathbf{x_2}, \dots, \mathbf{x_M}]$, one needs to compute, i.e., the probability of the sequence of feature vectors given the model of the ith speaker. We abbreviate this as $p(\mathbf{X}/L_i)$. The standard practice in speech literature is to assume independence of feature vectors with a certain degree of confidence [31] whereby the unconditional probability may be expressed in a product form as shown below:

$$p(\mathbf{X}/L_i) = p(\mathbf{x_1}, \mathbf{x_2}, \dots, \mathbf{x_M}/S_i) = \prod_{k=1}^{M} p(\mathbf{x_k}/L_i). \tag{19}$$

By the Bayes theorem,

$$g_{L_i}(\mathbf{x}) = \prod_{k=1}^{M} \frac{p(L_i/\mathbf{x_k})P_{av}(\mathbf{x_k})}{p(L_i)}. \tag{20}$$

The factor $P_{av}(\mathbf{x})$ (i.e., the average probability) in (8) cancels out of the likelihood ratio as discussed earlier and hence will be ignored. For computational and analytical simplicity (especially in case of maximum-likelihood estimation (MLE) approach [14, 33]), a log-likelihood function is employed, and since logarithm is a monotonic function, this will not affect the decision process.

$$\log\{g_{L_i}(\mathbf{x})\} = \sum_{k=1}^{M} \log\left[\frac{p(L_i/\mathbf{x_k})}{p(L_i)}\right]. \tag{21}$$

By using Taylor's series, a linear approximation of around is. Thus, we can approximate $\log\{g_{L_i}(\mathbf{x})\}$ as

$$\log\{g_{L_i}(\mathbf{x})\} \approx \sum_{k=1}^{M} \left(\frac{p(L_i/\mathbf{x_k})}{p(L_i)} - 1\right). \tag{22}$$

Since $x - 1$ goes to -1 as x as goes to zero and $\log(x)$ goes to $-\infty$, using $x-1$ is approximately equivalent to replacing $\log(x)$ by $\max(\log(x), -1)$. The approximation, $\max(\log(x), -1)$ is equivalent to ensuring that the probability is not allowed to go below a certain value. The discriminant function with all the above approximations and after normalization is

$$\log\{g_{L_i}(\mathbf{x})\} \approx \frac{1}{M}\sum_{k=1}^{M}\frac{p(L_i/\mathbf{x_k})}{p(L_i)} = \frac{1}{M}\sum_{k=1}^{M}\mathbf{w}_{L_i}^T p(\mathbf{x}_k)$$

$$= \mathbf{w}_{L_i}^T\left(\frac{1}{M}\sum_{k=1}^{M}p(\mathbf{x}_k)\right) = \mathbf{w}_{L_i}^T\bar{p}, \quad (23)$$

where $\bar{p} = \frac{1}{M}\sum_{k=1}^{M}p(\mathbf{x}_k)$. Thus, we first compute from the test utterance and then $\mathbf{w}_{L_i}^T\bar{p}$. Thus scoring for each language model is accomplished by one inner product a low-complexity operation. We have dropped the term -1 since a constant offset will be eliminated in a likelihood function. The process of normalization in (9) is normalized by the number of frames to ensure that a constant threshold can be used in recognition task. Finally, for optimal Bayes identification with equal class priors (e.g., LID task), we choose so that $i^* = \arg\max_i\{g_{L_i}(\mathbf{x})\}$. This scoring method can be extended to any linear approximation space. The key point that allows the simplification of the scoring is that the quantity $\frac{1}{M}\sum_{k=1}^{M}\mathbf{w}_{L_i}^T p(\mathbf{x}_k)$ can be written as $\mathbf{w}_{L_i}^T\left(\frac{1}{M}\sum_{k=1}^{M}p(\mathbf{x}_k)\right)$. This property is dependent only on the fact that we are using a linear basis, i.e., in this case the components of $p(\mathbf{x})$. Based on (9) Campbell has derived Naive a Posteriori Sequence (NAPS) kernel which greatly simplifies the computational complexities in SVM which will be discussed very briefly in next subsection [7].

3 MPN as Designing Neural Network using RBF

We can conceive the proposed technique of MPN for LID system in terms of a neural network which consists of an input layer, two hidden layers and an output layer (as shown in Fig. 6). The input layer consists of a sequence of M input testing feature vectors (each of D-dimensional), i.e., $\{\mathbf{x}_1, \ldots, \mathbf{x}_M\}$ (called as sensory units) given to the first hidden layer (or first network) which maps this to a vector of polynomial basis terms, i.e., $p_d(\mathbf{x})$ of polynomial basis terms depending upon the degree (d) of the network (i.e., second degree, third degree, ..., dth degree, etc.). Note that $p_d(\mathbf{x})$ is de-noted as in Sect. 2 for generalized discussion. The second hidden layer computes the inner product kernel of vector $p_d(\mathbf{x})$ with ith language model (stored during machine learning) to give $\mathbf{w}_{L_i}^T p_d(\mathbf{x})$. The elements of language model can be conceived as synaptic weights in neural network which are to be determined using MSE learning criterion. Since there is a sequence of M testing feature vectors, the total score (i.e., the output layer) for a particular language is computed as an average score over all input test feature vectors, i.e., $s_i = (1/M) \sum_{k=1}^{M} \mathbf{w}_{L_i}^T p_d(\mathbf{x}_k)$. Note that here we do not use a sigmoid function as activation function which is common in higher order neural networks [17]. However, the monomial terns, viz., $1, x_1, x_2, \ldots, x_D, x_1 x_2, \ldots, x_1 x_D, \ldots, x_1^2, x_2^2, \ldots, x_D^2, \ldots$ in the vector can be conceived as activation functions [14].

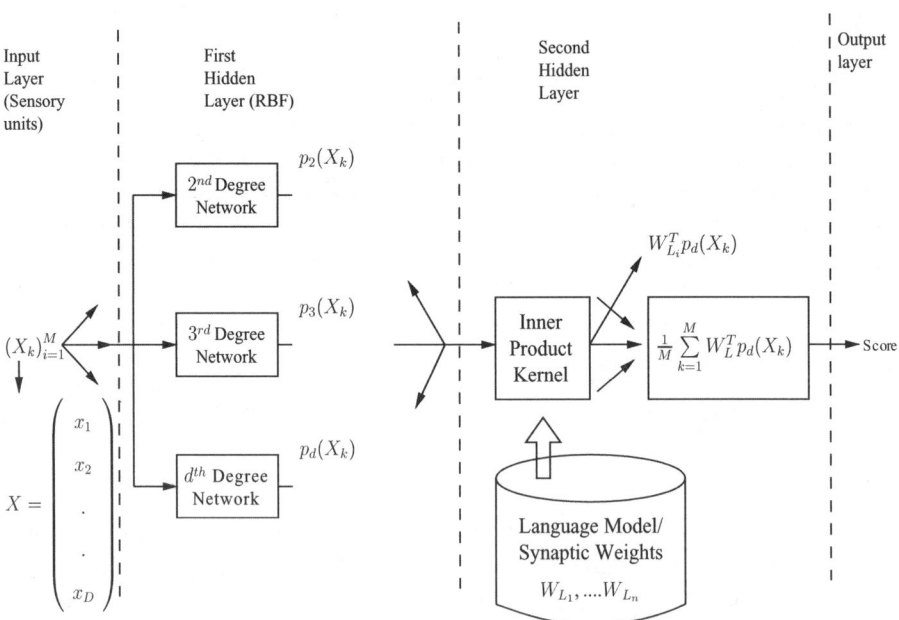

Fig. 6. Interpretation of MPN in terms of neural network structure

The scoring or more formally generalized discriminant function $\mathbf{w}_L^T p_d(\mathbf{x})$ is not linear in but it is linear in $p_d(\mathbf{x})$. Such mapping of input feature space to higher dimensional space help in capturing the language-specific information hidden in different components/directions of $p_d(\mathbf{x})$. This approach of classifier design can be interpreted as designing a neural network by viewing it as a curve-fitting (approximation) problem in a high-dimensional space [19]. According to this viewpoint, learning is equivalent to finding a surface in multi-dimensional space that provides a best fit to training data, with the criterion for "best fit" being measured in some statistical sense such as MSE. Such an interpretation is indeed the motivation for introducing the use of RBF into the neural network literature. Broomhead and Lowe were the first to exploit the use of RBF in the design of neural network [5]. The RBF network acts as "basis functions" in hidden layer (e.g., polynomial in the present work) which transforms original input feature space from input layer to a high-dimensional feature space (i.e., the mapping is non-linear) [4]. The output of RBF is mapped to output layer, i.e., score for a language using inner product kernel (i.e., the mapping is linear). Mathematical justification to this rationale can be traced back to the Cover's theorem in his early paper [12]. According to the Cover's theorem on separability of patterns, an input space made up of non-linearly separable patterns may be transformed into a feature space where the patterns are linearly separable with high probability, provided the transformation is non-linear and the dimensionality of the feature space is high enough [12, 19, 37]. Figure 7 shows the graphical illustration of these concepts.

The inner product kernel discussed above can be conceived as NAPS kernel (Sect. 2.5) since scoring assumes statistical independence of observations

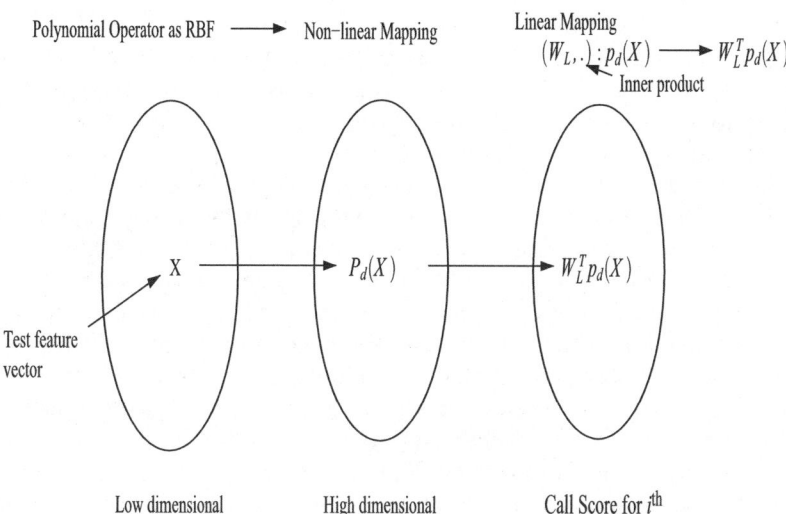

Fig. 7. Graphical illustration of MPN as designing a neural network using RBF

and training approximates the a posteriori probabilities in Bayesian framework [7]. The value of the NAPS kernel can be interpreted as scoring using a polynomial network on the sequence $\{x_1, \ldots, x_M\}$ with the MSE model trained using training feature vectors. Several observations could be made about the NAPS kernel. First, NAPS kernel is used to reduce scoring complexities using Cholesky decomposition. Second, since NAPS kernel explicitly performs expansion to "feature space", we can simplify the output of SVM, i.e., once we train the SVM, we can collapse all the support vectors down into a single model, which greatly simplifies the (soft) output of SVM. Third, although the NAPS kernel is reminiscent of Mahalanobis distance, it is distinct. No assumption of equal covariance matrices for different classes is made for the new kernel – the kernel covariance matrix is the mixture of individual class covariances. Also, the kernel is not a distance measure [7].

4 Setup

In this section brief detail of setup used for this study, viz., data collection and corpus design and different speech features is given.

4.1 Data Collection and Corpus Design

Database of 240 subjects (60 in each of Marathi, Hindi, Urdu and Oriya) is prepared from the different states of India, viz., Maharashtra, Uttar Pradesh, West Bengal and Orissa with the help of a voice activated tape recorder (Sanyo model no. M-1110C and Aiwa model no. JS299) with microphone input, a close talking microphone (viz., Frontech and Intex). The data is recorded on the Sony high fidelity voice and music recording cassettes (C-90HFB). A list consisting of five questions, isolated words, digits, combination-lock phrases, read sentences and a contextual speech of considerable duration was prepared. The contextual speech consisted of description of nature or memorable events, etc. of community or family life of the subject. The data was recorded with 10 repetitions except for the contextual speech. During recording of the contextual speech, the interviewer asked some questions to subject in order to motivate him or her to speak on his or her chosen topic. This also helps the subject to overcome the initial nervousness and come to his or her natural mode so that the acoustic characteristics of his or her language are tracked precisely. During recording, both the interviewer and subject were able to see and interact with each other. The subject's voice and interviewer's voice were recorded on the same track. Once the magnetic tape was played into the computer, the subject's voice was played again to check the wrong editing. The interviewer's voice was deleted from the speech file so that there will not be any biasing for a particular language. The automatic silence detector was employed to remove the silence periods in the speech recordings to get only the language model for the subject's voice and not the background noise and silence interval. Also, each subject's voice is normalized by the peak value so that the speech amplitude level is constant for all the subjects in a particular language. Finally, the corpus is designed into training segments of 30, 60, 90 and 120 s durations and testing segments of 1, 3, 5, 7, 10, 12 and 15 s durations in order to find the performance of the LID system for various training and testing durations. Table 1 shows the details of the corpus. Other details of the experimental setup and data collection are given in [30]. Following are salient features of our corpus:

- Wide varieties of acoustic environments were considered during recording ranging from office roads train, to noisy workstations, etc. which added realistic acoustic noise (e.g., crosstalk, burst, spurious noise activity, traffic noise, etc.) to the data. This is the most important feature of our corpus.
- Speech units including specific vocabularies, e.g., isolated words, digits, combination-lock phrases, read sentences, question–answer session, and contextual speech/spontaneous speech of considerable duration with varying topics were considered in the recording which added realistic situations in the speech data.
- Data is not recorded in closed booth (research lab/office), where the speaker may not feel free to give his/her speech data.
- Speakers of wide ranging ages (15–80 years) and a variety of educational backgrounds (from illiterate to university post graduates) have been considered in the corpus.
- Database is prepared from non-paid subjects and hence their natural mode of excitement was not altered and only those subjects who were ready to volunteer for this study were considered.

Following are some of our practical experiences during recording:

- The presence of observer which may be interviewer, recording equipment or any other tool of measurement affects the natural acoustical characteristics for a particular language (i.e., Labov's observer's paradox [11]).
- Subjects unconsciously talk louder in front of microphone was observed during recording (i.e., Lombard effect [23]).
- Some initial resistance was experienced from a small section of native speakers in Marathi for recording of Hindi language.
- Speakers occasionally became bored or distracted, and lowered their voice intensity or turned their heads away from the microphone.
- There was stammering, laughter, throat clearing, tittering and poor articulation. All these cases were recorded in normal fashion.

4.2 Speech Features Used

In this paper, Linear Prediction Coefficients (LPC), Linear Prediction Cepstral Coefficients (LPCC) [1, 2] and Mel frequency Cepstral Coefficients (MFCC) [13] are used as 12-dimensional spectral feature vectors for all the LID experiments. In next subsections, their computational details are discussed.

LP-Based Features

The combined effect of glottal pulse, vocal tract and radiation at lips can be modelled by a simple filter function, for a speech signal. A quasi-periodic impulse train is assumed for the voiced part and a random noise as the input for the unvoiced part at the output speech. The gain factor accounts for the intensity (assuming a linear system) [1]. Combining the glottal pulse, vocal tract and radiation yields a single all-pole transfer function given by

$$H(z) = \frac{G}{1 - \sum\limits_{k=}^{p} a_k z^{-k}}, \tag{24}$$

where are called as LPC and they are computed by using autocorrelation method. LPC are known to be dependent on text material used for the recording [1] and hence in some sense the syllables, phonemes in speech and in turn are supposed to carry language-specific information. Given that all the poles are inside the unit circle and the gain is 1, the causal LPCC of is given by [29]:

$$LPCC(n) = \frac{1}{n} \sum\limits_{i=1}^{p} |r_i|^n \cos(\theta_i n), n > 0, \ with \ z_i = r_i \exp(j\theta_i), \tag{25}$$

where are the poles of LP transfer function. LPCC is known to decouple the vocal tract spectrum with source spectrum. In general, the first 0–5 ms information in LPCC corresponds to vocal tract information (since pitch duration is normally in the range of 5–10 ms) and hence LPCC captures vocal tract information more dominantly than LPC. Moreover, the most important reason for dominant use of LP analysis in speech literature is that it carries directly information about time-varying vocal tract area function [30]. Since phoneme spectra in a particular language are produced with significant contribution of vocal tract spectral envelope and they are also language-specific, LPCC can track language-specific information better than LPC.

MFCC

Features for a particular speech sound in a language, the human perception process responds with better frequency resolution to lower frequency range and relatively low frequency resolution in high frequency range with the help of human ear. To mimic this process MFCC is developed. For computing MFCC, we warp the speech spectrum into Mel frequency scale. This Mel frequency warping is done by multiplying the magnitude of speech spectrum for a preprocessed frame by magnitude of triangular filters in Mel filterbank followed by log-compression of sub-band energies and finally DCT. Davis and Mermelstein proposed one such filterbank to simulate this in 1980 for speech recognition application [30]. The frequency spacing of the filters used in Mel

Table 3. Database description for LID system

Item	Details
No. of subjects	240 (60 in each of Marathi, Hindi, Urdu and Oriya)
No. of sessions	1
Data type	Speech
Sampling rate	22,050 Hz
Sampling format	1-channel, 16-bit resolution
Type of speech	Read sentences, isolated words and digits, combination-lock phrases, questions, contextual speech of considerable duration
Application	Text-independent language identification system
Training language	Marathi, Hindi, Urdu, Oriya
Testing language	Marathi, Hindi, Urdu, Oriya
No. of repetitions	10 except for contextual speech
Training segments (s)	30, 60, 90, 120
Test segments (s)	1, 3, 5, 7, 10, 12, 15
Microphone	Close talking microphone
Recording Equipment	Sanyo Voice Activated System (VAS: M-1110C), Aiwa (JS299), Panasonic magnetic tape recorders
Magnetic tape	Sony high-fidelity (HF) voice and music recording cassettes
Channels	EP to EP Wire
Acoustic environment	Home/slums/college/remote villages/roads

filterbank is kept as linear up to 1 kHz and logarithmic after 1 kHz. The frequency spacing is designed to simulate the subjective spectrum from physical spectrum to emphasize the human perception process. Thus, MFCC can be a potential feature to identify phonetically distinct languages (because for phonetically similar languages there will be confusion in MFCC due to its dependence of human perception process for hearing).

5 Experimental Results

Feature analysis was performed using a 23.2 ms duration frame with an overlap of 50%. Hamming window was applied to each frame and subsequently; each frame was pre-emphasized with a high-pass filter (1-0.97z-1). No data reduction techniques were applied. These feature vectors are fed to the polynomial network for building language model. The network builds up model for each language for different training speech durations such as 30, 60, 90, and 120 s by averaging polynomial coefficients of the feature vectors of 60 subjects for each language class. During testing phase, 12-dimensional LPC,

Table 4. Average success rates (%) for two languages (M and H) with second-order approximation

FS/TR	30 s	60 s	90 s	120 s
LPC	44.76	58.45	60.11	57.97
LPCC	60.59	56.31	57.38	54.76
MFCC	62.97	67.02	68.09	67.97

Table 5. Average success rates (%) for two languages (H and U) with second-order approximation

FS/TR	30 s	60 s	90 s	120 s
LPC	5	19.4	22.02	31.19
LPCC	37.38	43.57	46.07	43.81
MFCC	21.42	22.97	23.57	23.69

Table 6. Average success rates (%) for three languages (M, H and U) with second-order approximation

FS/TR	30 s	60 s	90 s	120 s
LPC	28.17	37.61	37.69	36.03
LPCC	71.74	69.6	71.66	67.54
MFCC	65.15	64.12	68.01	66.74

Table 7. Average success rates (%) for four languages (M, H, U and O) with second-order approximation

FS/TR	30 s	60 s	90 s	120 s
LPC	22.49	28.8	28.5	26.84
LPCC	53.33	50.83	52.61	52.31
MFCC	51.78	51.96	53.27	54.87

Table 8. Confusion matrix with second-order approximation for LPC (TR = 120 s and TE = 15 s) with four languages

Ident./act.	M	H	U	O
M	53.33	43.33	0	3.33
H	36.66	55	3.33	5
U	63.33	36.66	0	0
O	51.66	48.33	0	0

Table 9. Confusion Matrix with second-order approximation for LPCC (TR = 120 s and TE = 15 s) with four languages

Ident./act.	M	H	U	O
M	13.33	61.66	5	20
H	8.33	78.33	6.66	6.66
U	0	25	75	0
O	5	13.33	41.66	40

Table 10. Confusion Matrix with second-order approximation for MFCC (TR = 120 s and TE = 15 s) with four languages

Ident./act.	M	H	U	O
M	78.33	6.66	5	10
H	46.66	33.33	20	0
U	8.33	1.66	90	0
O	61.66	0	15	23.33

LPCC and MFCC feature vectors were extracted per frame from the testing speech and score for each unknown language is computed against each of the stored language models. Finally, test utterance is assigned to a language whose score gives maximum value. Results are shown in Tables 4–7 as average success rates (i.e., average over 1, 3, 5, 7, 10, 12, and 15 s testing speech durations and success rates refers the number of correctly identified speakers in % for a particular language) for different training (TR) durations taken two or three or four languages at a time and with different feature sets (FS). Tables 8–10 show confusion matrices (diagonal elements indicate % correct identification in a particular linguistic group and off-diagonal elements show the misidentification) for four languages with different feature sets (FS). In Tables 8–10, ACT represents the actual language of the speaker and IDENT represents the identified language of an unknown speaker. The performance of a confusion matrix is evaluated based on its diagonal and off-diagonal entries meaning a confusion matrix will be ideal, i.e., all the testing samples are correctly identified to their respective classes; if all the off-diagonal elements are zero and diagonal elements are 100. Any deviation from this will judge the relative performance of the confusion matrix). Thus the confusion matrix indicates the effectiveness of the proposed MPN model in capturing language-specific information with the help spectral features. Some of the observations from the results are as follows:

- For two class LID problem of Marathi and Hindi, MFCC performed well compared to LPC and LPCC. This may be due to the fact that MFCC is based on human perception process and hence is better able to discriminate between Marathi and Hindi whereas LPC and LPCC are known to be dependent on pronunciation of text material used for data collection [1]. Hindi data is recorded with subjects having Marathi as native language and confusion in lexicon (either from his or her native language or non-native language) used by the subjects cannot be ruled out and hence chances of misclassification amongst the classes increase (Table 4).

- For the two class LID problem of Hindi and Urdu, LPCC performed better than MFCC. This may be due to the fact that the sounds of Hindi and Urdu phonemes are very much similar in perception and hence MFCC features may fail sometimes in discriminating between these two classes as discussed in subsection MFCC and perform poor than LPCC (Table 5).

- For the two class LID problem of Hindi and Urdu, LPCC performed better than LPC as well. This may be due to the fact that LPC is dependent on recording text material and due to high similarity in phonemes for Hindi and Urdu, LPC fails to track language-specific information whereas LPCC captures gross (due to virtue of machine learning using MPN) vocal tract information of 60 speakers in a language class, it captures phoneme spectra of a language more dominantly than LPC and performs better than LPC (as discussed in subsection LP-Based Features) (Table 5). In fact, this is the reason for better performance of LPCC than LPC in all the cases of experiments reported in Tables 4–7.

- Success rates are relatively low for the two class problem of Marathi and Hindi as compared to Hindi and Urdu. This may be due to the similarity in perceiving the sounds of Hindi and Urdu as compared to that of Marathi and Hindi (Tables 2 and 3).

- For 3 and 4 class LID problems, both LPCC and MFCC performed almost equally well (Tables 4 and 5).

- Confusion matrices performed well for MFCC as compared to LP-based features for a majority of the cases (Tables 6–8).

Comparison of results with other existing techniques for LID will not be very accurate to a certain extent. This may be due to the fact that use of different experimental setups used for data collection and corpus design and the most important factors such as educational background, socio-economic status, speaking rate and intelligence will affect features for speech and hence the language spoken in an utterance.

6 Summary and Conclusion

In this paper, a language identification system is presented using new MPN-based approach to classifier design. The novelty of the proposed method consists of building up language models for each language using the normalized mean of the spectral training feature vectors for all the speakers in a particular language with discriminatively trained MPN having MSE learning criterion. Major contributions of the paper are as follows:

- Introduction of new classifier design technique, viz., MPN for LID problem.
- Formulation of LID problem using MPN in Bayesian framework, i.e., statistical interpretation of scoring.
- We believe that the first time MPN is interpreted as designing a neural network with RBF.
- Finally, results on LID tasks are reported with corpus prepared in real life settings, i.e., data with realistic acoustic noise and experimental conditions.

Acknowledgments

The authors of this paper would like to thank authorities of IIT Kharagpur and Dr. BCREC Durgapur for their kind support to carry out this research work. They also thank Prof. B. Yegnanarayana of IIIT Hyderabad for his valuable comments and suggestions about our work during WISP-06 at IIT Madras.

References

1. Atal, B. S., Hanuaer, S. L.: Speech analysis and synthesis by linear prediction of the speech wave. J. Acoust. Soc. Am. 50 (1971) 637–655
2. Atal, B. S.: Effectiveness of linear prediction of the speech wave for automatic speaker identification and verification. J. Acoust. Soc. Am. 55 (1974) 1304–1312
3. Bennani, Y., Gallinari, P.: On the use of TDNN-extracted features information in talker identification. Proc. Int. Conf. Acoustics, Speech, and Signal Processing, ICASSP (1991) 385–388
4. Bishop, C. M.: Neural Networks for Pattern Recognition. Oxford University Press, Oxford (1995)

5. Broomhead, D. S., Lowe, D.: Multivariate functional interpolation and adaptive networks. Complex Syst. 2 (1988) 321–355

6. Campbell, W. M., Torkkola, K., Balakrishnan, S. V.: Dimension reduction techniques for training polynomial networks. Proc. 17th Int. Conf. Machine Learning, (2000) 119–126

7. Campbell, W. M.: A sequence kernel and its application to speaker recognition. Advances in Neural Information Processing Systems, Dietrich, T. G., Becker, S., Ghahramani, Z. (eds.), MIT, Cambridge, MA, 14 (2002)

8. Campbell, W. M., Assaleh, K. T., Broun, C. C.: Speaker recognition with polynomial classifiers. IEEE Trans. Speech Audio Processing 10 (2002) 205–212

9. Campbell, W. M., Singer, E., Torres-Carrasquillo, P. A., Reynolds, D. A.: Language recognition with support vector machines. Proc. Odyssey: The Speaker and Language Recognition Workshop in Toledo, Spain, ISCA (2004) 41–44

10. Clarkson, T. G., Christoulou, C. C., Guan, Y., Gorse, D., Romano-Critchley D. A., Taylor, J. G.: Speaker identification for security systems using reinforcement-trained pRAM neural network architectures. IEEE Trans. Syst. Man Cybern. C Appl. Rev. 31 (2001) 65–76

11. Clopper, C. G., et al.: A multi-talker dialect corpus of spoken American English: An initial report. Research on Spoken Language Processing, Progress Report, Bloomington, IN: Speech Research Laboratory, Indiana University, 24 (2000) 409–413

12. Cover, T. M.: Geometrical and statistical properties of systems of linear inequalities with applications in pattern recognition. IEEE Trans. Electronic Computers 14 (1965) 326–334

13. Davis, S. B., Mermelstein, P.: Comparison of parametric representations for monosyllabic word recognition in continuously spoken sentences. IEEE Trans. Acoust. 28 (1980) 357–366

14. Duda, R. O., Hart, P. E., Stork, D. G.: Pattern Classification and Scene Analysis, 2nd edition. Wiley Interscience, New York (2001)

15. Fakunaga, K.: Introduction to Statistical Pattern Recognition. New York: Academic (1990)

16. Farrell, K. R., Mammone, R. J., Assaleh, K. T.: Speaker recognition using neural networks and conventional classifiers. IEEE Trans. Speech Audio Processing 2 (1994) 194–205

17. Giles, C. L., Maxwell, T.: Learning, invariance and generalizations in high-order neural networks. Appl. Opt. 26 (1987) 4972–4978

18. Gish, H., Schmidt, M.: Text-independent speaker identification. IEEE Signal Processing Mag. (1994) 18–32

19. Haykin, S.: Neural Networks: A Comprehensive Foundation. Macmillan, New York (2004)

20. Hazen, T. J., Zue, V. W.: Automatic language identification using a segment-based approach. Proc. Eurospeech 2 (1993) 1303–1306

21. Higgins, A. L., Bahler, L., Porter, J.: Speaker verification using randomized phrase prompting. Digital Signal Processing 1 (1991) 89–106

22. Li, K-P.: Automatic language identification using syllabic spectral features. Proc. Int. Conf. Acoustics, Speech, and Signal Processing, ICASSP, 1 (1994) 297–300

23. Lombard, E.: Le Signe de l'Elevation de la Voix. Ann. Maladies Oreille, Larynx, Nez, Pharynx, 37 (1911) 101–119

24. Mary, L., Yegnanarayana, B.: Autoassociative neural network models for language identification. Int. Conf. on Intelligent Sensing and Information Processing, ICISIP (2004) 317–320
25. Muthusamy, Y. K., Barnard, E., and Cole, R. A.: Reviewing automatic language identification. IEEE Signal Processing Mag. 11 (1994) 3341
26. O'Shaughnessy, D.: Speech Communications: Human and Machine, 2nd edition. Universities Press, Hyderabad (2001)
27. Oglesby, J., Mason, J. S.: Optimization of neural networks for speaker identification. Proc. Int. Conf. Acoustics, Speech, and Signal Processing, ICASSP (1990) 261–264
28. Oglesby, J., Mason, J. S.: Radial basis function networks for speaker recognition. Proc. Int. Conf. Acoustics, Speech, and Signal Processing, ICASSP (1991) 393–396
29. Oppenheim, A. V., Schafer, R. W.: Discrete-Time Signal Processing. Prentice-Hall, Englewood Cliffs, NJ (1989)
30. Patil, H. A.: Speaker Recognition in Indian Languages: A Feature Based Approach. Ph.D. thesis, Department of Electrical Engineering, IIT Kharagpur, India (2005)
31. Rabiner, L. R., Juang, B. H.: Fundamentals of Speech Recognition. Prentice Hall, Englewood Cliffs, NJ (1993)
32. Ramachandran, R. P., Farrell, K. R., Ramachandran, R., Mammone, R. J.: Speaker recognition-General classifier approaches and data fusion methods. Pattern Recognit. 35 (2002) 2801–2821
33. Reynolds, D. A., Rose R. C.: Robust text-independent speaker identification using Gaussian mixture models. IEEE Trans. Speech Audio Processing 3 (1995) 72–83
34. Rosenberg, A. E., DeLong, J., Lee, C. H., Juang, B. H., Soong, F. K.: The use of cohort normalized scores for speaker verification. Proc. Int. Conf. Spoken Language Processing, ICSLP, 2 (1992) 599–602
35. Rosenberg, A. E., Parthasarathy, S.: Speaker background models for connected digit password speaker verification. Proc. Int. Conf. Acoustics, Speech, and Signal Processing, ICASSP (1996) 81–84
36. Rudasi, L., Zahorian, S. A.: Text-independent talker identification with neural networks. Proc. Int. Conf. Acoustics, Speech, and Signal Processing, ICASSP (1991) 389–392
37. Satish, S. D., Sekhar C. C.: Kernel based clustering and vector quantization for speech recognition. IEEE Workshop on Machine Learning and Signal Processing (2004) 315–324
38. Schurmann, J.: Pattern Classification. Wiley, New York (1996)
39. Singer, E., Torres-Carrasquillo, P. A., Gleason, T. P., Campbell, W. M., Reynolds, D. A.: Acoustic, phonetic, and discriminative approaches to automatic language identification. Proc. Eurospeech in Geneva, Switzerland, ISCA (2003) 1345–1348
40. Specht, D. F.: Generation of polynomial discriminant functions for pattern recognition. IEEE Trans. Electronic Computers 16 (1967) 308–319
41. Wu, C.-H., Chiu, Y.-H., Shia, C.-J., Lin C.-Y.: Automatic Segmentation and Identification of Mixed-Language Speech Using Delta BIC and LSA-Based GMMs. IEEE Trans. Audio, Speech, and Language Processing 14 (2006) 266–276

42. Wang, L., Chi, H.: Capture interspeaker information with a neural network for speaker identification. IEEE Trans. Neural Netw. 13 (2002) 436–445
43. Xiang, B., Berger, T.: Efficient text-independent speaker verification with structural Gaussian mixture models and neural network. IEEE Trans. Speech Audio Processing 11 (2003) 447–456
44. Yegnanarayana, B.: Artificial Neural Networks. Prentice-Hall of India, New Delhi (1999)
45. Yegnanarayana, B., Kishore, S. P.: AANN: An alternative to GMM for pattern recognition. Neural Netw. 15 (2002) 459–469
46. Zissman, M. A.: Comparison of four approaches to automatic language identification of telephone speech. IEEE Trans. Speech Audio Processing 4 (1986) 31–44

Speech/Non-Speech Classification in Hearing Aids Driven by Tailored Neural Networks

Enrique Alexandre, Lucas Cuadra, and Manuel Rosa-Zurera, and Francisco López-Ferreras

Department of Signal Theory and Communications, University of Alcalá, 28805 – Alcalá de Henares, Spain, enrique.alexandre@uah.es

Summary. This chapter explores the feasibility of using some kind of tailored neural networks to automatically classify sounds into either speech or non-speech in hearing aids. These classes have been preliminary selected aiming at focusing on speech intelligibility and user's comfort. Hearing aids in the market have important constraints in terms of computational complexity and battery life, and thus a set of trade-offs have to be considered. Tailoring the neural network requires a balance consisting in reducing the computational demands (that is the number of neurons) without degrading the classification performance. Special emphasis will be placed on designing the size and complexity of the multilayer perceptron constructed by a growing method. The number of simple operations will be evaluated, to ensure that it is lower than the maximum sustained by the computational resources of the hearing aid.

1 Introduction

Approximately 13% of the population in the highly developed countries suffer from considerable hearing losses whose negative consequences could be mitigated by using some kind of hearing aid [1]. These hearing losses strongly affect the speech communication and disqualify most of hearing-impaired people from holding a normal life. Unavoidably, hearing impairment grows to be more prevalent with age. For instance, almost 50% of those elder over the age of 75 are hard of hearing. Taking into account that the population of most highly developed countries is quickly ageing, and that hearing impairment has an important impact on communication, our society faces a grave health and social problem. This compels scientists and engineers to develop more advanced digital hearing aids, and, what is of key importance, to make them comfortable and easy to handle, especially for elderly people.

Although modern digital hearing aids exhibit the ability to be designed and programmed to compensate most of the particular hearing losses any patient suffers from, sometimes this goal becomes difficult to reach because it involves a balance between the adequate compensation of the hearing loss and

E. Alexandre et al.: *Speech/Non-Speech Classification in Hearing Aids Driven by Tailored Neural Networks*, Studies in Computational Intelligence (SCI) **83**, 145–163 (2008)
www.springerlink.com   © Springer-Verlag Berlin Heidelberg 2008

the user's comfort and subjective preferences. To understand the importance of this assertion, it is worth mentioning that, lamentably, only 20% of hearing impaired people who could be benefited from a hearing aid purchase it. Furthermore, about 25% of them do not wear their hearing aid because of irritating and unpleasant problems related to the background noise appearing elsewhere in their everyday life.

What is the reason underlying these facts? Apart from economical motivations related to the price of the most advanced hearing aids (which may provide higher fidelity of sound, greater overall amplification, directional sound detection, dynamic compression and frequency-specific amplification), there is still a crucial, scientific and technological problem that has not been solved effectively and comfortably at low-cost: the automatic adaptation to the changing acoustic environment.

As shown in Fig. 1, this problem arises from the fact that hearing aid users face a variety of different acoustic environments in everyday life, for example, conversation in quiet, or speech in a noisy crowded cafe, traffic, music, etc.

Regrettably, although hearing aid users expect authentic, significant benefits in each of the mentioned situations, this is not the case. This is because most hearing aids are usually designed and programmed for only one listening environment. Note that, in spite of these devices exhibit the numerous aforementioned capabilities, these are only beneficial in certain hearing situations whereas they are detrimental in other situations. For instance, when entering a crowded bar the hearing aid user hears a suddenly, uncomfortable, irritating, and amplified noise. The problem becomes accentuated because,

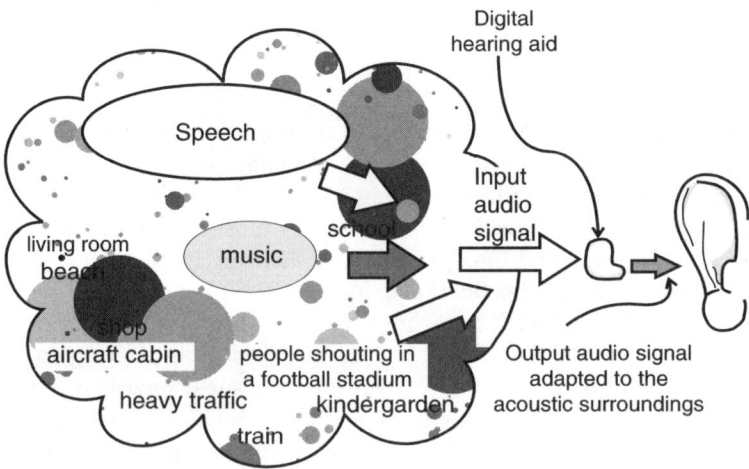

Fig. 1. Artistic representation of the variety of sounds a hearing aid user faces in his/her everyday life. Note that, apart from the speech signal, the target of our approach, a number of different sound sources has been represented such us traffic, school, people shouting in a football stadium, etc.

understanding speech in background noise is much more difficult for hearing impaired people than for normal hearing listeners: many times the patient can "hear" but not understand the speech signal.

With respect to the variety of sound environments the patient has to face on in his/her normal life, it has been shown that hearing aid users generally prefer to have a number of amplification schemes fitted to different listening conditions [2,3]. To satisfy such need there are two basic approaches in digital hearing aids. On the one hand, some modern hearing aids generally allow the user to manually select between a variety of programs (different frequency responses or other processing options such as compression methods, directional microphone, feedback canceller, etc.) depending on the listening conditions. The user has therefore to recognize the acoustic environment and choose the program that best fits this situation by using a switch on the hearing instrument or some kind of remote control. Nevertheless this approach exceeds the abilities of most hearing aid users, in particular for the smallest in-the-canal (ITC) or completely-in-the-canal (CIC) hearing aids, practically invisible to the casual observer. The second approach, which could significantly improve the usability of the hearing aid, is that the hearing aid itself selects the most appropriate program. This is the approach we consider in this work.

It seems to be apparent from the previous paragraphs the need for hearing aids that can be automatically fitted according to the preferences of the user for various listening conditions. As a matter of fact, in a field study with hearing-impaired subjects, it was observed that the automatic switching mode of the instrument was deemed useful by a majority of test subjects, even if its performance was not perfect [1]. Nevertheless, at present, many of the medium-cost hearing aids in the market cannot automatically adapt to the changing acoustical environment the user daily faces on.

Just regarding this necessity, the *purpose* of this chapter is to explore the feasibility of some kind of tailored neural networks (NN) for driving a particular sort of sound classification in digital hearing aids which aims at improving the speech intelligibility [4]. As a first glance one can expect that, in general, a neural network requires a so high computational burden that makes it unfeasible for being implemented on a hearing aid. The underlying reason is that most of hearing aids in the market have very strong constraints in terms of computational capacity, memory and battery, which, for the sake of clarity and for making this chapter self-contained, we have summarized in Sect. 2, placing particular emphasis on the restrictions imposed by the digital signal processing (DSP) the hearing aids are based on. Thus the approach we explore in this chapter seems to be a challenging goal. Although this is true, nevertheless, if a number of design considerations are taken into account, it is possible to construct a neural network tailored for running on a hearing aid and able to classify the sound into either speech or non-speech. The reasons for which this two classes has been selected in the detriment of an approach with more classes will appear clearer in Sect. 3. One of the aforementioned design trade-offs is based on the proper selection of the information entering

the NN. This adequate information is contained in a number of features that
will be calculated in Sect. 4. The tailoring of the NN also requires a balance
consisting in reducing the computational demands (that is the number of
neurons) without degrading the performance perceived by the user. These
trade-offs, involving a number of concepts related to the *training* of the neural
network, the use of Bayesian *regularization*, the tailoring of the NN *size*, and
the *growing* algorithm used to construct the NN, will be studied in Sect. 5.
Finally the performance of the design system will be evaluated.

2 A Brief Overview of the Hardware Limitations of Hearing Aids

As mentioned in Sect. 1, digital hearing aids suffer from important constraints
and, in this respect, Fig. 2 will assist us in introducing the key concepts
that strongly affect the design of our NN-based classifier. The typical digital
hearing aid illustrated in this figure consists basically of

- A microphone to convert the input sound into a electric signal
- A number of electronic blocks aiming at compensate the hearing losses.
  Among other signal processing stages, a proper sound amplifier will

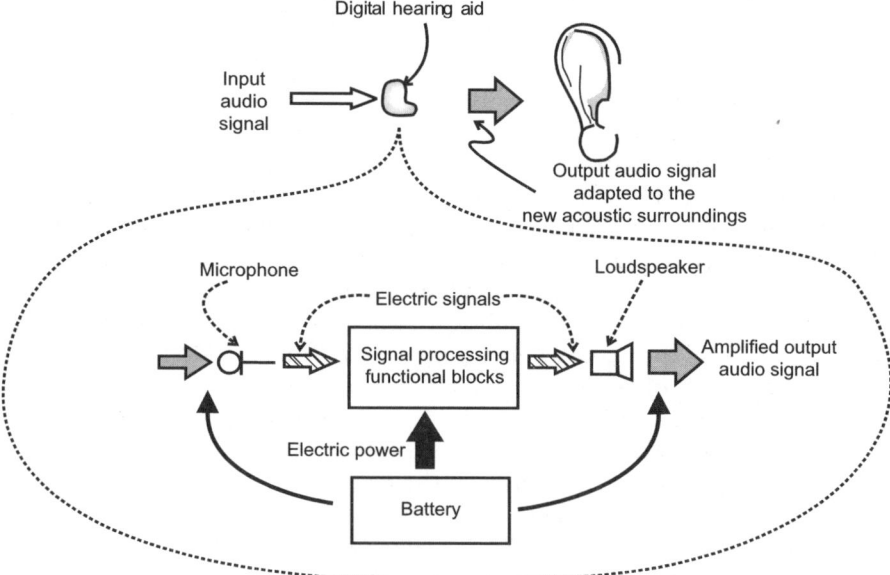

Fig. 2. Simplified diagram of the the structure of a typical hearing aid. The microphone transforms the input sound into a electric signal that can be processed by
the signal processing functional blocks in order to compensate the hearing losses the
patient suffer from. The loudspeaker converts this modified and amplified electrical
signal back to sound

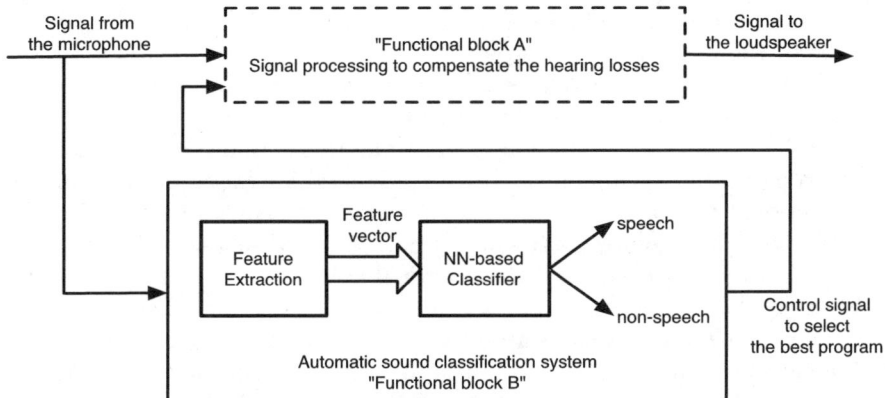

Fig. 3. Conceptual representation of the functional blocks to be implemented on the hearing aid. The block labeled "A" groups all the signal processing operations leading to compensate the particular hearing losses the patient suffer from. The other block, labeled "B", is related to the set of signal processing steps aiming at classifying the sound and, by means of a control signal, selecting the more appropriate program for user's comfort

increase the strength of the signal we are interested on (speech, in this particular application)
- A small loudspeaker to convert the electrical signal back to sound,
- A coupler to the ear canal usually called "ear mold"
- A battery to supply electric power to the electronic devices that compose the hearing instrument.

The signal processing, electronic stages represented in Fig. 2 are based on a DSP integrated circuit (IC). Figure 3 shows the two main functional blocks that must be implemented on the DSP. The first one, labeled "functional block A", corresponds to the set of signal processing stages aiming to compensate the hearing losses. The second one ("functional block B") is the classification system itself, whose implementation by means of a tailored neural network is the focus of this chapter. Note that the aforementioned classifying system consists conceptually of two basic parts:

- A feature extraction stage in order to properly characterize the signal to be classified. This part will be studied in more detail in Sect. 4
- A NN-based, two classes (speech/non-speech) classifier (Sect. 5).

The aforementioned restrictions arise mainly from the tiny size of the hearing aid (specially for the ITC and CIC models), which, as illustrated in Fig. 2, must contain not only the electronic equipment but also the battery. In fact the DSP usually integrates the A/D and D/A converters, the filterbank, RAM, ROM and EPROM memory, input/output ports, and the core. The immediate consequence is that the hearing aid has to work at very low clock frequencies in order to minimize the power consumption and thus maximize the life of the hearing aid batteries. Note that the power consumption must be low enough to ensure that neither the size of the battery pack nor the frequency of battery changes will annoy the user.

Another important issue to note regarding this is that a considerable part of the computational capabilities are already used for running the algorithms of "functional group A" aiming to compensate the acoustic losses. Therefore we are constrained to use the remaining part to implement the NN-based classifier. Roughly speaking, the computational power available does not exceed 3 MIPS (million instructions per second), with only 32 kB of internal memory. The time/frequency decomposition is performed by using an integrated weighted overlap-add (WOLA) filter bank, with 64 frequency bands. It must be considered that only the filter bank requires about 33% of the computation time of the DSP. Another 33% is needed by the so-called functional block A, only around 1 MIPS being thus free for our classifier.

3 Why a Speech/Non-Speech Classifier?

As previously mentioned in Sect. 1, the NN-based classifier we have selected has only two classes: speech and non-speech, this latter denoting any other signal different from speech.

Perhaps the reader may wonder why we restrict this process to only these two particular classes and why we have not considered a classier with, as the one illustrated in Fig. 4a, more classes. In this figure we have represented four classes that, as a first look, should allow the system to adapt more flexibly to a given acoustical environment. Apart from the speech-in-quiet and speech-in-noise classes, others involving noisy environments have been assumed. The class labeled "stationary noise" refers to situations containing monotonous noises such as, for instance, an air-craft cabin, while the "non-stationary" one denotes other non-monotonous noises (for example, people shouting in a crowded football stadium). Although will be explained later on, we can say in advance that, when interested in robustly classifying for improving the speech intelligibility and comfort, the crucial task is classify, as accurately as possible, the sound into either speech or non-speech. The system represented in Fig. 4b suggests that emphasis is placed on detecting speech to the detriment of other acoustic signals that are considered by the user as noise with no information. To understand this it is convenient to imagine two situations that we have considered as being representative enough of the problem at hand:

- *Situation 1*: The hearing aid user is in an acoustical environment consisting in speech in a quiet situation, for instance, at a conference. If the system erroneously classifies the speech-in-quiet signal as non-speech, then the hearing aid will likely decide that it is not worth amplifying such a signal, and thus will reduce the gain. The consequence is that the

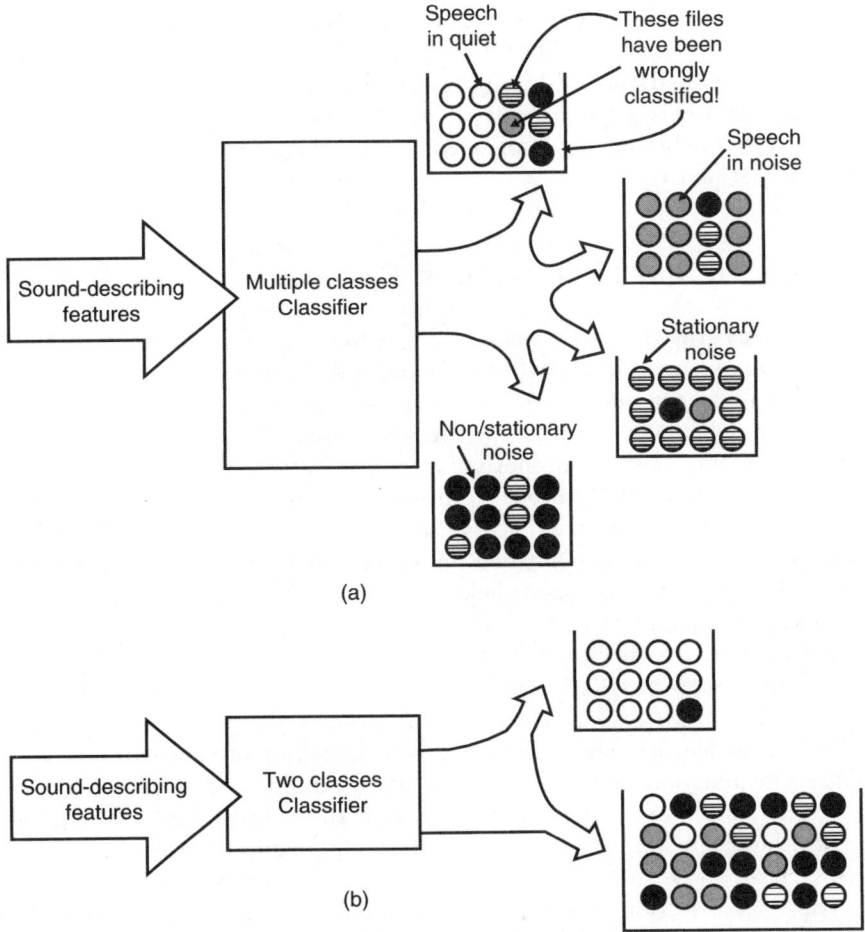

Fig. 4. (a) Conceptual representation of a more general classifier containing four classes. Sound-describing features should assist the system in classifying inputting sounds into any of the four classes illustrated. The class labeled "stationary noise" refers to environments containing monotonous noises such as, for instance, an aircraft cabin, while the "non-stationary" denotes other non-monotonous noises. The figure also illustrates the unavoidable existence of errors in the classification process. (b) Two-class system to illustrate that emphasis is placed on detect speech at the expense of other signals considered as noise from the conversational point of view

hearing-impaired person will loose all the information contained in the speech fragment. This degrades the usability of hearing aids.

- *Situation 2*: The patient is now embedded in a noisy place such us, for instance, a traffic jam. If the system wrongly classifies the sound as speech, then the hearing aid will decide that it is worth amplifying the signal. The immediate consequence is that the user hears a sudden, unpleasant and irritating amplified noise. This degrades the comfort.

This scenario illustrates that the *crucial* task is the discrimination between speech and non-speech [5]. Note that, within the stated framework, this latter class can be named "noise" because we lay particular emphasis on conversation and, from this point of view, it may contain any other signal different from speech, including not only environmental noise but also music, both instrumental and vocal.

4 Selection and Extraction of Features

As previously stated, these processes aim at finding and extracting the kind of information of the audio signal that should assist the classifier in distinguishing between the two aforementioned classes. The performance of the system is strongly dependent on what features are used. As a matter of fact, the classifier may not perform properly if its input features do not contain all this essential, adequate information. Additionally, and because of the aforementioned computational restrictions we have mentioned in Sect. 2, the features not only have to properly described the sound (allowing its subsequent classification), but also have to make efficient use of the DSP resources (i.e., reduced computational demand).

With regard to this issue, a point of the greatest impact in the design is that the feature extraction process can be the most time-consuming task in the classification system. It is very important therefore to carefully design and select the features that will be calculated, taking into account the limited facilities available in the DSP. An illustrative example of the consequences of these constraints can be found in the fact that the strong computational limits restrict the frequency resolution of the time/frequency transforms which are necessary for those that make use of spectral information.

Thus, the selection of an appropriate number of features becomes one of the key topics involved in tailoring the NN-based classifier. Note that this trade-off consists in reducing the number of features while maintaining the probability of correct classification and presumably the speech intelligibility perceived by the user. This, as illustrated in Fig. 5, is equivalent to diminish the dimension of a vector composed by the selected features. We have named this feature vector **F**.

4.1 The Feature Vector

In the problem at hand, classifying the audio signal entering the hearing aid as belonging to either speech or non-speech requires to extract features containing the relevant information that should assist the classifier in distinguishing between the two aforementioned classes. For illustrative purposes we have grouped these two tasks into the so-called Functional Group B illustrated in Fig. 3. For classifying the input sound it is necessary to carry out a number of signal processing steps that we summarized below and illustrated in Fig. 5.

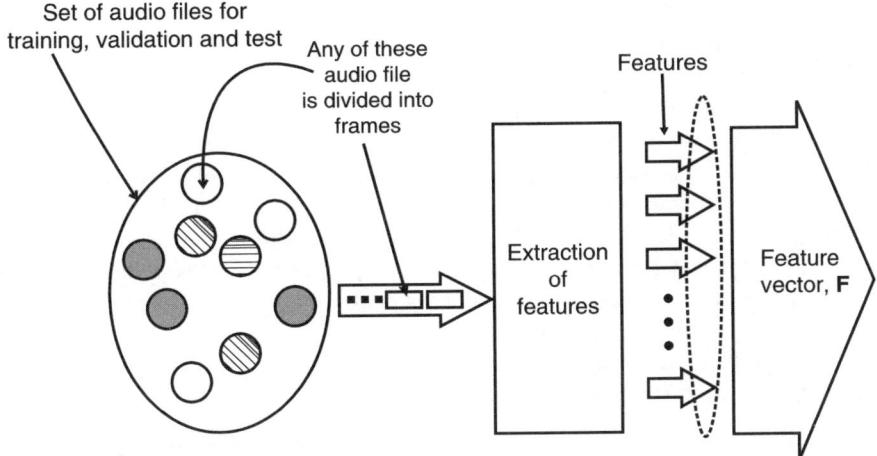

Fig. 5. Simplified diagram illustrating the feature extraction process

First of all, the input audio signal to be classified, $X(t)$, which will be assumed as a *stochastic process* is segmented into frames, $X_i(t)$, $i = 1, \ldots, r$, r being the number of frames into which the signal is divided. For computing the features, which will be used to characterize any frame $X_i(t)$, it is necessary for it to be sampled: $X_i(t_k)$, $k = 1, \ldots, p$, p being the number of samples per frame. Since each frame $X_i(t)$ is a windowed stochastic process, any of its samples, $X_i(t_k)$, is a *random variable* that, for simplicity, we label X_{ik}. Thus, for each audio frame, $X_i(t)$, the following random vector is obtained: $\mathbf{X_i} = [X_{i1}, \ldots, X_{ip}]$. This is just the initial information the system uses to calculate all the features describing frame $X_i(t)$.

With respect to the aforementioned features, let $\mathscr{F} = \{f_1, \ldots, f_{n_f}\}$ be the set that contains all the available features, n_f being the number of features. Any feature f_k can be assumed, in the most general case, as a complex function of p complex variables, $f_k : \mathbb{C}^p \longrightarrow \mathbb{C}$. Since frame $X_i(t)$ has been shown to be a random-variable vector, $\mathbf{X_i} = [X_{i1}, \ldots, X_{ip}]$, then any feature $f_k \in \mathscr{F}$ applied on it, $f_k(\mathbf{X_i})$, is thus a function of p random variables, $f_k(X_{i1}, \ldots, X_{ip})$, and, consequently, a random variable. In order to simplify the notation, the random variable $f_k(X_{i1}, \ldots, X_{ip})$ will be labeled f_{ki}. Finally, for completing the characterization of the input audio signal, the aforementioned sequence of processes has to be applied onto all the r frames into which the input audio signal has been segmented.

Let us imagine now that we are interested in how properly feature $f_k \in \mathscr{F}$ describes the input signal. One of the results that provides the previously described sequence of processes is the random data vector $[f_{k1}, \ldots, f_{kr}] \equiv \mathbf{F_k}$. The elements of this vector are the results obtained when feature f_k is applied on each of the r frames into which the input audio signal has been segmented to be processed. The random vector $\mathbf{F_k}$ can be characterized for example by estimating its mean value, $\hat{E}[\mathbf{F_k}]$, and its variance, $\hat{\sigma}^2[\mathbf{F_k}]$.

This statistical characterization can be done for all the available features $f_k \in \mathscr{F}$. The feature extraction algorithm generates the following feature vector: $\mathbf{F} = [\hat{E}[\mathbf{F}_1], \hat{\sigma}^2[\mathbf{F}_1], \dots, \hat{E}[\mathbf{F}_{n_f}], \hat{\sigma}^2[\mathbf{F}_{n_f}]]$, its dimension being $dim(\mathbf{F}) = 2n_f = m$. This is just the signal-describing vector that should feed the classifier. For the sake of clarity, we write it formally it as $\mathbf{F} = [F_1, \dots, F_{n_f}]$.

Taking this data structure in mind, the immediate goal is, given $\mathscr{F} = \{f_1, \dots, f_{n_f}\}$, the set of n_f available, general-purpose features, to select a subset $\mathscr{G} \subseteq \mathscr{F}$ that minimizes the mean error probability of correct classification, $\bar{P}_e$.

Please note that finding the solution requires, for any candidate feature subset $\mathscr{G}$, the following sequence of operations:

1. Feeding the complete system with input audio files from a proper database, which will be described in Sect. 6.
2. Calculating the feature vector $\mathbf{G}$ for the complete file, as previously explained.
3. Classifying by using a NN.
4. Calculate the mean squared error (MSE) between the classifier output ($\hat{o}$) and the correct output (o), labeled MSE $(\hat{o}, o)$, and the probability of correct classification P_{CC}.

4.2 The Selected Features

There is a number of interesting features that could potentially exhibit different behavior for speech than for non-speech and thus may assist the system to classify the signal into either speech or non-speech. Some experiments [6] and the results of this chapter suggest that the features listed below provide a high discriminating capability for the problem of speech/non-speech classification. We have briefly described below for making the chapter as self-content as possible. For doing so and for the sake of clarity we have labeled the audio frame to be characterized $X_i(t)$ and the output of the WOLA filter bank $\chi_i(t)$. The features we have found to be the more appropriate for the problem at hand are thus:

- *Spectral centroid.* The spectral centroid of frame $X_i(t)$ can be associated with the measure of brightness of the sound, and is obtained by evaluating the center of gravity of the spectrum. Its centroid can be calculated by making use of the formula:

$$Centroid_i = \frac{\sum_{k=1}^{N} |\chi_i(k)| \cdot k}{\sum_{k=1}^{N} |\chi_i(k)|}, \tag{1}$$

where $\chi_i[k]$ represents the kth frequency bin of the spectrum at frame i, and N is the number of samples.

- *Voice2white.* This parameter, proposed in [7], is a measure of the energy inside the typical speech band (300–4,000 Hz) respect to the whole energy of the signal.
- *Spectral Flux.* It is associated with the amount of spectral local changes, and is defined as follows:

$$Flux_t = \sum_{k=1}^{N} (|\chi_t(k)| - |\chi_{t-1}(k)|)^2 . \tag{2}$$

- *Short Time Energy (STE).* It is defined as the mean energy of the signal within each analysis frame (N samples):

$$STE_t = \frac{1}{N} \sum_{k=1}^{N} |\chi_t(k)|^2. \tag{3}$$

As mentioned before some preliminary results [6] had suggested that these features provide a high discriminating capability for the problem of speech/non-speech classification. One may wonder what happens if the number N of frequency bands is reduced. Is the overall performance degraded? To what extent? Figure 6 represents the influence of the number N of bands on the mean error probability for the Fisher linear discriminant (grey columns) and for the k−NN classifier (white columns).

To complete this section it is worth mentioning that other widely-used features such as the Mel-Frequency Cepstral Coefficients (MFCC) or the linear prediction coefficients (LPC) have not be considered because of their excessively high computational cost.

A batch of preliminary experiments has allowed us to select a feature set composed of the mean value and variance estimators of the four previously

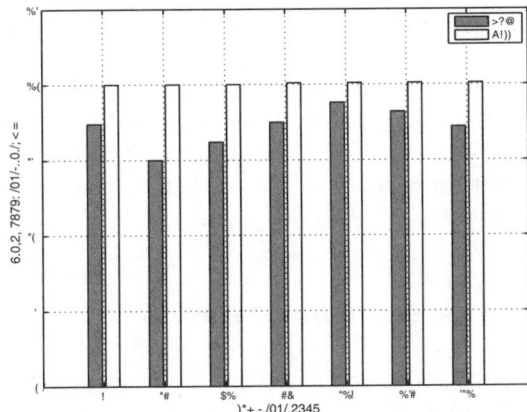

Fig. 6. Probability of error of the k-NN and FLD as a function of the number of frequency bands N. Grey columns correspond to the results of the FLD, while the white ones correspond to the k-NN

Table 1. Probability of error corresponding to the k-NN, FLD, and a MLP as a function of the number of features selected

Features	1 (%)	2 (%)	3 (%)	4 (%)
k-NN	34.4	28.3	34.4	33.7
FLD	31.9	22.0	20.8	18.7
MLP	21.1	9.9	8.5	8.1

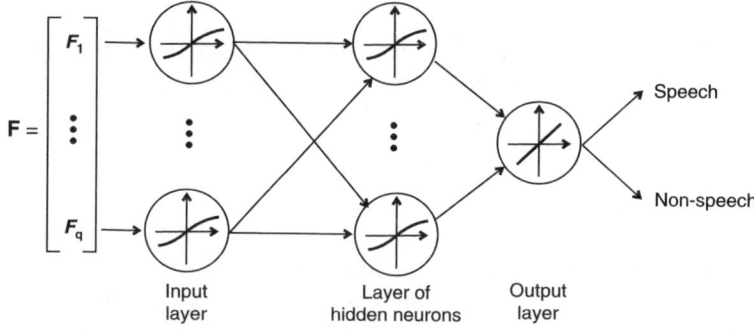

Fig. 7. Multilayer perceptron consisting in three layers of neural networks. q represents the number of features selected in a general case, that is, the dimension of the feature vector **F**. The figure also illustrates the activation functions considered as more appropriate for the problem at hand: logsig for the hidden layer neurons, and a linear one for the output neuron

described features. In this case with only eight potential features, it is possible to employ the brute force method, since there are only 255 possible combinations. Table 1 lists the mean error probability corresponding to the k-NN algorithm, a FLD, and a multilayer perceptron (MLP) as a function of the number of features selected. The way we designed the MLP is explained in the next sections.

Note that the results achieved using the MLP are always better than those obtained with the other algorithms. This motivated our election of this kind of classifiers for our system. Next sections will deal with the details of the implementation and the discussion of the computational complexity associated to this kind of structures.

5 Neural Network Implementation

Figure 7 will assist us in introducing the notation and the main concepts involved in the design trade-offs for tailoring our NN-based, speech/non-speech classifier. These concepts are related to the *training* of the neural network (Sect. 5.1), the use of Bayesian *regularization* (Sect. 5.2), the tailoring of the NN *size* (Sect. 5.3), and the *growing* algorithm used to construct the

NN (Sect. 5.4). With respect to this, Fig. 7 shows a simple MLP consisting of three layers of neurons (input, hidden, and output layers), interconnected by links with adjustable *weights*. The number of input and output layers seems to be clear: the input neurons represent the components of the feature vector; the two-state signal emitted by the output neuron will indicate what class is selected, either speech or non-speech. Nevertheless, what is the number of hidden neurons? This is related to the adjustment of the complexity of the network [8]. If too many free weights are used, the capability to generalize will be poor; on the contrary if too few parameters are considered, the training data cannot be learned satisfactorily. In this way adjusting the *weight vector* **w** becomes a fundamental topic whose main issues we explore in the following subsections.

5.1 Training of the Neural Network

In spite of the gradient descent method exhibits the advantage of its simplicity, nevertheless we have finally adopted the strategy of using the Levenberg–Marquardt method for reasons that will be appeared clear in the following paragraphs.

Gradient descent, as mentioned, is one of the simplest network training algorithms [9]. The algorithm starts usually initializing the weight vector, denoted by $\mathbf{w}^{(0)}$, with random values. Then, this weight vector is updated iteratively, moving a short distance in the direction of the greatest rate of decrease of the error. $\mathbf{w}^{(i)}$ refers to the weight vector at epoch number i. Denoting the error function for a given weight vector $\mathbf{w}^{(i)}$ by $E(\mathbf{w}^{(i)})$, the weight vector for the epoch $i + 1$ is thus given by:

$$\mathbf{w}^{(i+1)} = \mathbf{w}^{(i)} - \eta\nabla E(\mathbf{w}^{(i)}) = \mathbf{w}^{(i)} - \eta\mathbf{J}^{(i)}, \tag{4}$$

where $\mathbf{J}^{(i)}$ represents the Jacobian of the error function evaluated in $\mathbf{w}^{(i)}$, given by:

$$\mathbf{J}^{(i)} = \begin{pmatrix} \frac{\partial e(\mathbf{w}^{(i)}, \mathbf{x}_1)}{\partial w_1} & \cdots & \frac{\partial e(\mathbf{w}^{(i)}, \mathbf{x}_1)}{\partial w_P} \\ \vdots & \ddots & \vdots \\ \frac{\partial e(\mathbf{w}^{(i)}, \mathbf{x}_N)}{\partial w_1} & \cdots & \frac{\partial e(\mathbf{w}^{(i)}, \mathbf{x}_N)}{\partial w_P} \end{pmatrix}. \tag{5}$$

Note that the gradient is re-evaluated at each step. The parameter η is called the learning rate, and is of great importance for the convergence of the algorithm. If η is small, the learning process will be slow, and it may take a long time. On the contrary, if η is too high, the learning process will be much faster, but the algorithm may not converge, and the system could become unstable.

The Newton method can be seen as an evolution of the gradient descent where information from the second derivative is considered. The expression for the weight update is given by:

$$\mathbf{w}^{(i+1)} = \mathbf{w}^{(i)} - \mathbf{H}^{(i)^{-1}}\nabla e(\mathbf{w}^{(i)}), \tag{6}$$

where $\mathbf{H}^{(i)}$ represents the Hessian matrix of the error function $e(\mathbf{w})$ evaluated in $\mathbf{w}^{(i)}$, given by:

$$\mathbf{H}^{(i)} = \begin{pmatrix} \frac{\partial^2 e(\mathbf{w}^{(i)},\mathbf{x}_1)}{\partial w_1^2} & \cdots & \frac{\partial^2 e(\mathbf{w}^{(i)},\mathbf{x}_1)}{\partial w_1 \partial w_P} \\ \vdots & \ddots & \vdots \\ \frac{\partial^2 e(\mathbf{w}^{(i)},\mathbf{x}_N)}{\partial w_P \partial w_1} & \cdots & \frac{\partial^2 e(\mathbf{w}^{(i)},\mathbf{x}_N)}{\partial w_P^2} \end{pmatrix}. \tag{7}$$

This method converges in only one iteration when the error surface is quadratic, and is in general, much more efficient than the gradient descent for two main reasons: first, there is no need to adapt any constants, and second, because the direction in which the weights vary is more efficient than for the gradient descent.

The main problem of the Newton method is the calculation of the Hessian matrix and its inverse. The Gauss–Newton method is a simplification for the case when the error function is a sum of squared errors. In this case, the hessian matrix can be approximated by a function of the jacobian ($\mathbf{J}^T\mathbf{J}$), and the gradient of the error function can be approached by a function of the jacobian and the error function ($\mathbf{J}^T\mathbf{e}$). The expression for the calculation of the network weights is thus given by:

$$\mathbf{w}^{(i+1)} = \mathbf{w}^i - \left(\mathbf{J}^{(i)^T}\mathbf{J}^{(i)} + \alpha\mathbf{I}\right)^{-1}\mathbf{J}^{(i)^T}\mathbf{e}^{(i)}, \tag{8}$$

where the parameter α is chosen to ensure that the matrix $\mathbf{J}^{(i)^T}\mathbf{J}^{(i)} + \alpha\mathbf{I}$ is positively defined. That is, the parameter α compensates the most negative eigenvalue in case the matrix is not positively defined. Note that if $\alpha = 0$, (8) leads to the one of the Newton method, while if α is very high the method behaves like the gradient descent with a small learning rate. In the Levenberg–Marquardt algorithm [10] this parameter varies depending on the value of the error function, being smaller when the error function decreases and higher when it increases. For the reasons stated above, it was decided to train the networks using the Levenberg–Marquardt algorithm.

Once the Levenberg–Marquardt training method has been selected, the key point of adjusting the size and complexity of the NN arises: How to obtain good generalization capability with a "small" NN size? This objective motivates the use of regularization techniques. In particular, Bayesian regularization exhibits a good performance, the next section being devoted to this issue.

5.2 Use of Regularization

The technique of Bayesian regularization is based on the minimization of the following error function:

$$E(\mathbf{w}) = \alpha\mathbf{w}^T\mathbf{w} + \beta E_D, \tag{9}$$

where E_D is the sum of squared errors. If $\alpha \ll \beta$, then the training algorithm will make the errors smaller. If $\alpha \gg \beta$, training will emphasize weight size reduction. The bayesian regularization algorithm tries to find the optimal values of α and β by means of maximizing the joint probability function of α and β assuming a uniform density for them [11]. This leads to the following expressions for α and β:

$$\alpha^{(i+1)} = \frac{P - \frac{2\alpha^{(i)}}{tr(\mathbf{H}^{(i+1)})}}{2\mathbf{w}^{(i+1)^T}\mathbf{w}^{(i+1)}} = \frac{\gamma}{2\mathbf{w}^{(i+1)^T}\mathbf{w}^{(i+1)}}, \tag{10}$$

$$\beta^{(i+1)} = \frac{N - P + \frac{2\alpha^{(i)}}{tr(\mathbf{H}^{(i+1)})}}{2E_D^{(i+1)}} = \frac{N - \gamma}{2E_D^{(i+1)}}. \tag{11}$$

The term γ is called the effective number of parameters, P is the total number of parameters in the network, and N the total number of training vectors. The main advantage of using regularization techniques is that the generalization capabilities of the classifier are improved, and it is possible to obtain better results with smaller networks, since the regularization algorithm itself prunes those neurons that are not strictly necessary. From this, all our experiments will be done using the Levenberg–Marquardt algorithm with Bayesian regularization.

5.3 Network Size Considerations

One argument against the feasibility of neural networks for being used on DSP-based hearing aids consist in the, a priori, high computational complexity, which, among other topics, is related to the network size. However it is worth exploring its implementation because, as pointed out in [1] and [4], neural networks are able to achieve very good results in terms of probability of error when compared to other popular algorithms such as a rule-based classifier, the FLD, the minimum distance classifier, the k-Nearest Neighbor algorithm, or a Bayes classifier.

The "negative" facet, as mentioned, could arise from the fact that the computational complexity of a neural network is the highest of all those classifiers. This complexity depends on the number of weights that need to be adapted, and consequently on the number of neurons which compose the neural network. In particular, the number of simple operations required by a neural network to produce one output is given by:

$$N_{op} = W(2L + 2M + 1) + 2M - 1, \tag{12}$$

where W, L and M are the number of hidden, input and output neurons, respectively. Note that L equals the dimension of the feature vector.

From this, it can be observed that one way to achieve this goal is, as commented before, to decrease the number of input features (that is the dimensionality of the feature vector inputting the network), and thus the number of input neurons. Note again the key importance of the selection of a reduced number of appropriate features.

On the other hand, it will be also necessary to reduce the number of neurons in the hidden layer. As a rule of thumb, a number of hidden neurons equal to the logarithm of the number of training patterns has been empirically shown to be appropriate [12]. A number of algorithms have been proposed in the literature to reduce the number of weights to be considered, and with this, the size of the neural network. The purpose of these algorithms is to estimate the importance of each weight and eliminate those having the least importance. The optimal brain damage (OBD) algorithm and its descendent, optimal brain surgeon (OBS), use a second-order approximation to predict how the training error depends upon a weight, and eliminate the weight that leads to the smallest increase in the training error [8]. The approach we have considered more appropriate makes use of the growing algorithm we describe below.

5.4 Growing Algorithm

A simpler approach to reduce the number of hidden neurons, which however involves exhaustive search through a restricted class of network architectures, is to make use of growing and pruning algorithms [13, 14].

In this work, a growing algorithm has been used, inspired in [13]. This algorithm progressively constructs the NN by adding hidden neurons and re-training the network until a minimum of the validation error is achieved.

Figure 8 illustrates this algorithm. When it is decided that a new neuron must be added, the following process is carried out:

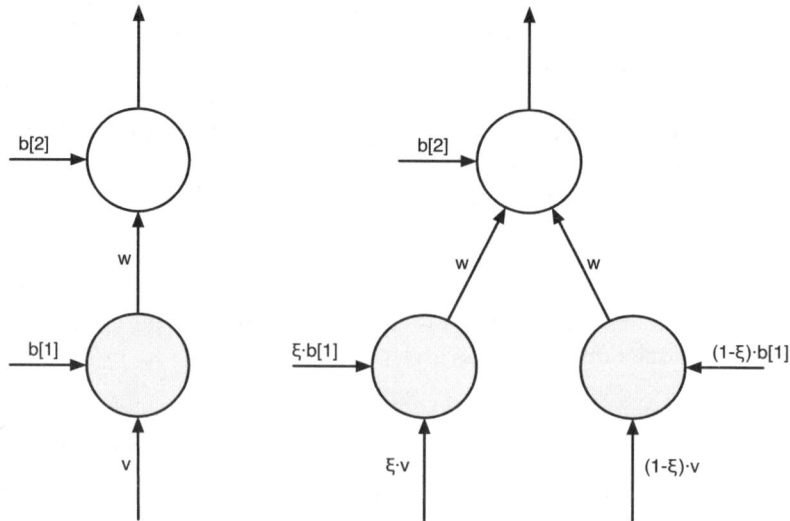

Fig. 8. Schematic representation of the growing algorithm

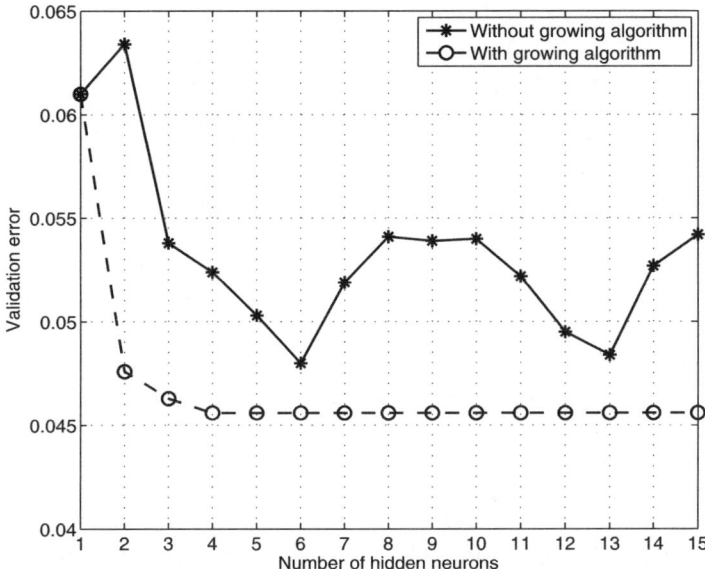

Fig. 9. Validation error as a function of the number of hidden neurons obtained with (*dashed line*) and without (*solid line*) the growing algorithm

1. The neuron with the highest weight at its output is selected. This will be considered as the "parent" neuron. Let it be the shaded one in Fig. 8.
2. The bias ($b[1]$) and the input weights (v) of the parent neuron are divided to obtain the bias and input weights of the new neuron. A factor of ξ, with $0 < \xi < 1$ is used.
3. The output weight of the parent neuron is copied to the new neuron.

Figure 9 shows the behavior of this algorithm. In this figure we have represented the validation error as a function of the number of hidden neurons. The dashed, monotonously decreasing line represents the validation error when the growing algorithm is used. The solid one corresponds to the case in which the algorithm is not considered. As it can be observed, the growing algorithm allows a smoother behavior of the system, being able to obtain also a lower validation error.

6 Experimental Works

With respect to our experimental work, we summarize below important details related to the database used (Sect. 6.1), numerical results considered as representative enough of the feasibility of our method (Sect. 6.2) and an strategy for selecting training patterns by using genetic algorithms (Sect. 6.3).

6.1 Database Used

As mentioned before, our goal is the implementation of an efficient, resource-saving, automatic sound classification system for DSP-based hearing aids. The sound database we have used for the experiments consist of 2,936 files, with a length of 2.5 s each. The sampling frequency is 22,050 Hz with 16 bits per sample. The files correspond to the following categories: speech in quiet (509 files), speech in stationary noise (727 files), speech in non-stationary noise (728 files), stationary noise (486 files) and non-stationary noise (486 files). We have taken into account a variety of noise sources, including those corresponding to the following diverse environments: aircraft, bus, cafe, car, kindergarten, living room, nature, school, shop, sports, traffic, train, and train station. Music files, both vocal and instrumental, have also been assumed as noise sources. The files with speech in noise that we have explored exhibit different signal to noise ratios (SNRs) ranging from 0 to 10 dB.

For training, validation and test it is necessary for the database to be divided into three different sets. These sets contain 1,074 files (35%) for training, 405 (15%) for validation, and 1,457 (50%) for testing. This division has been made randomly and ensuring that the relative proportion of files of each category is preserved for each set.

6.2 Numerical Results

In the batches of experiments we have put into practice, the neural networks have been independently trained 20 times, and the best realization in terms of probability of correct classification for the validation set has been selected. All the results shown here refer to the test set.

Table 2 shows the results obtained with the proposed algorithm for different number of hidden neurons (W) and input features (L). Each value represents the best case in terms of validation error obtained for the pair (W, L). Note that the best result is achieved with four input features and five hidden neurons, with a probability of error equal to 8.1%. The four features corresponding to this minimum error are: variance of the spectral centroid, mean of the Voice2White and mean and variance of the STE.

Table 2. Error probability as a function of the number of hidden neurons (W) and input features (L)

	W = 1 (%)	W = 2 (%)	W = 3 (%)	W = 4 (%)	W = 5 (%)	W = 6 (%)	W = 7 (%)	W = 8 (%)	W = 9 (%)	W = 10 (%)
L = 1	21.1	21.8	22.3	22.7	22.0	22.0	21.9	22.0	22.0	22.0
L = 2	14.0	12.4	10.2	10.3	10.1	10.3	9.9	9.9	10.0	10.2
L = 3	13.2	9.2	9.4	9.5	9.2	9.2	9.1	8.6	8.5	8.5
L = 4	12.3	9.5	8.9	8.6	8.1	8.2	8.3	8.4	8.4	8.4

It is also important to highlight that good results (9.5%) can be obtained with significantly smaller networks (four features and two hidden neurons). The feature set for this suboptimal case is the same as before.

6.3 Selection of the Training Patterns

The training database plays a crucial role in order to prevent overfitting [9]. While an excessively complex NN may provide very good results in classifying the training samples, however, it may be not capable of performing this on new acoustic signals. Therefore it will be necessary to adjust its complexity: not so simple that it cannot distinguish between the two classes, yet no so complex as to fail in classifying novel patterns. The goal is to make the NN able to generalize and classify as better as possible the acoustic signals the hearing aid user faces in everyday life.

It is very important therefore to have a good training database, which should be representative enough of each of the classes that are being considered. A typical problem is the presence of outlayers in the training set. These are patterns that are not representative of the class they belong to, and that could interfere in the learning process of the network. To solve this problem, it was decided to use a genetic algorithm [15] to select the training patterns in order to minimize the error on the validation set.

The main problem here is that a genetic algorithm with a fitness function driven by a neural network implies a huge computational complexity, since for each generation a number of neural networks equal to the number of individuals per generation must be trained. In our case, and given the size of the search space, a population of 5,000 individuals was considered. Thus, to simplify the problem, it was decided to use as the fitness function the probability of error on the validation set returned by a classifier based on the FLD. Then, with the training set returned by the genetic algorithm, a neural network was trained.

The probability of error achieved with the four features using the FLD is equal to 15.9%. After the selection of the training patterns, this probability drops to 12.6%. If the neural network with only two hidden neurons is trained using this set, the probability of error on the test set changes from 9.5 to 8.6%. The training set was reduced from 943 patterns to only 434.

7 Considerations on Real Time Implementation

In our implementation, a sampling frequency of 22,050 Hz is considered, with a frame length of 64 samples, which corresponds to 2.9 µs. This implies a rate of 344 frames per second. Since our DSP has a total computational power of approximately 3 MIPS, around 2 MIPS being already used, only 1 MIPS is thus available for the classification algorithm. This means that roughly speaking, a total number of 2,900 instructions is available per frame.

Once the performance of the proposed system has been stated, it is interesting to analyze the computational complexity of this implementation. First, each one of the considered features will be analyzed in terms of the number of simple operations required by the DSP to calculate them. Note that, as commented before, the hearing aid DSP implements a WOLA filter bank with 64 bands. This calculation will therefore not be considered in this section.

- STE: It requires 63 sums and 1 division.
- Total energy: It requires 63 sums.
- Centroid: It requires 64 products, 63 sums and 1 division.
- Voice2White: It requires 21 sums and 1 division.

This makes a total of 277 simple operations. Nevertheless, to complete the calculation of the selected features it is necessary also to compute their mean and variance. For doing so the following estimators are considered:

$$\hat{\mu} = \frac{1}{N} \sum_i X_i, \tag{13}$$

$$\hat{\sigma}^2 = \frac{1}{N} \sum_i (X_i - \hat{\mu})^2 = \frac{1}{N} \sum_i X_i^2 - \hat{\mu}^2. \tag{14}$$

The mean value requires thus 2 sums and 1 division per frame, and the standard deviation 3 sums, 1 product and 1 division. With all this, for each frame, and considering the four selected features, the number of simple operations required to compute them is equal to $277 + 2 \cdot 3 + 2 \cdot 5 = 293$.

Equation (12) allowed us to estimate the number of simple operations required by a MLP. A more conservative estimation is:

$$N_{op} = W(2L + 2M + 1) + 2M - 1 + 20W. \tag{15}$$

Note that this is (12) with an additional number of operations equal to $20W$. This is necessary to compute the logsig activation function (the logarithm is tabbed in the DSP). For the considered case, where $L = 4$, $W = 2$ and $M = 1$, the number of simple operations needed per frame is given by $23 + 40 = 43$. This makes a total of $293 + 43 = 336$ operations needed by the whole classification algorithm per frame.

The summarized results suggest that, when properly tailored, the proposed MLP can be feasible implemented on the DSP aiming at classifying into the two classes of interest. For other applications that presumably may require more computational demands we propose the following approach illustrated in Fig. 10.

As shown, the input audio signal is divided into frames of t_{frame} seconds. The calculation of the vector containing the previously selected features takes a time $t_{feat} < t_{frame}$. The neural network classifies over a number M ($M = 3$ in the figure) of feature vectors corresponding to M frames. In order to compose the input to the neural network, M feature vectors are combined, for

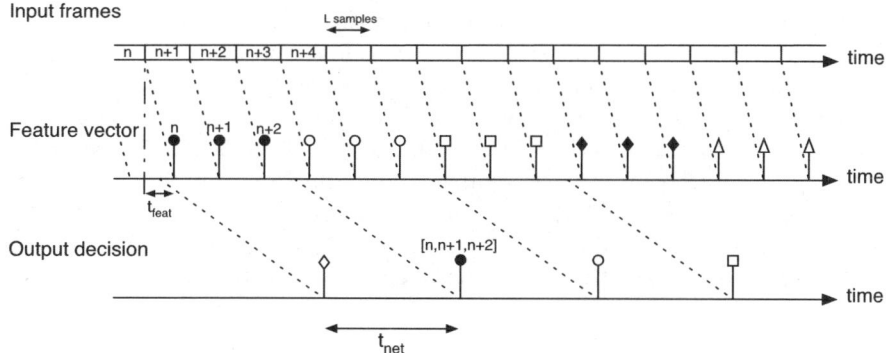

Fig. 10. Scheme to illustrate the delayed decision mechanism implemented on the hearing aid. Different symbols have been used to show the groups of frames used to obtain each output decision

instance by calculating their mean value. Assuming that it takes to the neural network t_{net} seconds to return a decision for a given set of input features, the number of frames that can be processed during that time would be $M = \lfloor t_{net}/t_{frame} \rfloor$, where $\lfloor \cdot \rfloor$ denotes rounding towards the nearest lower integer. This reaches a two-side goal: on the one side, CPU time per frame dedicated to the NN calculation is reduced; on the other, it avoids spurious decisions. This algorithm can be seen as an inertial system, which will prevent the hearing aid from spuriously switching among the different programs. Since the decisions are taken upon a number of input frames, the system will be more robust against transients.

8 Concluding Remarks

In this chapter we have explored the feasibility of using some kind of tailored neural networks to automatically classify sounds into either speech or non-speech in hearing aids. These classes have been preliminary selected aiming at focusing on speech intelligibility and user's comfort. Hearing aids in the market have important constraints in terms of computational complexity and battery life, a set of trade-offs have to be considered. Tailoring the neural network requires a balance consisting in reducing the computational demands (that is the number of neurons) without degrading the classification performance.

Especial emphasis has been place on design the size and complexity of the MLP constructed by a growing method. The necessary reduction in size can be achieved in two ways. On the one hand, by reducing the number of input neurons. This is equivalent to diminish the number of signal-describing features used to characterize the sounds. The second one is related to the number of neurons in the hidden layer. After training by using the Levenberg–Marquardt and Bayesian regularization it has been found that the best three-layer perceptron, which exhibits the best performance, has four input neurons, five hidden neurons, and one output neuron. Its error probability has been found to be 8.1%. Using a smaller neural network, with only two hidden neurons and the same input features, the achieved probability of error is 9.5%. A lower probability of error (8.6%) can be obtained with the same number of features and hidden neurons by applying a genetic algorithm to automatically select the elements of the training set. The number of simple operations has been also proved to be lower than the maximum sustained by the computational resources of the hearing aid.

References

1. Büchler, M.: Algorithms for sound classification in hearing instruments. PhD thesis, Swiss Federal Institute of Technology, Zurich (2002)
2. Keidser, G.: The relationships between listening conditions and alterative amplification schemes for multiple memory hearing aids. Ear Hear **16** (1995) 575–586
3. Keidser, G.: Selecting different amplification for different listening conditions. Journal of the American Academy of Audiology **7** (1996) 92–104
4. Alexandre, E., Cuadra, L., Perez, L., Rosa-Zurera, M., Lopez-Ferreras, F.: Automatic sound classification for improving speech intelligibility in hearing aids using a layered structure. In: Lecture Notes in Computer Science. Springer Verlag, Berlin (2006)
5. Alexandre, E., Alvarez, L., Cuadra, L., Rosa, M., Lopez, F.: Automatic sound classification algorithm for hearing aids using a hierarchical taxonomy. In: Dobrucki, A., Petrovsky, A., Skarbek, W., eds.: New Trends in Audio and Video. Volume 1. Politechnika Bialostocka (2006)
6. Alexandre, E., Rosa, M., Lopez, F.: Application of Fisher Linear Discriminant Analysis to Speech/Music Classification. In: proceedings of EUROCON 2005 - The International Conference on "Computer as a tool", Belgrade, Serbia and Montenegro, Volume: 2, ISBN: 1-4244-0049-X
7. Guaus, E., Batlle, E.: A non-linear rhythm-based style classification for broadcast speech-music discrimination. In: AES 116th Convention (2004)
8. Duda, R.O., Hart, P.E., Stork, D.G.: Pattern Classification. Wiley, New York (2001)
9. Bishop, C.M.: Neural Networks for Pattern Recognition. Oxford University Press, Oxford (1995)
10. Hagan, M., Menhaj, M.: Training feedforward networks with the marquardt algorithm. IEEE Transactions on Neural Networks **5**(6) (1994) 989–993
11. Foresee, F.D., Hagan, M.T.: Gauss-newton approximation to bayesian learning. Proceedings of the 1997 International Joint Conference on Neural Networks (1997) 1930–1935
12. Wanas, N., Auda, G., Kamel, M.S., Karray, F.: On the optimal number of hidden nodes in a neural network. In: IEEE Canadian Conference on Electrical and Computer Engineering (1998) 918–921

13. Mathia, K.: A variable structure learning algorithm for multilayer perceptrons. Intelligent Engineering Systems Through Artificial Neural Networks **14** (2004) 93–100
14. Pearce, B.Y.: Implementation of a variable structure neural network. Master's Thesis, Auburn University (2001)
15. Goldberg, D.: Genetic Algorithms in Search, Optimization and Machine Learning. Addison-Wesley, Reading, MA (1989)

Audio Signal Processing

Preeti Rao

Department of Electrical Engineering, Indian Institute of Technology Bombay,
India, prao@ee.iitb.ac.in

Summary. In this chapter, we review the basic methods for audio signal processing, mainly from the point of view of audio classification. General properties of audio signals are discussed followed by a description of time-frequency representations for audio. Features useful for classification are reviewed. In addition, a discussion on prominent examples of audio classification systems with particular emphasis on feature extraction is provided.

1 Introduction

Our sense of hearing provides us rich information about our environment with respect to the locations and characteristics of sound producing objects. For example, we can effortlessly assimilate the sounds of birds twittering outside the window and traffic moving in the distance while following the lyrics of a song over the radio sung with multi-instrument accompaniment. The human auditory system is able to process the complex sound mixture reaching our ears and form high-level abstractions of the environment by the analysis and grouping of measured sensory inputs. The process of achieving the segregation and identification of sources from the received composite acoustic signal is known as auditory scene analysis. It is easy to imagine that the machine realization of this functionality (sound source separation and classification) would be very useful in applications such as speech recognition in noise, automatic music transcription and multimedia data search and retrieval. In all cases the audio signal must be processed based on signal models, which may be drawn from sound production as well as sound perception and cognition. While production models are an integral part of speech processing systems, general audio processing is still limited to rather basic signal models due to the diverse and wide-ranging nature of audio signals.

Important technological applications of digital audio signal processing are audio data compression, synthesis of audio effects and audio classification. While audio compression has been the most prominent application of digital

P. Rao: *Audio Signal Processing*, Studies in Computational Intelligence (SCI) **83**, 169–185 (2008)
www.springerlink.com   © Springer-Verlag Berlin Heidelberg 2008

audio processing in the recent past, the burgeoning importance of multimedia content management is seeing growing applications of signal processing in audio segmentation and classification. Audio classification is a part of the larger problem of audiovisual data handling with important applications in digital libraries, professional media production, education, entertainment and surveillance. Speech and speaker recognition can be considered classic problems in audio retrieval and have received decades of research attention. On the other hand, the rapidly growing archives of digital music on the internet are now drawing attention to wider problems of nonlinear browsing and retrieval using more natural ways of interacting with multimedia data including, most prominently, music. Since audio records (unlike images) can be listened to only sequentially, good indexing is valuable for effective retrieval. Listening to audio clips can actually help to navigate audiovisual material more easily than the viewing of video scenes. Audio classification is also useful as a front end to audio compression systems where the efficiency of coding and transmission is facilitated by matching the compression method to the audio type, as for example, speech or music.

In this chapter, we review the basic methods for signal processing of audio, mainly from the point of view of audio classification. General properties of audio signals are discussed followed by a description of time-frequency representations for audio. Features useful for classification are reviewed followed by a discussion of prominent examples of audio classification systems with particular emphasis on feature extraction.

2 Audio Signal Characteristics

Audible sound arises from pressure variations in the air falling on the ear drum. The human auditory system is responsive to sounds in the frequency range of 20 Hz to 20 kHz as long as the intensity lies above the frequency dependent "threshold of hearing". The audible intensity range is approximately 120 dB which represents the range between the rustle of leaves and boom of an aircraft take-off. Fig. 1 displays the human auditory field in the frequency-intensity plane. The sound captured by a microphone is a time waveform of the air pressure variation at the location of the microphone in the sound field. A digital audio signal is obtained by the suitable sampling and quantization of the electrical output of the microphone. Although any sampling frequency above 40 kHz would be adequate to capture the full range of audible frequencies, a widely used sampling rate is 44,100 Hz, which arose from the historical need to synchronize audio with video data. "CD quality" refers to 44.1 kHz sampled audio digitized to 16-bit word length.

Sound signals can be very broadly categorized into environmental sounds, artificial sounds, speech and music. A large class of interesting sounds is time-varying in nature with information coded in the form of temporal sequences of atomic sound events. For example, speech can be viewed as a sequence of

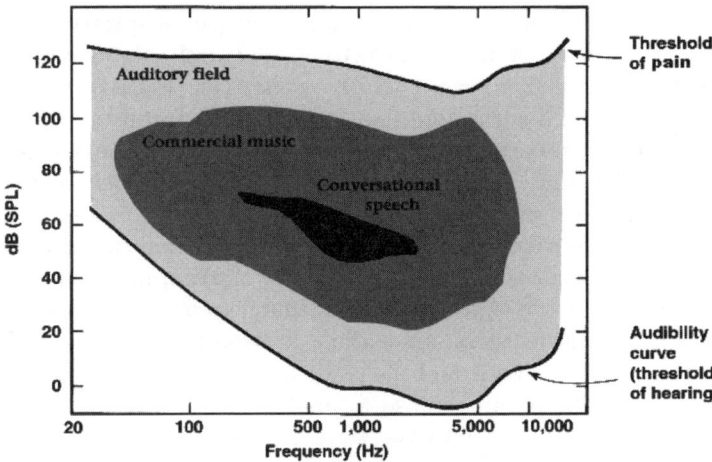

Fig. 1. The auditory field in the frequency-intensity plane. The sound pressure level is measured in dB with respect to the standard reference pressure level of 20 microPascals

phones, and music as the evolving pattern of notes. An atomic sound event, or a single gestalt, can be a complex acoustical signal described by a specific set of temporal and spectral properties. Examples of atomic sound events include short sounds such as a door slam, and longer uniform texture sounds such as the constant patter of rain. The temporal properties of an audio event refer to the duration of the sound and any amplitude modulations including the rise and fall of the waveform amplitude envelope. The spectral properties of the sound relate to its frequency components and their relative strengths. Audio waveforms can be periodic or aperiodic. Except for the simple sinusoid, periodic audio waveforms are complex tones comprising of a fundamental frequency and a series of overtones or multiples of the fundamental frequency. The relative amplitudes and phases of the frequency components influence the sound "colour" or timbre. Aperiodic waveforms, on the other hand, can be made up of non-harmonically related sine tones or frequency shaped noise. In general, a sound can exhibit both tone-like and noise-like spectral properties and these influence its perceived quality. Speech is characterized by alternations of tonal and noisy regions with tone durations corresponding to vowel segments occurring at a more regular syllabic rate. Music, on the other hand, being a melodic sequence of notes is highly tonal for the most part with both fundamental frequency and duration varying over a wide range.

Sound signals are basically physical stimuli that are processed by the auditory system to evoke psychological sensations in the brain. It is appropriate that the salient acoustical properties of a sound be the ones that are important to the human perception and recognition of the sound. Hearing perception has been studied since 1870, the time of Helmholtz. Sounds are described in terms of the perceptual attributes of pitch, loudness, subjective duration and timbre. The human auditory system is known to carry out the frequency analysis of sounds to feed the higher level cognitive functions. Each of the subjective sensations is correlated with more than one spectral property (e.g. tonal content) or temporal property (e.g. attack of a note struck on an instrument) of the sound. Since both spectral and temporal properties are relevant to the perception and cognition of sound, it is only appropriate to consider the representation of audio signals in terms of a joint description in time and frequency.

While audio signals are non stationary by nature, audio signal analysis usually assumes that the signal properties change relatively slowly with time. Signal parameters, or features, are estimated from the analysis of short windowed segments of the signal, and the analysis is repeated at uniformly spaced intervals of time. The parameters so estimated generally represent the signal characteristics corresponding to the time center of the windowed segment. This method of estimating the parameters of a time-varying signal is known as "short-time analysis" and the parameters so obtained are referred to as the "short-time" parameters. Signal parameters may relate to an underlying signal model. Speech signals, for example, are approximated by the well-known source-filter model of speech production. The source-filter model is also applicable to the sound production mechanism of certain musical instruments where the source refers to a vibrating object, such as a string, and the filter to the resonating body of the instrument. Music due to its wide definition, however, is more generally modelled based on observed signal characteristics as the sum of elementary components such as continuous sinusoidal tracks, transients and noise.

3 Audio Signal Representations

The acoustic properties of sound events can be visualized in a time-frequency "image" of the acoustic signal so much so that the contributing sources can often be separated by applying gestalt grouping rules in the visual domain. Human auditory perception starts with the frequency analysis of the sound in the cochlea. The time-frequency representation of sound is therefore a natural starting point for machine-based segmentation and classification. In this section we review two important audio signal representations that help to visualize the spectro-temporal properties of sound, the spectrogram and an auditory representation. While the former is based on adapting the Fourier transform to time-varying signal analysis, the latter incorporates the knowledge of hearing perception to emphasize perceptually salient characteristics of the signal.

3.1 Spectrogram

The spectral analysis of an acoustical signal is obtained by its Fourier transform which produces a pair of real-valued functions of frequency, called the amplitude (or magnitude) spectrum and the phase spectrum. To track the time-varying characteristics of the signal, Fourier transform spectra of overlapping windowed segments are computed at short successive intervals. Time-domain waveforms of real world signals perceived as similar sounding actually show a lot of variability due to the variable phase relations between frequency components. The short-time phase spectrum is not considered as perceptually significant as the corresponding magnitude or power spectrum and is omitted in the signal representation [1]. From the running magnitude spectra, a graphic display of the time-frequency content of the signal, or spectrogram, is produced.

Figure 2 shows the waveform of a typical music signal comprised of several distinct acoustical events as listed in Table 1. We note that some of the events overlap in time. The waveform gives an indication of the onset and the rate of decay of the amplitude envelope of the non-overlapping events. The spectrogram (computed with a 40 ms analysis window at intervals of 10 ms) provides a far more informative view of the signal. We observe uniformly-spaced horizontal dark stripes indicative of the steady harmonic components of the piano notes. The frequency spacing of the harmonics is consistent with the relative pitches of the three piano notes. The piano notes' higher harmonics are seen to decay fast while the low harmonics are more persistent even as the overall amplitude envelope decays. The percussive (low tom and cymbal crash) sounds are marked by a grainy and scattered spectral structure with a few weak inharmonic tones. The initial duration of the first piano strike is dominated by high frequency spectral content from the preceding cymbal crash as it decays. In the final portion of the spectrogram, we can now clearly detect the simultaneous presence of piano note and percussion sequence.

The spectrogram by means of its time-frequency analysis displays the spectro-temporal properties of acoustic events that may overlap in time and frequency. The choice of the analysis window duration dictates the trade-off between the frequency resolution of steady-state content versus the time resolution of rapidly time-varying events or transients.

Table 1. A description of the audio events corresponding to Fig. 2

Time duration (s)	Nature of sound event
0.00–0.15	Low Tom (Percussive stroke)
0.15–1.4	Cymbal crash (percussive stroke)
1.4–2.4	Piano note (Low pitch)
2.4–3.4	Piano note (High pitch)
3.4–4.5	Piano note (Middle pitch) occurring simultaneously as the low tom and cymbal crash (from 0 to 1.4 s)

3.2 Auditory Representations

Auditory representations attempt to describe the acoustic signal by capturing the most important aspects of the way the sound is perceived by humans. Compared with the spectrogram, the perceptually salient aspects of the audio signal are more directly evident. For instance, the spectrogram displays the time-frequency components according to their physical intensity levels while the human ear's sensitivity to the different components is actually influenced by several auditory phenomena. Prominent among these are the widely differing sensitivities of hearing in the low, mid and high frequency regions in terms of the threshold of audibility (see Fig. 1), the nonlinear scaling of the perceived loudness with the physically measured signal intensity and the decreasing frequency resolution, with increasing frequency, across the audible range [2]. Further, due to a phenomenon known as auditory masking, strong signal components suppress the audibility of relatively weak signal components in their time-frequency vicinity. The various auditory phenomena are explained based on the knowledge of ear physiology combined with perception models derived from psychoacoustical listening experiments [2].

An auditory representation is typically obtained as the output of a computational auditory model and represents a physical quantity at some stage of the auditory pathway. Computational auditory models simulate the outer, middle and inner ear functions of transforming acoustic energy into a neural code in the auditory nerve. The computational models are thus based on the approximate stages of auditory processing with the model parameters fitted to explain empirical data from psychoacoustical experiments. For instance, cochlear models simulate the band-pass filtering action of the basilar membrane and the subsequent firing activity of the hair-cell neurons as a function of place along the cochlea. Several distinct computational models have been proposed over the years [3, 4]. However a common and prominent functional block of the cochlear models is the integration of sound intensity over a finite frequency region by each of a bank of overlapping, band-pass filters. Each of these "critical bandwidth" filters corresponds to a cochlear channel whose output is processed at the higher levels of the auditory pathway relatively independently of other channels. The filter bandwidths increase roughly logarithmically with center frequency indicating the non-uniform frequency resolution of hearing. The output of the cochlear model is a kind of "auditory image" that forms the basis for cognitive feature extraction by the brain. This is supported by the psychoacoustic observation that the subjective sensations that we experience from the interaction of spectral components that fall within a cochlear channel are distinctly different from the interaction of components falling in separate channels [2].

In the Auditory Image Model of Patterson [3], the multichannel output of the cochlea is simulated by a gammatone filterbank followed by a bank of adaptive thresholding units simulating neural transduction. Fourth-order gammatone filters have frequency responses that can closely approximate psychoacoustically measured auditory filter shapes. The filter bandwidths follow the measured critical bandwidths ($25\,\mathrm{Hz} + 10\%$ of the center frequency) and center frequencies distributed along the frequency axis in proportion to their bandwidths. The filter outputs are subjected to rectification, compression and low-pass filtering before the adaptive thresholding. The generated auditory image is perceptually more accurate than the spectrogram in terms of the time and frequency resolutions of signal components and their perceived loudness levels.

4 Audio Features for Classification

While the spectrogram and auditory signal representations discussed in the previous section are good for visualization of audio content, they have a high dimensionality which makes them unsuitable for direct application to classification. Ideally, we would like to extract low-dimensional features from these representations (or even directly from the acoustical signal) which retain only the important distinctive characteristics of the intended audio classes. Reduced-dimension, decorrelated spectral vectors obtained using a linear transformation of a spectrogram have been proposed in MPEG-7, the audio-visual content description standard [5, 6].

A more popular approach to feature design is to use explicit knowledge about the salient signal characteristics either in terms of signal production or perception. The goal is to find features that are invariant to irrelevant transformations and have good discriminative power across the classes. Feature extraction, an important signal processing task, is the process of computing the numerical representation from the acoustical signal that can be used to characterize the audio segment. Classification algorithms typically use labeled training examples to partition the feature space into regions so that feature vectors falling in the same region come from the same class. A well-designed set of features for a given audio categorization task would make for robust classification with reasonable amounts of training data. Audio signal classification is a subset of the larger problem of auditory scene analysis. When the audio stream contains many different, but non-simultaneous, events from different classes, segmentation of the stream to separate class-specific events can be achieved by observing the transitions in feature values as expected at segment boundaries. However when signals from the different classes (sources) overlap in time, stream segregation is a considerably more difficult task [7].

Research on audio classification over the years has given rise to a rich library of computational features which may be broadly categorized into physical features and perceptual features. Physical features are directly related to the measurable properties of the acoustical signal and are not linked with human perception. Perceptual features, on the other hand, relate to the subjective perception of the sound, and therefore must be computed using

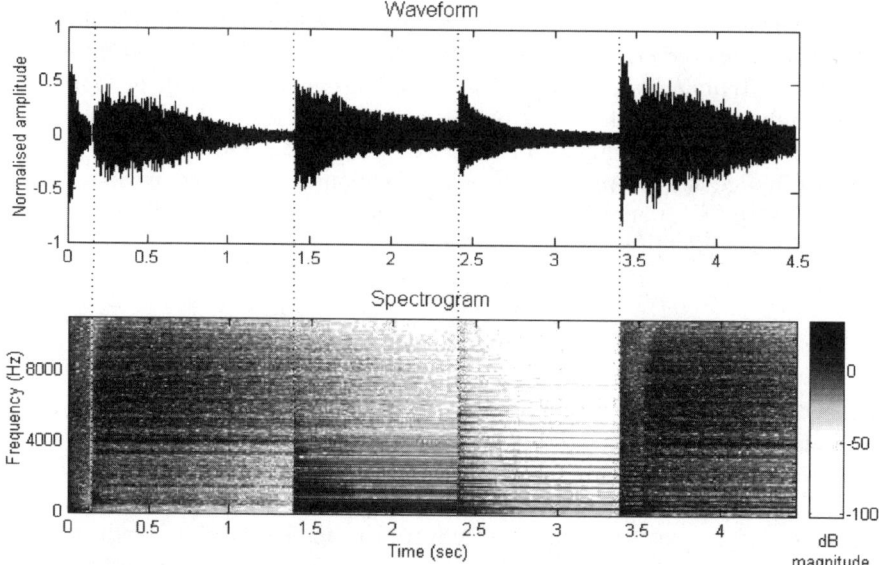

Fig. 2. (a) Waveform, and (b) spectrogram of the audio segment described in Table 1. The *vertical dotted lines* indicate the starting instants of new events. The spectrogram relative intensity scale appears at lower right

auditory models. Features may further be classified as static or dynamic features. Static features provide a snapshot of the characteristics of the audio signal at an instant in time as obtained from a short-time analysis of a data segment. The longer-term temporal variation of the static features is represented by the dynamic features and provides for improved classification. Figure 3 shows the structure of such a feature extraction framework [8]. At the lowest level are the analysis frames, each representing windowed data of typical duration 10 to 40 ms. The windows overlap so that frame durations can be significantly smaller, usually corresponding to a frame rate of 100 frames per second. Each audio frame is processed to obtain one or more static features. The features may be a homogenous set, like spectral components, or a more heterogenous set. That is, the frame-level feature vector corresponds to a set of features extracted from a single windowed audio segment centered at the frame instant. Next the temporal evolution of frame-level features is observed across a larger segment known as a texture window to extract suitable dynamic features or feature summaries. It has been shown that the grouping of frames to form a texture window improves classification due to the availability of important statistical variation information [9, 10]. However increasing the texture window length beyond 1 s does not improve classification any further. Texture window durations typically range from 500 ms to 1 s. This implies a latency or delay of up to 1 s in the audio classification task.

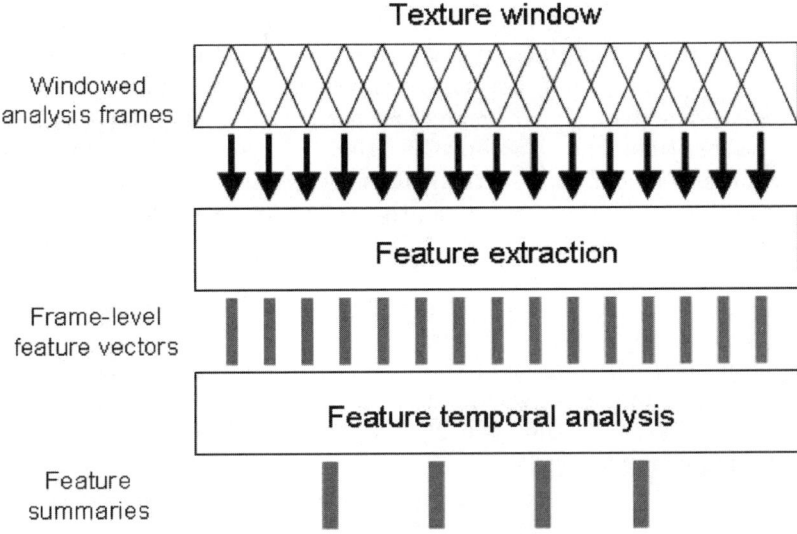

Fig. 3. Audio feature extraction procedure (adapted from [8])

4.1 Physical Features

Physical features are low-level signal parameters that capture particular aspects of the temporal or spectral properties of the signal. Although some of the features are perceptually motivated, we classify them as physical features since they are computed directly from the audio waveform amplitudes or the corresponding short-time spectral values. Widely applied physical features are discussed next. In the following equations, the subindex "r" indicates the current frame so that $x_r[n]$ are the samples of the N-length data segment (possibly multiplied by a window function) corresponding to the current frame. We then have for the analysis of the rth frame,

$$\left\{ \begin{matrix} x_r[n] \\ n = 1 \ldots N \end{matrix} \right\} \longrightarrow \left\{ \begin{matrix} X_r[k] \text{ at freq. f[k]} \\ k = 1 \ldots N \end{matrix} \right\} \tag{1}$$

Zero-Crossing Rate

The Zero-Crossing Rate (ZCR) measures the number of times the signal waveform changes sign in the course of the current frame and is given by

$$ZCR_r = \frac{1}{2} \sum_{n=1}^{N} | \, sign(x_r(n)) - sign(x_{r-1}(n)) \, | \tag{2}$$

where,

$$sign(x) = \left\{ \begin{matrix} 1, & x \geq 0; \\ -1, & x < 0. \end{matrix} \right.$$

For several applications, the ZCR provides spectral information at a low cost. For narrowband signals (e.g. a sinusoid), the ZCR is directly related to the fundamental frequency. For more complex signals, the ZCR correlates well with the average frequency of the major energy concentration. For speech signals, the short-time ZCR takes on values that fluctuate rapidly between voiced and unvoiced segments due to their differing spectral energy concentrations. For music signals, on the other hand, the ZCR is more stable across extended time durations.

Short-Time Energy

It is the mean squared value of the waveform values in the data frame and represents the temporal envelope of the signal. More than its actual magnitude, its variation over time can be a strong indicator of underlying signal content. It is computed as

$$E_r = \frac{1}{N} \sum_{n=1}^{N} \mid x_r(n) \mid^2 \tag{3}$$

Band-Level Energy

It refers to the energy within a specified frequency region of the signal spectrum. It can be computed by the appropriately weighted summation of the power spectrum as given by

$$E_r = \frac{1}{N} \sum_{k=1}^{\frac{N}{2}} (X_r[k]W[k])^2 \tag{4}$$

$W[k]$ is a weighting function with non-zero values over only a finite range of bin indices "k" corresponding to the frequency band of interest. Sudden transitions in the band-level energy indicate a change in the spectral energy distribution, or timbre, of the signal, and aid in audio segmentation. Generally log transformations of energy are used to improve the spread and represent (the perceptually more relevant) relative differences.

Spectral Centroid

It is the center of gravity of the magnitude spectrum. It is a gross indicator of spectral shape. The spectral centroid frequency location is high when the high frequency content is greater.

$$C_r = \frac{\sum_{k=1}^{\frac{N}{2}} f[k] \mid X_r[k] \mid}{\sum_{k=1}^{\frac{N}{2}} \mid X_r[k] \mid} \tag{5}$$

Since moving the major energy concentration of a signal towards higher frequencies makes it sound brighter, the spectral centroid has a strong correlation to the subjective sensation of brightness of a sound [10].

Spectral Roll-off

It is another common descriptor of gross spectral shape. The roll-off is given by

$$R_r = f[k] \tag{6}$$

where K is the largest bin that fulfills

$$\sum_{k=1}^{K} | X_r[k] | \le 0.85 \sum_{k=1}^{\frac{N}{2}} | X_r[k] | \tag{7}$$

That is, the roll-off is the frequency below which 85% of accumulated spectral magnitude is concentrated. Like the centroid, it takes on higher values for right-skewed spectra.

Spectral Flux

It is given by the frame-to-frame squared difference of the spectral magnitude vector summed across frequency as

$$F_r = \sum_{k=1}^{\frac{N}{2}} (| X_r[k] | - | X_{r-1}[k] |)^2 \tag{8}$$

It provides a measure of the local spectral rate of change. A high value of spectral flux indicates a sudden change in spectral magnitudes and therefore a possible segment boundary at the rth frame.

Fundamental Frequency (F_0)

It is computed by measuring the periodicity of the time-domain waveform. It may also be estimated from the signal spectrum as the frequency of the first harmonic or as the spacing between harmonics of the periodic signal. For real musical instruments and the human voice, F_0 estimation is a non-trivial problem due to (i) period-to-period variations of the waveform, and (ii) the fact that the fundamental frequency component may be weak relative to the other harmonics. The latter causes the detected period to be prone to doubling and halving errors. Time-domain periodicity can be estimated from the signal autocorrelation function (ACF) given by

$$R(\tau) = \frac{1}{N} \sum_{n=0}^{N-1} (x_r[n]x_r[n+\tau]) \tag{9}$$

The ACF, $R(\tau)$, will exhibit local maxima at the pitch period and its multiples. The fundamental frequency of the signal is estimated as the inverse of the lag "τ" that corresponds to the maximum of $R(\tau)$ within a predefined range. By favouring short lags over longer ones, fundamental frequency multiples are avoided. The normalized value of the lag at the estimated period represents the strength of the signal periodicity and is referred to as the harmonicity coefficient.

Mel-Frequency Cepstral Coefficients (MFCC)

MFCC are perceptually motivated features that provide a compact representation of the short-time spectrum envelope. MFCC have long been applied in speech recognition and, much more recently, to music [11]. To compute the MFCC, the windowed audio data frame is transformed by a DFT. Next, a Mel-scale filterbank is applied in the frequency domain and the power within each sub-band is computed by squaring and summing the spectral magnitudes within bands. The Mel-frequency scale, a perceptual scale like the critical band scale, is linear below 1 kHz and logarithmic above this frequency. Finally the logarithm of the bandwise power values are taken and decorrelated by applying a DCT to obtain the cepstral coefficients. The log transformation serves to deconvolve multiplicative components of the spectrum such as the source and filter transfer function. The decorrelation results in most of the energy being concentrated in a few cepstral coefficients. For instance, in 16 kHz sampled speech, 13 low-order MFCCs are adequate to represent the spectral envelope across phonemes.

A related feature is the cepstral residual computed as the difference between the signal spectrum and the spectrum reconstructed from the prominent low-order cepstral coefficients. The cepstral residual thus provides a measure of the fit of the cepstrally smoothed spectrum to the spectrum.

Feature Summaries

All the features described so far were short-time parameters computed at the frame rate from windowed segments of audio of duration no longer than 40 ms, the assumed interval of audio signal stationarity. An equally important cue to signal identity is the temporal pattern of changing signal properties observed across a sufficiently long interval. Local temporal changes may be described by the time derivatives of the features known as delta-features. Texture windows, as indicated in Fig. 3, enable descriptions of the long-term characteristics of the signal in terms of statistical measures of the time variation of each feature. Feature average and feature variance over the texture window serve as a coarse summary of the temporal variation. A more detailed description of a feature's variation with time is provided by the frequency spectrum of its temporal trajectory. The energy within a specific frequency band of this spectrum is termed a "modulation energy" of the feature [8]. For example, the short-time energy feature of speech signals shows high modulation energy in a band around 4 Hz due to the syllabic rate of normal speech utterances (that is, approximately 4 syllables are uttered per second leading to 4 local maxima in the short-time energy per second).

4.2 Perceptual Features

The human recognition of sound is based on the perceptual attributes of the sound. When a good source model is not available, perceptual features provide an alternative basis for segmentation and classification. The psychological sensations evoked by a sound can be broadly categorized as loudness, pitch and timbre. Loudness and pitch can be ordered on a magnitude scale of low to high. Timbre, on the other hand, is a more composite sensation with several dimensions that serves to distinguish different sounds of identical loudness and pitch. A computational model of the auditory system is used to obtain numerical representations of short-time perceptual parameters from the audio waveform segment. Loudness and pitch together with their temporal fluctuations are common perceptual features and are briefly reviewed here.

Loudness

Loudness is a sensation of signal strength. As would be expected it is correlated with the sound intensity, but it is also dependent on the duration and the spectrum of the sound. In physiological terms, the perceived loudness is determined by the sum total of the auditory neural activity elicited by the sound. Loudness scales nonlinearly with sound intensity. Corresponding to this, loudness computation models obtain loudness by summing the contributions of critical band filters raised to a compressive power [12]. Salient aspects of loudness perception captured by loudness models are the nonlinear scaling of loudness with intensity, frequency dependence of loudness and the additivity of loudness across spectrally separated components.

Pitch

Although pitch is a perceptual attribute, it is closely correlated with the physical attribute of fundamental frequency (F_0). Subjective pitch changes are related to the logarithm of F0 so that a constant pitch change in music refers to a constant ratio of fundamental frequencies. Most pitch detection algorithms (PDAs) extract F_0 from the acoustic signal, i.e. they are based on measuring the periodicity of the signal via the repetition rate of specific temporal features, or by detecting the harmonic structure of its spectrum. Auditorily motivated PDAs use a cochlear filterbank to decompose the signal and then separately estimate the periodicity of each channel via the ACF [13]. Due to the higher channel bandwidths in the high frequency region, several higher harmonics get combined in the same channel and the periodicity detected then corresponds to that of the amplitude envelope beating at the fundamental frequency. The perceptual PDAs try to emulate the ear's robustness to interference-corrupted signals, as well as to slightly anharmonic signals, which still produce a strong sensation of pitch. A challenging problem for PDAs is the pitch detection of a voice when multiple sound sources are present as occurs in polyphonic music.

As in the case of the physical features, temporal trajectories of pitch and loudness over texture window durations can provide important cues to sound source homogeneity and recognition. Modulation energies of bandpass filtered audio signals, corresponding to the auditory gammatone filterbank, in the 20–40 Hz range are correlated with the perceived roughness of the sound while modulation energies in the 3–15 Hz range are indicative of speech syllabic rates [8].

5 Audio Classification Systems

We review a few prominent examples of audio classification systems. Speech and music dominate multimedia applications and form the major classes of interest. As mentioned earlier, the proper design of the feature set considering the intended audio categories is crucial to the classification task. Features are chosen based on the knowledge of the salient signal characteristics either in terms of production or perception. It is also possible to select features from a large set of possible features based on exhaustive comparative evaluations in classification experiments. Once the features are extracted, standard machine learning techniques to design the classifier. Widely used classifiers include statistical pattern recognition algorithms such as the k nearest neighbours, Gaussian classifier, Gaussian Mixture Model (GMM) classifiers and neural networks [14]. Much of the effort in designing a classifier is spent collecting and preparing the training data. The range of sounds in the training set should reflect the scope of the sound category. For example, car horn sounds would include a variety of car horns held continuously and also as short hits in quick succession. The model extraction algorithm adapts to the scope of the data and thus a narrower range of examples produces a more specialized classifier.

5.1 Speech-Music Discrimination

Speech-music discrimination is considered a particularly important task for intelligent multimedia information processing. Mixed speech/music audio streams, typical of entertainment audio, are partitioned into homogenous segments from which non-speech segments are separated. The separation would be useful for purposes such as automatic speech recognition and text alignment in soundtracks, or even simply to automatically search for specific content such as news reports among radio broadcast channels. Several studies have addressed the problem of robustly distinguishing speech from music based on features computed from the acoustic signals in a pattern recognition framework. Some of the efforts have applied well-known features from statistical speech recognition such as LSFs and MFCC based on the expectation that their potential for the accurate characterization of speech sounds would help distinguish speech from music [11,15]. Taking the speech recognition approach further, Williams and Ellis [16] use a hybrid connectionist-HMM speech recogniser to obtain the posterior probabilities of 50 phone classes from a temporal window of 100 ms of feature vectors. Viewing the recogniser as a system of highly tuned detectors for speech-like signal events, we see that the phone posterior probabilities will behave differently for speech and music signals. Various features summarizing the posterior phone probability array are shown to be suitable for the speech-music discrimination task.

A knowledge-based approach to feature selection was adopted by Scheirer and Slaney [17], who evaluated a set of 13 features in various trained-classifier paradigms. The training data, with about 20 min of audio corresponding to each category, was designed to represent as broad a class of signals as possible. Thus the speech data consisted of several male and female speakers in various background noise and channel conditions, and the music data contained various styles (pop, jazz, classical, country, etc.) including vocal music. Scheirer and Slaney [17] evaluated several of the physical features, described in Sect. 4.1, together with the corresponding feature variances over a one-second texture window. Prominent among the features used were the spectral shape measures and the 4 Hz modulation energy. Also included were the cepstral residual energy and, a new feature, the pulse metric. Feature variances were found to be particularly important in distinguishing music from speech. Speech is marked by strongly contrasting acoustic properties arising from the voiced and unvoiced phone classes. In contrast to unvoiced segments and speech pauses, voiced frames are of high energy and have predominantly low frequency content. This leads to large variations in ZCR, as well as in spectral shape measures such as centroid and roll-off, as voiced and unvoiced regions alternate within speech segments. The cepstral residual energy too takes on relatively high values for voiced regions due to the presence of pitch pulses. Further the spectral flux varies between near-zero values during steady vowel regions to high values during phone transitions while that for music is more steady. Speech segments also have a number of quiet or low energy frames which makes the short-time energy distribution across the segment more left-skewed for speech as compared to that for music. The pulse metric (or "rhythmicness") feature is designed to detect music marked by strong beats (e.g. techno, rock). A strong beat leads to broadband rhythmic modulation in the signal as a whole. Rhythmicness is computed by observing the onsets in different frequency channels of the signal spectrum through bandpass filtered envelopes. There were no perceptual features in the evaluated feature set. The system performed well (with about 4% error rate), but not nearly as well as a human listener. Classifiers such as k-nearest neighbours and GMM were tested and performed similarly on the same set of features suggesting that the type of classifier and corresponding parameter settings was not crucial for the given topology of the feature space.

Later work [18] noted that music dominated by vocals posed a problem to conventional speech-music discrimination due to its strong speech-like characteristics. For instance, MFCC and ZCR show no significant differences between speech and singing. Dynamic features prove more useful. The 4 Hz modulation rate, being related to the syllabic rate of normal speaking, does well but is not sufficient by itself. The coefficient of harmonicity together with its 4 Hz modulation energy better captures the strong voiced-unvoiced temporal variations of speech and helps to distinguish it from singing. Zhang and Kuo [19] use the shape of the harmonic trajectories ("spectral peak tracks") to distinguish singing from speech. Singing is marked by relatively long durations of continuous harmonic tracks with prominent ripples in the higher harmonics due to pitch modulations by the singer. In speech, harmonic tracks are steady or slowly sloping during the course of voiced segments, interrupted by unvoiced consonants and by silence. Speech utterances have language-specific basic intonation patterns or pitch movements for sentence clauses.

5.2 Audio Segmentation and Classification

Audiovisual data, such as movies or television broadcasts, are more easily navigated using the accompanying audio rather than by observing visual clips. Audio clips provide easily interpretable information on the nature of the associated scene such as for instance, explosions and shots during scenes of violence where the associated video itself may be fairly varied. Spoken dialogues can help to demarcate semantically similar material in the video while a continuous background music would help hold a group of seemingly disparate visual scenes together. Zhang and Kuo [19] proposed a method for the automatic segmentation and annotation of audiovisual data based on audio content analysis. The audio record is assumed to comprise of the following non-simultaneously occurring sound classes: silence, sounds with and without music background including the sub-categories of harmonic and inharmonic environmental sounds (e.g. touch tones, doorbell, footsteps, explosions). Abrupt changes in the short-time physical features of energy, zero-crossing rate and fundamental frequency are used to locate segment boundaries between the distinct sound classes. The same short-time features, combined with their temporal trajectories over longer texture windows, are subsequently used to identify the class of each segment. To improve the speech-music distinction, spectral peaks detected in each frame are linked to obtain continuous spectral peak tracks. While both speech and music are characterized by continuous harmonic tracks, those of speech correspond to lower fundamental frequencies and are shorter in duration due to the interruptions from the occurrence of unvoiced phones and silences.

Wold et al. [20] in a pioneering work addressed the task of finding similar sounds in a database with a large variety of sounds coarsely categorized as musical instruments, machines, animals, speech and sounds in nature. The individual sounds ranged in duration from 1 to 15 s. Temporal trajectories of short-time perceptual features such as loudness, pitch and brightness were examined for sudden transitions to detect class boundaries and achieve the temporal segmentation of the audio into distinct classes. The classification itself was based on the salient perceptual features of each class. For instance, tones from the same instrument share the same quality of sound, or timbre. Therefore the similarity of such sounds must be judged by descriptors of temporal and spectral envelope while ignoring pitch, duration and loudness level. The overall system uses the short-time features of pitch, amplitude, brightness and bandwidth, and their statistics (mean, variance and autocorrelation coefficients) over the whole duration of the sound.

5.3 Music Information Retrieval

Automatically extracting musical information from audio data is important to the task of structuring and organizing the huge amounts of music available in digital form on the Web. Currently, music classification and searching depends entirely upon textual meta-data (title of the piece, composer, players, instruments, etc.). Developing features that can be extracted automatically from recorded audio for describing musical content would be very useful for music classification and subsequent retrieval based on various user-specified similarity criteria. The chief attributes of a piece of music are its timbral texture, pitch content and rhythmic content. The timbral texture is related to the instrumentation of the music and can be a basis for similarity between music drawn from the same period, culture or geographical region. Estimating the timbral texture would also help to detect a specified solo instrument in the audio record. The pitch and rhythm aspects are linked to the symbolic transcription or the "score" of the music independent of the instruments playing. The pitch and duration relationships between successive notes make up the melody of the music. The rhythm describes the timing relation between musical events within the piece including the patterns formed by the accent and duration of the notes. The detection of note onsets based on low-level audio features is a crucial component of rhythm detection. Automatic music transcription, or the conversion of the sound signal to readable musical notation, is the starting point for retrieval of music based on melodic similarity. As a tool, it is valuable also to music teaching and musicological research. A combination of low-level features for each of higher-level musical attributes viz. timbre, rhythmic structure and pitch content are used in identification of musical style, or genre [9, 10]. We next discuss the various features used for describing musical attributes for specific applications in music retrieval.

Timbral texture of music is described by features similar to those used in speech and speaker recognition. Musical instruments are modeled as resonators with periodic excitation. For example, in a wind instrument, a vibrated reed delivers puffs of air to a cylindrical bore resonator. The fundamental frequency of the excitation determines the perceived pitch of the note while the spectral resonances, shaping the harmonic spectrum envelope, are characteristic of the instrument type and shape. MFCC have been very successful in characterizing vocal tract resonances for vowel recognition, and this has prompted their use in instrument identification. Means and variances of the first few cepstral coefficients (excluding the DC coefficient) are utilized to capture the gross shape of the spectrum [9]. Other useful spectral envelope descriptors are the means and variances over a texture window of the spectral centroid, roll-off, flux and zero-crossing rate. The log of the attack time (duration between the start of the note and the time at which it reaches its maximum value) together with the energy-weighted temporal centroid are important temporal descriptors of instrument timbre especially for percussive instruments [21]. Given the very large number of different instruments, a hierarchical approach to classification is sometimes taken based on instrument taxonomy.

A different, and possibly more significant, basis for musical similarity is the melody. Melodic similarity is an important aspect in music copyright and detection of plagiarism. It is observed that in the human recall of music previously heard, the melody or tune is by far the most well preserved aspect [22]. This suggests that a natural way for a user to query a music database is to perform the query by singing or humming a fragment of the melody. "Query-by-humming" systems allow just this. Users often prefer humming the tune in a neutral syllable (such as "la" or "ta") to singing the lyrics. Retrieving music based on a hummed melody fragment then reduces to matching the melodic contour extracted from the query with pre-stored contours corresponding to the database. The melodic contour is a mid-level data representation derived from low-level audio features. The mid-level representation selected defines a trade-off between retrieval accuracy and robustness to user errors. Typically, the melody is represented as a sequence of discrete-valued note pitches and durations. A critical component of query signal processing then is audio segmentation into distinct notes, generally considered a challenging problem [23]. Transitions in short-term energy or in band-level energies derived from either auditory or acoustic-phonetic motivated frequency bands have been investigated to detect vocal note onsets [23,24]. A pitch label is assigned to a detected note based on suitable averaging of frame-level pitch estimates across the note duration. Since people remember the melodic contour (or the shape of the temporal trajectory of pitch) rather than exact pitches, the note pitches and durations are converted to relative pitch and duration intervals. In the absence of an adequate cognitive model for melodic similarity, modifications of the basic string edit distance measure are applied to template matching in the database search [26].

The instrument identification and melody recognition methods discussed so far implicitly assumed that the music is monophonic. In polyphonic music, where sound events corresponding to different notes from one or more instruments overlap in time, pitch and note onset detection become considerably more difficult. Based on the observation that humans can decompose complex sound mixtures to perceive individual characteristics such as pitch, auditory models are being actively researched for polyphonic music transcription [13, 25].

6 Summary

The world's ever growing archives of multimedia data pose huge challenges for digital content management and retrieval. While we now have the ability to search text quite effectively, other multimedia data such as audio and video remain opaque to search engines except through a narrow window provided by the possibly attached textual metadata. Segmentation and classification of raw audio based on audio content analysis constitutes an important component of audiovisual data handling. Speech and music constitute the commercially most significant components of audio multimedia. While speech and speaker recognition are relatively mature fields, music information retrieval is a new and growing research area with applications in music searching based on various user-specified criteria such as style, instrumentation and melodic similarity. In this chapter we have focused on signal processing methods for the segmentation and classification of audio signals. Assigning class labels to sounds or audio segments is aided by an understanding of either source or signal characteristics. The general characteristics of audio signals have been discussed, followed by a review of time-frequency representations that help in the visualization of the acoustic spectro-temporal properties. Since the important attributes of an audio signal are its salient features as perceived by the human ear, auditory models can play a substantial role in the effective representation of audio. Two chief components of a classification system are the feature extraction module and the classifier itself. The goal is to find features that are invariant to irrelevant transformations and have good discriminative power across the classes. Feature extraction, an important signal processing task, is the process of computing a numerical representation from the acoustical signal that can be used to characterize the audio segment. Important audio features, representative of the rich library developed by research in audio classification over the years, have been reviewed. The acoustic and perceptual correlates of the individual features have been discussed, providing a foundation for feature selection in specific applications.

Local transitions in feature values provide the basis for audio segmentation. Sound tracks could, for instance, be comprised of alternating sequences of silence, spoken dialog and music needing to be separated into homogenous segments before classification. Classification algorithms typically use labeled training examples to partition the feature space into regions so that feature vectors falling in the same region come from the same class. Statistical pattern classifiers and neural networks are widely used. Since the methods are well documented in the general machine learning literature, they have not been discussed here. A few important audio classification tasks have been presented together with a discussion of some widely cited proposed solutions. The problems presented include speech-music discrimination, audiovisual scene segmentation and classification based on audio features, and music retrieval applications such as genre or style identification, instrument identification and query by humming. Active research continues on all these problems for various database compilations within the common framework of feature extraction followed by pattern classification. Significant progress has been made and improvements in classification accuracy continue to be reported from minor modifications to the basic low-level features or in the classification method employed. However it is believed that such an approach has reached a "glass ceiling", and that really large gains in the performance of practical audio search and retrieval systems can come only with a deeper understanding of human sound perception and cognition [27]. Achieving the separation of the components of complex sound mixtures, and finding objective distance measures that predict subjective judgments on the similarity of sounds are specific tasks that would benefit greatly from better auditory system modelling.

References

1. Oppenheim A V, Lim J S (1981) The Importance of Phase in Signals. Proc of the IEEE 69(5):529–550
2. Moore B C J (2003) An Introduction to the Psychology of Hearing. Academic, San Diego
3. Patterson R D (2000) Auditory Images: How Complex Sounds Are Represented in the Auditory System. J Acoust Soc Japan (E) 21(4)
4. Lyon R F, Dyer L (1986) Experiments with a Computational Model of the Cochlea. Proc of the International Conference on Acoustics, Speech and Signal Processing (ICASSP)
5. Martinez J M (2002) Standards - MPEG-7 overview of MPEG-7 description tools, part 2. IEEE Multimedia 9(3):83–93
6. Xiong Z, Radhakrishnan R, Divakaran A, Huang T (2003) Comparing MFCC and MPEG-7 Audio Features for Feature Extraction, Maximum Likelihood HMM and Entropic Prior HMM for Sports Audio Classification. Proc of the International Conference on Multimedia and Expo (ICME)
7. Wang L, Brown G (2006) Computational Auditory Scene Analysis: Principles, Algorithms and Applications. Wiley-IEEE, New York
8. McKinney M F, Breebaart J (2003) Features for Audio and Music Classification. Proc of the International Symposium on Music Information Retrieval (ISMIR)
9. Tzanetakis G, Cook P (2002) Musical Genre Classification of Audio Signals. IEEE Trans Speech Audio Process 10(5):293–302
10. Burred J J, Lerch A (2004) Hierarchical Automatic Audio Signal Classification. J of Audio Eng Soc 52(7/8):724–739

11. Logan B (2000) Mel frequency Cepstral Coefficients for Music Modeling. Proc of the International Symposium on Music Information Retrieval (ISMIR)
12. Zwicker E, Scharf B (1965) A Model of Loudness Summation. Psychol Rev 72:3–26
13. Klapuri A P (2005) A Perceptually Motivated Multiple-F0 Estimation Method for Polyphonic Music Signals. IEEE Workshop on Applications of Signal Processing to Audio and Acoustics (WASPA)
14. Duda R, Hart P, Stork D (2000) Pattern Classification. Wiley, New York
15. El-Maleh K, Klein M, Petrucci G, Kabal P (2000) Speech/Music Discrimination for Multimedia Applications. Proc of the International Conference on Acoustics, Speech and Signal Processing (ICASSP)
16. Williams G, Ellis D (1999) Speech/Music Discrimination based on Posterior Probability Features. Proc of Eurospeech
17. Scheirer E, Slaney M (1997) Construction and Evaluation of a Robust Multi-feature Speech/Music Discriminator. Proc of the International Conference on Acoustics, Speech and Signal Processing (ICASSP)
18. Chou W, Gu L (2001) Robust Singing Detection in Speech/Music Discriminator Design. Proc of the International Conference on Acoustics, Speech and Signal Processing (ICASSP)
19. Zhang T, Kuo C C J (2001) Audio Content Analysis for Online AudioVisual Data Segmentation and Classification. IEEE Trans on Speech and Audio Processing 9(4):441–457
20. Wold E, Blum T, Keisler D, Wheaton J (1996) Content-based Classification, Search and Retrieval of Audio. IEEE Multimedia 3(3):27–36
21. Peeters G, McAdams S, Herrera P (2000) Instrument Sound Description in the Context of MPEG-7. Proc of the International Computer Music Conference (ICMC)
22. Dowling W J (1978) Scale and Contour: Two Components of a Theory of Memory for Melodies. Psychol Rev 85:342–389
23. Pradeep P, Joshi M, Hariharan S, Dutta-Roy S, Rao P (2007) Sung Note Segmentation for a Query-by-Humming System. Proc of the International Workshop on Artificial Intelligence and Music (Music-AI) in IJCAI
24. Klapuri A P (1999) Sound Onset Detection by Applying Psychoacoustic Knowledge. Proc of the International Conference on Acoustics, Speech and Signal Processing (ICASSP)
25. de Cheveigne A, Kawahara H (1999) Multiple Period Estimation and Pitch Perception Model. Speech Communication 27:175–185
26. Uitdenbogerd A, Zobel J (1999) Melodic Matching Techniques for Large Music Databases. Proc of the 7th ACM International Conference on Multimedia (Part 1)
27. Aucouturier J J, Pachet F (2004) Improving Timbre Similarity: How High is the Sky. J Negat Result Speech Audio Sci 1(1)

Democratic Liquid State Machines for Music Recognition

Leo Pape[1], Jornt de Gruijl[2], and Marco Wiering[2]

[1] Department of Physical Geography, Faculty of Geosciences, IMAU,
Utrecht University; P.O. Box 80.115, 3508 TC Utrecht, Netherlands,
l.pape@geo.uu.nl
[2] Intelligent Systems Group, Department of Information and Computing Sciences,
Utrecht University, P.O. Box 80.089, 3508 TB Utrecht, Netherlands,
jornt.degruijl@phil.uu.nl, marco@cs.uu.nl

Summary. The liquid state machine (LSM) is a relatively new recurrent neural network (RNN) architecture for dealing with time-series classification problems. The LSM has some attractive properties such as a fast training speed compared with more traditional RNNs, its biological plausibility, and its ability to deal with highly non-linear dynamics. This paper presents the democratic LSM, an extension of the basic LSM that uses majority voting by combining two dimensions. First, instead of only giving the classification at the end of the time-series, multiple classifications after different time-periods are combined. Second, instead of using a single LSM, multiple ensembles are combined. The results show that the democratic LSM significantly outperforms the basic LSM and other methods on two music composer classification tasks where the goal is to separate Haydn/Mozart and Beethoven/Bach, and a music instrument classification problem where the goal is to distinguish between a flute and a bass guitar.

1 Introduction

Liquid state machines [1] and echo state networks [2] are relatively novel algorithms that can handle problems involving time-series data. These methods consist of a liquid (or dynamic reservoir) that receives time-series data as input information, which causes the liquid to make transitions to dynamical states dependent on the temporal information stream. After a specific time-period the liquid is read out by a trainable readout network (such as a perceptron or multilayer perceptron) that learns to map liquid states to a particular desired output. Earlier recurrent neural networks that could also be used for handling time-series data such as Elman or Jordan networks suffer from the problem of vanishing gradients which makes learning long time-dependencies a very complex task [3]. A more advanced recurrent neural network architecture is Long Short-Term Memory (LSTM) [4]. LSTM is ideally suited for

L. Pape et al.: *Democratic Liquid State Machines for Music Recognition*, Studies in Computational Intelligence (SCI) **83**, 191–211 (2008)
www.springerlink.com

remembering particular events, such as remembering whether some lights were on or off [5], but training the LSTM is more complicated than training a liquid state machine. The attractive property of a liquid state machine (LSM) is that the liquid which represents temporal dependencies does not have to be trained, which therefore makes the learning process fast and simple. Furthermore, the liquid state machine bears some similarities to biological neural networks and can therefore profit from new findings in neurobiology.

In this chapter we extend the liquid state machine for real-world time-series classification problems. Our democratic liquid state machine (DLSM) uses voting in two different complementary dimensions. The DLSM uses an ensemble of single-column LSMs and multiple readout periods to classify an input stream. Therefore instead of a single election on the classification, many votes are combined through majority voting. DLSM can therefore also be seen as an extension to bagging [6] LSMs for time-series classification problems, since multiple votes in the time-dimension are combined. We argue that this multiple voting technique can significantly boost the performance of LSMs, since they can exhibit a large variance and small bias. It is well-known that use of ensembles and majority voting significantly reduces the variance of a classifier [7].

In the present work we apply the LSM and its democratic extension on two different music recognition problems: composer identification and musical instrument classification. In the first problem the algorithm receives time-series input about the pitch and duration of notes of a particular classical music piece and the goal is to identify the composer of the piece. Here we will deal with two binary classification tasks. In the first task the goal is to distinguish between Beethoven and Bach as being the composer, and in another more complex task the goal is to distinguish between Haydn and Mozart. Furthermore, we will also study the performance of the LSM and the DLSM on a musical instrument classification problem where the goal is to distinguish between a flute and a bass guitar.

The rest of the chapter is organized as follows: In Sect. 2 we describe neural networks, the LSM, the DLSM and the implementation we used. Then in Sect. 3 we describe the experimental results on the classical music recognition tasks. In Sect. 4 we describe the results on the task of musical instrument classification, and we conclude in Sect. 5.

2 Democratic Liquid State Machines

2.1 Feedforward Neural Networks

Artificial neural networks are based on the low-level structure of the human brain, and can mimic the capacity of the brain to learn from observations. Neural networks have several advantages compared to other pattern recognition techniques. Neural networks can learn non-linear relationships, do not require any a priori knowledge, can generalize well to unseen data, are able to cope with severe noise, can deal with high-dimensional input spaces, and are fast in their usage (although not in training). A last important advantage of neural networks compared to algorithms based on computing distances between feature vectors is that no higher-level features need to be designed. Instead a neural network learns to extract higher level features from raw data itself. However, not all types of pattern recognition problems are solved equally well by neural networks.

A feedforward neural network or multilayer perceptron is a trainable differentiable function that maps an input x to an output y. It can be used both for regression and classification problems. A neural network uses a topology of neurons and weighted connections between the neurons. The output of a fully connected feedforward neural network with three layers (one hidden layer) given some input x is computed by first copying the input to the input layer of the neural network. After this the activations h_i of the hidden units are computed as follows:

$$h_i = f\left(\sum_j w_{ij}^h x_j + b_i^h\right), \tag{1}$$

where w_{ij}^h is the strength of the weighted connection between input j and hidden unit i, b_i^h is the bias of the hidden unit, and $f(.)$ is some differentiable activation function. Here we will use the commonly used sigmoid function: $f(z) = \frac{1}{1+\exp(-z)}$.

After the values for all hidden units are computed we can compute the value y_i of output unit i

$$y_i = \sum_j w_{ij}^o h_j + b_i^o, \tag{2}$$

where we use a linear activation function for the output units.

The weights w can be trained using gradient descent on the error function for a training example. The error function that is most commonly used is the squared error defined as $E = \sum_i (d_i - y_i)^2$, where y_i is the i-th output of the neural network and d_i is the i-th target output for the example. Now we can use the derivative of the error function to each weight to change the weight with a small step

$$\frac{\partial E}{\partial w_{ij}^o} = \frac{\partial E}{\partial y_i}\frac{\partial y_i}{\partial w_{ij}^o} = -(d_i - y_i)h_j. \tag{3}$$

The errors are all backpropagated to change the weights between inputs and hidden units

$$\frac{\partial E}{\partial w_{ij}^h} = \sum_k \frac{\partial E}{\partial y_k}\frac{\partial y_k}{\partial h_i}\frac{\partial h_i}{\partial w_{ij}^h} = \sum_k -(d_k - y_k)w_{ki}^o(1 - h_i)h_i x_j. \tag{4}$$

The learning rules then become

$$w_{ij} = w_{ij} - \alpha \frac{\partial E}{\partial w_{ij}}, \tag{5}$$

where α is known as the learning rate. The updates for the bias weights is very similar and are therefore not presented here.

2.2 Time-Series Classification with Neural Networks

At present there exist a lot of pattern recognition methods that operate on snapshots of data, but these methods cannot always be applied to time-series data. Time-series data consist of sequences of snapshots that are typically measured at successive times and are usually spaced at uniform intervals. Often successive patterns in time-series show dependencies. Because these dependencies can span over patterns at different moments in time, they are referred to as temporal dependencies. When information on the class of time-series data is contained in the temporal dependencies, a pattern recognition method should be able to deal with dependencies in the temporal dimension. In the case of artificial neural networks it is of course possible to increase the number of input units, such that several snapshots containing temporal dependencies can be provided to the network at once. It is however not always a priori known over how many time steps the temporal dependencies span, which makes it difficult to determine how many patterns should be presented to the network at once. Besides, in case the dependencies span a very long time, the network becomes very large, which results in low learning speeds and increased chance of overfitting. Whereas neural networks without recurrent connections cannot use any information that is present in the dependencies between inputs that are presented at different time points, recurrent connections allow a network to keep some sort of short-term memory on previously seen inputs. However, using the short-term memory that is based on the activations of the units in the past is very difficult.

The reason that short-term memory based on previous activities in the network is really short, is that it is very difficult to retain a stable level of activity in the network. Once a certain level of activity is reached, connections with a large weight cause the activation to increase exponentially, and connections with a small weight cause the activations to decrease exponentially. Non-linear activation functions of the units prevent the activities from growing without bound, but the problem is that it is very difficult to train the network to make use of this sort of memory [4]. For example in training algorithms based on error backpropagation, which are often used in non-recurrent neural networks, the contribution to the error of a particular weight situated lower in the network becomes more indirect. Consequently, when the error is propagated back over time, each propagation dilutes the error until far back in time it is too small to have any influence. This makes it very difficult to assess the importance of network states at times that lie far back in the past. In general, recurrent neural networks that are trained by backpropagation methods such as Realtime Recurrent Learning (RTRL) and Backpropagation Through Time (BPTT) [8, 9] cannot reliably use states of the network that lie more than about 10 time steps in the past. In the literature, this problem is often referred to as the problem of the vanishing gradient [3, 10–13].

2.3 The Liquid State Machine

Instead of training a recurrent neural network to learn to exploit memory over previous states, the liquid state machine [1] is based on the idea that due to their dynamics, recurrent neural networks can be employed as temporal filters. Just as a real liquid contains information in the form of waves and surface ripples on perturbations that were applied to it, the liquid filter of an LSM contains information on its present and past inputs at any moment. At certain times, the state of this liquid filter can be read out and presented to a separate readout mechanism. This readout mechanism can be trained to learn to extract useful information from the state of the liquid. The problems that arise when training recurrent neural networks disappear in this approach: the liquid itself is a recurrent neural network, but is not trained, while the readout mechanism does not need to implement any memory since the information on previous inputs is already reflected in the state of the liquid. Figure 1 shows the LSM architecture, connecting a liquid to a readout network.

Although the idea for the LSM originates from the dynamics of spiking recurrent neural networks, it can include any liquid that shows some kind of dynamical behavior. For a liquid to serve as a salient source of information on temporal dependencies in the input data, it should satisfy the separation property, which is the property to produce increasing divergence of internal states over time caused by different input patterns [1]. Furthermore, it should have

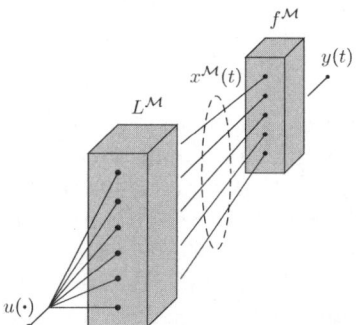

Fig. 1. The architecture of the liquid state machine. $u(\cdot)$ is the input, L^M the liquid, $x^M(t)$ the state of the liquid at time t, f^M the readout network, and $y(t)$ the output of the liquid state machine at time t

a fading memory to prevent the state trajectories from becoming chaotic [14]. Whether these conditions are satisfied for a certain liquid, depends on the complexity and dynamics of the liquid. In general, more complex liquids with more degrees of freedom of the processes that govern their behavior, produce more diverging internal states, but also have a greater risk of showing chaotic behavior. The way in which perturbations in the liquid are formed and propagated varies for different types of liquids and strongly determines how useful it is.

Recurrent neural networks with spiking units not only have a large variety of mechanisms and constants determining their interactions, but can also have local as well as non-local interactions. This means that they can effectively be used as high-dimensional liquid equivalents [15]. Maass et al. [1] used a simplified model of the cerebral cortex, in which the individual neurons are organized in columns. These columns are all very similar and can be used to perform many complex tasks in parallel. Not only are these heterogeneous column-like structures found in the human cortex; they are also present in the cortex of other species. Implementations of these columns in the form of spiking neural networks, have already shown to perform very well; even in noisy environments their dynamics remain reliable [16].

The internal state of the liquid at a certain time reflects the perturbations that were applied to it in the past. The capability of the readout mechanism to distinguish and transform the different internal states of the liquid to match a given target, or the approximation property, depends on the ability to adjust to the required task. The readout mechanism should be able to learn to define its own notion of equivalence of dynamical states within the liquid, despite the fact that the liquid may never revisit the same state. It is however argued by [1] and more elaborately by [17] that the LSM operates by the principle of projecting the input patterns in a high-dimensional space. That means that very simple and robust (linear) readout mechanisms should be able to perform very well, given enough degrees of freedom in the liquid from which the readout mechanism can extract individualized states.

2.4 Democratic Liquid State Machines

Maass et al. [1] show that the separation property of the LSM can be improved by adding more liquid columns. In a multiple column configuration the input patterns are presented to each individual column. Because the columns consist of different randomly initialized neural circuitries, each column produces its own representation of the input patterns. In case of a single liquid column, the intrinsic complexity of this column determines the separation property, whereas the complexity of a column in case of multiple liquid columns becomes less important. This allows for a trade-off between the intrinsic complexity of the microcircuitry and the number of liquid columns.

An important difference between other multiple classifier systems and the multiple column LSM as described by [1] is that although each liquid column extracts different features from the input patterns, no actual classification is being made at the level of the liquid. In the multiple column LSM the states of all liquid columns together are presented to a single readout mechanism. Since each liquid column has its own intrinsic structure, each column produces its own kind of equivalence in its dynamic states. When a single readout mechanism is used, it should learn to define a notion of equivalence of the dynamic states for *all* liquid columns. This imposes certain requirements on the size of the readout mechanism (in terms of connections) and the complexity of its training algorithm. In order to decrease the complexity of the readout mechanism and to use the advantages of a combination function over multiple classifiers, a separate readout mechanism can be used for each individual liquid column. In this configuration each readout mechanism learns its own notion of equivalence of dynamic states in a separate liquid column. The classification results of each liquid-readout pair are now equivalent to classification results of multiple classifier systems, and can be combined using multiple classifier combination functions as for example is done in bagging [6].

Improving performance using multiple classifiers does not only depend on the diversity between classifiers, but also on the function that is used to combine them. One of the simplest combination functions that has been studied thoroughly, is majority voting, which can be applied to classification tasks by counting the number of times each classifier assigns a certain class to an input pattern. The majority of the classification results is taken as the outcome of the majority voting combination function. In case more than one class gets the highest number of votes, there is no majority, and the outcome of the majority voting function is undecided. Benefits of majority voting have been studied for both odd and even numbers of classifiers, as well as dependent and independent errors [7, 18, 19]. Even without the assumption of a binary classification task, independent errors or equally performing classifiers majority voting can substantially increase performance.

Apart from using multiple classifiers that each involve their own notion of equivalence over input patterns, the LSM allows for multiple classification results over time. As stated in Sect. 2.3, the liquid state contains information on present and past perturbations at any moment. Because the liquid should involve some kind of 'forgetting behavior' to prevent the state trajectories from becoming chaotic, additional information can be gathered by using the state of the liquid at different moments instead of just the final liquid state. In case the length of the time-series of input patterns is relatively large compared to the time-constants of the liquid, multiple readout moments can be used. The readout mechanism produces a classification using the liquid state at each readout moment. In this fashion the input data are divided in parts, and a classification is performed on each part. In case the activity in the liquid is not reset at the beginning of each part of the input data, the liquid still satisfies the requirements of the LSM. As with multiple classifier systems, the individual classification results at each readout moment can be combined using a certain combination function. Although the different classification results are by no means independent, majority voting can still be used to boost performance over multiple results [7]. Majority voting over the results of the same classifier but for different input patterns at different moments, can be used, at least in theory, to improve performance of the LSM.

Combining majority voting over multiple classifiers with majority voting at multiple times results in a strategy in which the classification of a time-series pattern is not determined by a single election, but by the result of multiple elections performed by different classifiers, each having their own generalization strategy. LSMs that combine majority voting over spatial (multiple classifiers) and temporal (multiple classification moments) dimensions are what we call democratic liquid state machines (DLSMs).

2.5 Implementation of the Liquid State Machine

As stated in Sect. 2.3, spiking recurrent neural networks can effectively be used as high-dimensional liquid equivalents. A spiking neural network tries to capture the behavior of real biological neural networks, but does not necessary implement all features and interactions of its biological counterpart. One model that is commonly used for networks of spiking units, is the integrate-and-fire model [20]. This is a general model that implements certain properties of biological neurons, such as spikes and membrane potentials, and describes in basic terms that the membrane potential of a unit slowly leaks away, and that a unit fires if the membrane potential reaches a certain value. This model is rather simple to understand and implement, and can be described by giving the differential equations that govern the behavior of the membrane potentials, spikes and some other variables over time. Although it is common practice in the field of LSMs to use spiking neural networks with a continuous-time dynamics, in the present work a spiking neural network is used in which the states and variables of the units are computed in discrete time. Therefore the equations that describe the behavior of such units are not given in differential equations, but are described in terms of the updates of the membrane potentials and activities of the units, that are computed each time step t.

The units in the present work are a simplified model of their more realistic biological counterparts. Spikes are modeled by short pulses that are communicated between units. Action potentials in biological neurons are all very similar. Therefore all spikes are modeled to have the same value, so the output y_i of unit i is 1 if a unit fires and 0 if it does not fire. Incoming spikes affect the membrane potential of a unit. How much each incoming spike decreases or increases the membrane potential is determined by the synaptic strength, or weight w_{ij}, of the connection between unit j and i. In the spiking neural network, a certain proportion p^i of the weights is set to a negative value to provide for inhibitory connections, while the remaining weights are set to positive values, representing excitatory connections. Because in biological neurons the effect of an incoming signal lasts longer than the duration of the pulse itself and the membrane potential in absence of any input slowly returns to its resting potential, the accumulated input of a unit at each time step is added to a proportion of the membrane potential of the previous step. To keep the equations for the membrane potential simple, the resting potential for each unit is 0. The parameter that controls the speed at which the membrane potential returns to its resting potential is

$$\tau = \left(\frac{1}{2}\right)^{\frac{1}{t^h}}, \qquad (6)$$

where t^h is the number of time steps in which half of the membrane potential is left if no additional inputs are presented. The membrane potential μ_i, of unit i is computed each time step according to

$$\mu_i(t) = \tau \mu_i(t-1) + \sum_j w_{ij} y_j(t-1). \qquad (7)$$

Subsequently the output, or whether the unit fires or not, is computed by comparing it with a positive threshold value η

$$y_i(t) = \begin{cases} 1 \text{ if } \mu_i(t) \geq \eta \\ 0 \text{ otherwise} \end{cases}, \qquad (8)$$

where η has a value of 1 throughout all experiments. After sending an action potential, a biological neuron cannot fire again for a certain period, followed by a period in which it is more difficult to fire. This behavior is modeled in the artificial neurons by implementing an absolute and a relative refractory period. After sending an action potential, it is impossible for a unit to fire again for a number of t_i^a time steps, where t_i^a is the absolute refractory period of unit i. Combining this with (8), yields

$$y_i(t) = \begin{cases} 1 \text{ if } \mu_i(t) \geq \eta \text{ and } (t - t_i^f) > t_i^a \\ 0 \text{ otherwise} \end{cases}, \qquad (9)$$

where t_i^f is the last time unit i generated an action potential. When the absolute refractory period has ended, a relative refractory period is simulated by resetting the membrane potential μ_i to some negative value $-r_i$ which influences the relative refractory period of unit i, thus making it harder for the membrane potential to reach the positive threshold value η. To implement this relative refractory period, (7) is augmented with an additional term

$$\mu_i(t) = \begin{cases} \tau \mu_i(t-1) + \sum_j w_{ij} y_j(t-1) \text{ if } (t - t_i^f) > t_i^a \\ -r_i \qquad\qquad\qquad\qquad\qquad\qquad \text{otherwise} \end{cases}. \qquad (10)$$

In case there is no additional input presented to the unit, the negative membrane potential slowly returns to the resting potential, just as it would for a positive membrane potential in case no additional inputs are presented. Equations (9) and (10) sum up the whole of the dynamics of spiking units of which the spiking neural network consists.

The structure of the liquid is an n-dimensional space, in which each unit is located at the integer points. The units in this space are connected to each other according to a stochastic process such that units that are close to each other have a higher chance to be connected than units that lie farther from each other. To create such a network with mainly locally connected units, the chance $p_i^c(j)$ for each unit i that it has a connection to unit j is computed according to

$$p_i^c(j) = Ce^{(D(i,j)/\lambda)^2}, \tag{11}$$

where $D(i,j)$ is the Euclidean distance between units i and j. λ is the parameter that controls both the probability of a connection and the rate at which the probability that two units are connected decreases with increasing distance between the two. C is the parameter that controls just the probability that two units are connected, and can be used in combination with λ to scale the proportion between the rate at which the probability that unit i and j are connected decreases with increasing distance between i and j, and the actual probability of a connection. Maass et al. [1] found that the connectivity in biological settings can be approximated by setting the value of λ to 2, and C to a value between 0.1 and 0.4, depending on whether the concerning units are inhibitory or excitatory. However, that does not mean that other values for C and λ cannot yield a reasonable separation property of the liquid. In the present work λ is set to a value of 2 in all experiments, while the value of C was set depending on the task. The probability that a connection is inhibitory or excitatory is computed independent of the properties of the units it connects, so C was the same for all units.

When time-series data of input patterns are provided to the network, the state $y_i(t)$, of unit i in the network is computed at each discrete time step t, using the states of connected units at the previous time step. The incoming signals propagate through the network, and produce certain activities in the liquid units. The total activity of all units in the network at a certain time step contains information on the inputs that were provided during the previous time steps. It is not necessary to present the liquid state to the readout mechanism at each time step, because the liquid can preserve information on previously seen inputs over several time steps. The interval between successive readouts is denoted by t^v. At the beginning of each time-series pattern, the activity in the network is reset to zero. Because it takes some time before the activity has spread from the input units to the rest of the network, the collection of liquid states is started after a number of time steps. The time between reset and the first readout moment is called the warmup period t^w.

Because in a spiking neural network the units are not firing most of the time the activity in the network at the readout moments is sparse. That means that little information can be extracted from the activities at just a single time step. There are several ways to create representations of liquid states that overcome the problem of sparse activity. For example, the membrane potential μ can be used, which reflects more information about the activity in the network than just the unit's activity. Another method is to take the number of times a unit fires during a certain period as measure of activity in the network.

As stated in Sect. 2.3, the LSM operates by projecting input patterns in a high-dimensional space, which makes the learning process less complicated [21]. Due to the high-dimensionality of the liquid, the projections of the original input patterns become linearly separable. In general, this means that the more degrees of freedom the liquid has, usually the easier it becomes for the readout mechanism to learn to predict the right targets. Because no theory exists at present that tells for which problems and liquids the task becomes linearly separable, we cannot be sure whether this is attained for a certain amount of units and parameter settings of the spiking neural network. Therefore the readout mechanism in the present work is a multilayer perceptron [9], which can also learn non-linearly separable tasks.

At the end of intervals with length t^v the representations of the liquid states are presented to the multilayer perceptron. All units in the liquid other than the input units are used for readout, so the number of units in the first layer of the multilayer perceptron is equal to the number of units in the liquid minus the input units. The multilayer perceptron is trained by the supervised backpropagation algorithm [9], with first-order gradient descent as explained in Sect. 2.1. In order to compute the error, a representation of the desired output has to be created. In the classification tasks studied in the present work, each output unit is assigned to a class. The desired activity of the unit that represents the class that corresponds to an input pattern is 1, while the desired activity for all other output units is 0. In this fashion the output unit that represents the class that corresponds to an input pattern learns to have a high activity, while the other output units have a low activity. For testing purposes the unit with the highest activity determines the class that is assigned by the network.

In the implementation of the DLSM in the present work, the time-series of input patterns is presented to several spiking neural networks, where each network has its own multilayer perceptron that serves as readout mechanism. Each perceptron is trained to learn a mapping from the states of the spiking network to classification outputs represented in the activation of the output units. After the training of all perceptrons is finished, the performance on novel data can be determined by multiple elections over the classification results of the multilayer perceptrons. First, the outcome at each readout moment is determined by applying majority voting over all readout mechanisms. Next, majority voting is applied on the results of the previous step for all readout moments. The outcome of the elections determine the class of a time-series of input patterns.

3 Classical Music Recognition

3.1 Learning Task

The performance of the DLSM can be investigated using a classification task involving time-series data in which the classes are determined by temporal dependencies. A real-world pattern recognition task that satisfies this requirement is the classification of music based on musical content. A set of musical pieces can be divided into different classes based on several criteria. The criterion used here is the composer. Two different binary classification tasks were constructed: a task that is quite easily solved, and a far more difficult task (even for human experts). The first task consists of the classification of Johann Sebastian Bach and Ludwig van Beethoven, using music from the 'Well-Tempered Clavier' by Bach, and the 'Piano Sonatas' by Beethoven obtained from several sources on the internet. The second task is the classification of Joseph Haydn and Wolfgang Amadeus Mozart, using files from the 'String Quartets' of both composers. The difficulty of the second task is demonstrated by the current identification statistics, which indicate an average performance of 53% for self-reported novices and a maximum of 66% for self-reported experts [22]. The musical instruments for which the works are composed – keyboard instruments and string quartets – are the same in each of the two classification tasks, so recognition cannot be based on features that are specific for a certain musical instrument.

3.2 Input Representation

Whereas most pattern recognition systems dealing with sound first transform raw wave data into frequency bands, *pitch* representation rather than a representation involving frequency bands is used in the music classification task. Pitch representation is also used by the human brain when processing music. There are specific neurons that respond to specific pitches rather than frequencies, which are spatially ordered by pitch, not frequency, just as the keys on a piano [23, 24]. A musical representation that consists of information on the pitch of a note, its onset, its duration and its loudness, is the MIDI format. Because neural networks cannot deal with 'raw' MIDI files, the files were processed to time-series data with ordered pitch information.

A MIDI file consists of events. There are a large number of events that can be represented in such a file, most of which have to do with specific properties of electronic instruments, but are not relevant for the experiments. The events used here are: the onset and duration of a note, and its pitch. Each note is set to the same velocity value, and the beginning and duration of the notes are set to discrete time steps with intervals of $\frac{1}{32}$ note. This value is chosen because a note with a duration shorter than $\frac{1}{32}$ has little implication for the melodic line, and is more often used as ornament. Furthermore, all files are transposed to the same key, namely the key of C-major for pieces in a major scale, and

A-minor for pieces in a minor scale, such that classes cannot be determined based on key. In a MIDI file a range of 128 different pitches can be represented, but because the data sets contain files composed for certain instruments, notes that cannot be played or are rarely played on such instruments, are filtered out. After preprocessing the MIDI files, the results are represented in the form of time-series data, in which the time is divided into discrete time steps of $\frac{1}{32}$ note. Each input pattern in the time-series data consists of a set of 72 numbers (corresponding to the 72 most-used keys on a piano starting at f2), which have a value of 1 if the corresponding note is 'on' and 0 if the corresponding note is 'off'. A simplified version of the process of converting MIDI files to time-series data with ordered pitches is depicted in Fig. 2.

3.3 Setup

The time-series data extracted from the MIDI files consisting of information on the presence of each of the 72 notes are presented to the liquid columns. The structure of each liquid column is a three-dimensional space, in which

Fig. 2. Representation of MIDI file in staff (*top*), and part of the corresponding time-series data (*bottom*). The numbers within the *dashed box* represent an input pattern to the liquid filter at one time step

the units are located at the integer points. The time-series data are presented to the leftmost row with dimensions 12×6 to represent the 12-note scale property of the music over 6 octaves ($12 \times 6 = 72$ notes). Each unit in a horizontal row of 12 units corresponds to a note in the 12-note scale, and each vertical row to one octave. For all notes that are played in the MIDI file at a certain time step, the output of the corresponding units is set to 1, while the output of the other units is set to 0. To create a three-dimensional space, five additional rows of the same size are placed next to the input row. The resulting three-dimensional space of units has dimensions $12 \times 6 \times 6$. In the experiments on music classification by composer the liquid units are prevented from having self-recurrent connections.

The interval between successive readouts t^v, is set to 128 for all experiments. To give an idea about the duration of the music that is presented to the liquid during this period: 128 time steps correspond to $\frac{128}{32}$ notes in the MIDI file (= 4 bars in $\frac{4}{4}$ measure). Furthermore, the warmup period t^w is chosen to be half as long as the interval between successive readouts.

As stated in Sect. 2.5, little information can be extracted from the sparse activity in the network at a single time step. Therefore not the information whether a unit is firing or not at a single time step, but the number of times a unit fires during a certain readout period is used as measure of activity in the network

$$z_i(t) = \sum_{t'=t-t^s}^{t} y_i(t'), \qquad (12)$$

where t is the time step at which the liquid state is read, $z_i(t)$ is a measure of activity of unit i, and t^s is the number of time steps in the readout period. Next, the results of (12) are scaled between 0 and 1 by dividing it by the maximum number of times a unit can fire during the readout period

$$x_i(t) = \frac{z_i(t)t_i^a}{t^s}. \qquad (13)$$

The results of (13) for all units in the liquid other than the input units are presented to the readout mechanism. The readout mechanism for each liquid column is a three-layer perceptron. The number of units in the input layer is equal to the number of units in the spiking neural network other than the input units, which is $12 \times 6 \times 5 = 360$. In both classification tasks the number of classes is two, so there are two units in the output layer of the multilayer perceptron. The optimal number of units in the hidden layer is to be determined empirically. The training process of the multilayer perceptron is performed in batch-wise mode.

In the experiments the generalization capacity of an algorithm is tested, so the data are divided into a training set that is used for the training algorithm and a test set that is used to compute the performance of the selected method. Because the learning speed of several training algorithms can substantially be improved by using the same number of patterns in each class, the number of files in the training sets are such that an equal amount of classification moments occurs for both classes. In the Bach/Beethoven classification task, Book I of the 'Well-Tempered Clavier' and 20 movements of the 'Piano Sonatas' are used as training data, resulting in approximately the same number of classification moments for both classes. Book II of the 'Well-Tempered Clavier' and 30 movements of the 'Piano Sonatas' are used as test data. For the Haydn/Mozart classification task, approximately two-third of the movements of Mozart is used as training set, together with 64 movements of Haydn's 'String Quartets' which results in the same amount of classification moments for both classes. The remaining movements of both composers are used as test set.

Both the multilayer perceptron used as readout mechanism, and the spiking neural network used as liquid column have a number of adjustable parameters. For some parameters the values can be inferred from existing literature, as described in Sect. 2.5, while other parameters have to be attuned to the task at hand. Several experiments are performed to find the right parameter settings of the neural networks for the learning tasks. To make a quick comparison between the results of different parameter settings, the average percentages of correctly classified liquid states over a total of 20 LSMs are compared. Table 1 shows the values that were found during the experiments for a number of parameters, together with the values of the parameters that are inferred from existing literature as described in Sect. 2.5. The parameter values for the readout networks that were found during the experiments are given in Table 2. Note that the number of epochs needed to train the networks was different in both classification tasks because the number of patterns in the training set is not the same.

Table 1. Parameters for the spiking neural network

Parameter description	Symbol	Value
Liquid dimensions		$12 \times 6 \times 6$
Connectivity parameter	C	0.4
Connectivity parameter	λ	2
Proportion of inhibitory connections	p^i	0.35
Weight between unit i and j	w_{ij}	0.4
Firing threshold	η	1
Half-time of the membrane potential	t^h	4
Mean of Gaussian distribution of absolute refractory period	t^a	4
Standard deviation of Gaussian distribution of t^a		2
Relative refractory period	r_i	$\dfrac{\tau t_i^a}{10}$
Readout interval	t^v	128
Warmup period	t^w	$\frac{1}{2}t^v$
Readout summation steps in Bach/Beethoven task	t^s	100
Readout summation steps in Haydn/Mozart task	t^s	50

Table 2. Parameters for the multilayer perceptron

Parameter description	Value
Number of units in each layer	$360 \times 8 \times 2$
Initial weight interval	$[-0.3, 0.3]$
Initial bias interval	$[-0.1, 0.1]$
Batch size	10
Learning rate	0.00025
Number of epochs in Bach/Beethoven task	3,000
Number of epochs in Haydn/Mozart task	2,000

3.4 Experiments

A task that is difficult for humans does not always have to be difficult for computers as well, and vice versa, so to get an indication how difficult the composer identification task is for a machine learning algorithm a number of experiments were performed with some rather simple algorithms. In this fashion the performance of the DLSM can not only by compared to that of the LSM, but also to some less-sophisticated methods.

The first method that is tested is a very simple note-counting algorithm. All notes that are played are summed during the entire duration of the songs in the test set, and the result is compared to the average number of times these notes are used in the training set. The class with the smallest absolute difference is assigned as the outcome for each song. In Tables 3 and 4 the results for the note counting experiments are given.

The setup of the second method is more like the LSM, but instead of a spiking neural network as a liquid, readout samples are created by summing over the notes that are played within the same readout interval as the LSM, and scaling the results between 0 and 1. The other parameters have the same values as those that were found in the LSM and DLSM experiments, and all results are averaged over 20 repeated experiments. To see how this simple 'summation liquid' compares to an LSM with a spiking neural network liquid, majority voting over multiple readout moments (MV over time) and the spatial dimension (MV over multiple classifiers) was also tested. While it is straightforward to create different liquid columns for the multi-column LSM and the DLSM, summation can only be performed in one way, so only the initial weights and bias values of the readout networks were different for a comparison with the 'multiple liquid columns' experiments. The results are given in Tables 3 and 4. The 'average' columns gives the average over 20 experiments, the 'best' column the best out of 20 experiments, and the 'sd' column the standard deviation.

Next several experiments are performed to investigate the results of majority voting over different dimensions, with spiking neural networks as liquids. Just as in the previous setup the experiments for each parameter setting are repeated 20 times. In the first experiment (MV over time) majority voting is

Table 3. Results for the Bach/Beethoven classification task

Note counting	Correct		
All samples	71.3%		

Summation over time	Average (%)	Best (%)	SD
All samples	73.4	73.9	0.3
MV over time (correct)	88.3	90.8	1.3
MV over time (undecided)	1.7		0.7
MV over multiple classifiers (correct)	73.4	73.6	0.1
MV over multiple classifiers (undecided)	−		−
MV over multiple classifiers and time (correct)	87.8	89.8	0.9
MV over multiple classifiers and time (undecided)	1.2		0.4

Spiking neural network liquid	Average (%)	Best (%)	SD
LSM (average correct over all samples)	73.3	75.6	1.3
MV over time (correct)	90.2	93.5	2.1
MV over time (undecided)	2.4		1.5
MV over multiple liquid columns (correct)	77.6	79.1	0.8
MV over multiple liquid columns (undecided)	−		−
DLSM (correct)	**92.7**	**96.7**	**1.6**
DLSM (undecided)	1.5		1.0

Table 4. Results for the Haydn/Mozart classification task

Note counting	Correct		
All samples	54.9%		

Summation over time	Average (%)	Best (%)	SD
All samples	55.6	55.8	0.1
MV over time (correct)	57.3	59.6	1.2
MV over time (undecided)	5.1		1.6
MV over multiple classifiers (correct)	55.6	55.8	0.1
MV over multiple classifiers (undecided)	0.4		0.1
MV over multiple classifiers and time (correct)	57.0	57.8	0.9
MV over multiple classifiers and time (undecided)	5.9		0.9

Spiking neural network liquid	Average (%)	Best (%)	SD
LSM (average correct over all samples)	56.0	58.3	1.1
MV over time (correct)	55.8	66.1	4.6
MV over time (undecided)	5.5		2.3
MV over multiple liquid columns (correct)	57.4	58.9	1.0
MV over multiple liquid columns (undecided)	0.2		0.6
DLSM (correct)	**63.5**	**73.5**	**4.5**
DLSM (undecided)	7.1		2.6

applied over different readout moments of a single liquid column. In the second experiment (MV over multiple liquid columns) majority voting is applied over the time-averaged results of multiple liquid columns, and in the last experiment (DLSM) majority voting is applied over multiple liquid columns and over multiple readout moments. For majority voting over multiple liquid columns we tested multiple-column LSMs with increasing amounts of different liquid columns, up to 20 columns. As expected the performance of the multiple-column LSM improves when more liquid columns are added. The best results in the Bach/Beethoven classification task were obtained with 15 columns, while a DLSM with eight liquid columns achieved the best results in the Haydn/Mozart classification task. The results of the experiments for the optimal number of liquid columns are given in Tables 3 and 4. Note that the 'undecided' row for majority voting over multiple liquid columns is missing in Table 3, because of an odd number of liquid columns.

For both types of liquids (simple summation and the spiking neural network) we find in most experiments a significant increase in performance for majority voting over time. While for the DLSM additional performance increase can be obtained by adding different liquid columns, this is of course not the case for the summation liquid (there is only one way to compute a sum). Another striking result is that the performance of a summation liquid is almost the same as that of a single-column LSM. This might be caused by the fact that a relatively simple liquid was used here compared to the more sophisticated liquids used by others [1, 16], but the fact that even real water can effectively be applied as liquid column [25] indicates that this is not always the case. Whereas the results of the previous experiments hardly encourage the use of an LSM, the strength of this method becomes clear when majority voting is applied to multiple readout moments and multiple different classifiers as is done in the DLSM. The DLSM achieves a significantly ($p = 0.01$) higher performance than the single-column LSM and other less-sophisticated algorithms.

When the results of the different composer identification tasks in Tables 3 and 4 are compared, it becomes immediately clear that the Bach/Beethoven task is more easy to solve than the Haydn/Mozart task, at least for the algorithms used here. Because no identification statistics for human beings on the Bach/Beethoven classification task are currently known to the authors we cannot say anything with certainty about the difficulty of the Bach/Beethoven classification task. Although both tasks might be equally difficult for the untrained human ear, the identification statistics of the 'Haydn/Mozart String Quartet Quiz' show a maximum of 66% correct movements for self-reported experts [22], indicating that the task is very difficult even for the trained human ear. However, a direct comparison between the quiz and the results in Table 4 cannot be made because in the quiz no distinction is being made between train and test sets. The DLSM, with an average of 63.5% correct (and 7.1% undecided) on the test has not been presented with any file in the test set before, while in case of the 'Haydn/Mozart String Quartet Quiz' the contesters might have heard a piece before, for several times, or might even be able to play (reproduce) a piece themselves. Furthermore the 'Haydn/Mozart String Quartet Quiz' gives only averages, so we cannot compare our best DLSM (73.5%) to the best contester in the quiz. Although no fair comparison can be made between the performance of the DLSM and the human expert we can conclude that the performance of the DLSM comes close to the performance of human experts even under less favorable circumstances.

4 Musical Instrument Recognition

4.1 Learning Task

Apart from the composer classification task, we also test the DLSM architecture on the more basal task of timbre recognition. Musical instruments generate tones that can potentially span many octaves and thus cover a wide range of frequencies, with relations between frequency bands largely determining the timbre. While in the composer classification tasks timbre information was left out of the input data representation, in the present task timbre is the most important factor used to distinguish between musical instruments. For this task we use two musical instruments: the bass guitar and the flute. These instruments differ significantly in sound (even to the untrained ear) and have a relatively small tonal overlap, providing a convenient focus for this experiment.

In an experiment using nine wind instruments and a piano in a pairwise comparison classification paradigm, 98% of the piano-vs-other cases were classified correctly[3] [26]. These results were obtained using extensive preprocessing on the input data, the need for which is decreased dramatically when using reservoir computing variants. Although the aforementioned results indicate that this task is fairly easy, there is still room for improvement.

4.2 Data Set and Input Representation

The majority of the samples used here are made especially for this classification task. The data set consists of single tones, two-tone samples and partial scales. For the samples made for this research three different bass guitars and three different flutes (one of which was an alto flute to extend the range of the tonal overlap) that differ significantly in timbre are used. A small number of additional samples is taken from several CDs and can feature background

---

[3] The piano is the closest match found to a bass guitar in musical instrument classification literature. Since the difference between a piano and a wind instrument is not only great, but also qualitatively similar (string instrument vs wind instrument) to the difference between flute and bass, this seems a reasonable indication of the difficulty of the task.

noise such as roaring crowds and other instruments playing. The samples taken from CDs can also venture outside the tonal overlap.

All of the samples that are used in our timbre recognition experiments are 16-bit mono wave files with a 22 kHz sample rate. Most files are roughly one second in length, but some are longer. The data set consists of 234 wave files, with an equal number of bass guitar and flute fragments. Instead of presenting the samples to an LSM as raw wave data, the samples are transformed into the frequency domain by applying a Fast Fourier Transform (FFT). This yields an efficient representation of the input data and bears some similarities to audio processing in the human cochlea.

The length of the time window on which the FFT algorithm is applied is 5.8 ms, and the resolution of the FFT is set such that the raw wave data are converted into vectors of 64 frequency bands with their respective amplitudes. After applying the FFT the data is normalized between 0 and 1 for compatibility with the liquid's input units. The normalized FFT vectors are fed into the liquid's input units every 5 time steps, making 1 time step roughly equal to 1 ms and allowing the input to propagate through the liquid before the next is presented.

4.3 Setup

The model used for the musical instrument classification task is somewhat different from the one used for music recognition. For this task a two-dimensional rather than a three-dimensional liquid is sufficient. The liquid is organized in a grid of 64×5 units, including one layer of 64 input units, representing the 64-band frequency resolution of the FFT transform. The dynamics of the liquid used for the musical instrument recognition task are as described in (9) and (10), but no relative refractory period is implemented ($r = 0$). Readout periods are determined by cutting up samples into five fragments of roughly the same duration. Since most sound samples are approximately one second in length, readout periods would span some 200 ms, which translates to 40 input patterns.

Because the activity in the liquid is sparse, little information can be extracted from the activity in the network at one time step. Therefore not the information whether a unit fired at a single time step, but the membrane potential equivalent μ, is used as measure of activity in the network: $x_i(t) = \mu_i(t)$. The value of x_i for each non-input unit i in the liquid is presented to the readout mechanism.

For the readout function a three-layer perceptron is used. This neural network is trained using backpropagation with first-order gradient-descent. A special requirement has to be met in order to end the training period: the output for the correct class has to be bigger than the output for the incorrect class by at least a factor 1.5. If this requirement is not met after 2,000 iterations, training is also stopped. The number of units in the input layer is equal to the number of the non-input units in the liquid: $64 \times 4 = 256$.

In the musical instrument recognition task the number of classes is two, so the readout network has two output units.

To determine the parameter settings for the liquid and the readout mechanism for which the best generalization capacity is achieved, several experiments are performed. Each experiment is repeated 50 times with a different distribution of samples over the training and test sets. Of the 234 wave files in total, 34 are picked as a test set while the system is trained on the remaining 200 files. Both the test set and the training set consist of an equal amount of bass and flute samples. The best parameter settings for the liquid columns and the readout neural network as found in a number of experiments are given in Tables 5 and 6.

The optimal value of 50 units in the hidden layer of the multilayer perceptron is somewhat surprising. Networks with smaller numbers of hidden units are outperformed by this setting, as are those with larger numbers. Normally, such a large number would cause overfitting, but with the early stopping criterion – which causes the training to stop when the output for the correct outcome is at least 1.5 times as large as the incorrect classification – this problem is alleviated. The large number of hidden units combined with a rule that ensures generalization enables the system to take more features into account than it would otherwise be able to.

Table 5. Parameters for the spiking neural network

Parameter description	Symbol	Value
Liquid dimensions		64×5
Connectivity parameter	C	0.9
Connectivity parameter	λ	2
Proportion of inhibitory connections	p^i	0.1
Weight between unit i and j	w_{ij}	0.2
Firing threshold	η	1
Half-time of the membrane potential	t^h	4.5
Absolute refractory period	t^a	15
Relative refractory period	r	0
Readout interval (see text)	t^v	~ 40
Warmup period	t^w	t^v

Table 6. Parameters for the multilayer perceptron

Parameter description	Value
Number of units in each layer	$256 \times 50 \times 2$
Initial weight and bias intervals	$[-0.5, 0.5]$
Learning rate	0.1
High/low output value ratio	1.5
Maximum number of epochs	2,000

4.4 Experiments

Since the liquids that are used consist of a set of randomly connected nodes it is expected that the performance of individual LSMs will not be steady over the entire frequency range. We hypothesize that the DLSM having multiple different liquid columns, minimizes this problem and thus will outperform the single-column LSM.

A comparison between a single-column LSM in which the activity is read once per file, and several possibilities of applying majority voting over multiple readout moments and the results of multiple liquid columns is given in Table 7. For the results of both majority voting over multiple liquid columns and the DLSM, 10 combinations of a liquid and readout mechanism were tested. All experiments were repeated 10 times to get a more reliable average.

For comparison purposes, a simpler algorithm is also investigated. This model uses a readout network with 15 units in the hidden layer and a simple 64-band 'summation filter' replacing the liquid. The filter sums all the inputs per frequency band in a time span of 200 time steps, after which the readout function is called for classification and the filter is reset. This algorithm is also investigated in several majority voting settings. The results are given in Table 7.

For all setups, the classification accuracy for bass guitar samples is about the same as that for flute samples. Even though the simple summation algorithm performs well, it is outperformed by LSMs. And, as expected, LSMs in the DLSM setting structurally outperform single-column LSMs. The improved classification accuracy on individual readout moments has a marked effect on classification of entire audio files, regardless of the type of sample (e.g., from

Table 7. Results for the musical instrument classification task

Summation over time	Average (%)	Best (%)	SD
All samples	85.0	94.7	5.4
MV over time (correct)	94.6	100	4.0
MV over multiple classifiers (correct)	85.3	91.8	4.6
MV over multiple classifiers (undecided)	0.6		1.3
MV over multiple classifiers and time (correct)	95.9	100	3.4
MV over multiple classifiers and time (undecided)	0		0

Spiking neural network liquid	Average (%)	Best (%)	SD
LSM (average correct over all samples)	88.4	100	5.2
MV over time (correct)	95.9	100	4.1
MV over time (undecided)	2.4		1.5
MV over multiple liquid columns (correct)	91.9	100	4.8
MV over multiple liquid columns (undecided)	1.4		2.6
DLSM (correct)	**99.1**	**100**	**2.6**
DLSM (undecided)	0.3		0.3

CD, single tone, interval, etc.). The only times when the DLSM incorrectly classifies certain samples occur when there are no comparable samples in the training set.

The performance of individual LSMs is fairly good, but leaves enough room for improvement. As demonstrated in the experiments in this section applying majority voting variants can increase the classification accuracy. In the DLSM setting, a marked increase in classification accuracy is observed for all samples, effectively going from an average of 88% to an average of 99% correctly classified samples. This is comparable to results obtained by using extensive preprocessing of the input (98%, [26]) and exceeds the results that a simpler setup yields (96%, Sect. 4.4) significantly.

5 Conclusion

The experiments in the previous sections show that the DLSM outperforms a configuration in which majority voting is not applied over multiple liquid columns. The reason for this is that using ensembles of multiple classifiers can significantly boost performance, especially when there is a large variance between the individual classifiers. As the combination of a liquid column with a trained readout mechanism involves its own kind of equivalence over input patterns, the way in which such a combination decides the class of an input pattern is probably different for each liquid, and the variance of the error will also be high. The concept of majority voting over multiple classifiers is therefore highly applicable to the LSM.

Majority voting as applied in the temporal dimension in the DLSM minimizes the chance of judgment errors due to atypical liquid states. These can be caused by many things, such as a momentary lapse in the signal, a high amount of noise that drowns out the signal completely or a partial input pattern that resembles a property of another class. Since input samples will yield a sufficiently characteristic liquid state most of the time, using multiple readout moments reduces the chances of misidentification of an input pattern due to atypical periods in the input pattern. As becomes clear from the experiments, majority voting over the temporal dimension also seems to be a good strategy.

A possible improvement to the DLSM is to use liquids with different characteristics. As stated before, the DLSM operates by the principle that different liquid columns represent other aspects of the input data. The difference between liquid columns can not only be achieved by using columns that are initialized with different random seeds, but also by changing the parameters of the liquid columns. Maass et al. [1] for example, used multiple liquid columns with different values of the connection parameter λ, but it is also possible to change other parameters or even the network topology to increase the difference between liquid columns.

Since the liquids we used in the experiments are randomly generated, the separation between liquid states of patterns belonging to different classes is not optimized. We are currently investigating optimization techniques such as reinforcement learning to optimize the separation property, which could reduce the needed size of the liquid and increase generalization performance.

References

1. Maass, W., Natschläger, T., Markram, H.: Real-time computing without stable states: a new framework for neural computation based on perturbations. Neural Computation **14** (2002) 2531–2560
2. Jaeger, H.: The 'echo state' approach to analyzing and training recurrent neural networks. GMD report **148** (2001)
3. Hochreiter, S.: Recurrent neural net learning and vanishing gradient. International Journal Of Uncertainity, Fuzziness and Knowledge-Based Systems **6** (1998) 107–116
4. Hochreiter, S., Schmidhuber, J.: Long short-term memory. Neural Computation **9** (1997) 1735–1780
5. Bakker, B., Zhumatiy, V., Gruener, G., Schmidhuber, J.: A robot that reinforcement-learns to identify and memorize important previous observations. In: Proceedings of the 2003 IEEE/RSJ International Conference on Intelligent Robots and Systems (IROS2003) (2003) 430–435
6. Breiman, L.: Bagging predictors. Machine Learning **24** (1996) 123–140
7. Ruta, D., Gabrys, B.: A theoretical analysis of the limits of majority voting errors for multiple classifier systems. Technical Report 11, ISSN 1461-6122, Department of Computing and Information Systems, University of Paisley (2000)
8. Williams, R., Zipser, D.: A learning algorithm for continually running fully recurrent neural networks. Neural Computation **1** (1989) 270–280
9. Rumelhart, D., Hinton, G., Williams, R.: Learning internal representations by error propagation. Parallel Distributed Processing **1** (1986) 318–362
10. Bengio, Y., Simard, P., Frasconi, P.: Learning long-term dependencies with gradient descent is difficult. IEEE Transactions on Neural Networks **5** (1994) 157–166
11. Pearlmutter, B.: Gradient calculations for dynamic recurrent neural networks: a survey. IEEE Transactions on Neural Networks **6** (1995) 1212–1228
12. Williams, R., Zipser, D.: Gradient-based learning algorithms for recurrent networks and their computational complexity. In Chauvin, Y., Rumelhart, D., eds.: Back-propagation: theory, architectures and applications. Lawrence Erlbaum, Hillsdale, NJ (1995) 433–486
13. Lin, T., Horne, B., Tino, P., Giles, C.: Learning long-term dependencies is not as difficult with NARX networks. In: Touretzky, D., Mozer, M., Hasselmo, M., eds.: Advances in Neural Information Processing Systems, Volume 8. MIT, Cambridge, MA (1996) 577–583
14. Natschläger, T., Bertschinger, N., Legenstein, R.: At the edge of chaos: Real-time computations and self-organized criticality in recurrent neural networks. In: Proceedings of Neural Information Processing Systems, Volume 17 (2004) 145–152

15. Gupta, A., Wang, Y., Markram, H.: Organizing principles for a diversity of GABAergic interneurons and synapses in the neocortex. Science **287** (2000) 273–278

16. Vreeken, J.: On real-world temporal pattern recognition using liquid state machines. Master's thesis, Utrecht University, University of Zürich (2004)

17. Häusler, S., Markram, H., Maass, W.: Perspectives of the high dimensional dynamics of neural microcircuits from the point of view of low dimensional readouts. Complexity (Special Issue on Complex Adaptive Systems) **8** (2003) 39–50

18. Narasimhamurthy, A.: Theoretical bounds of majority voting performance for a binary classification problem. IEEE Transactions on Pattern Analysis and Machine Intelligence **27** (2005) 1988–1995

19. Lam, L., Suen, C.Y.: Application of majority voting to pattern recognition: an analysis of its behaviour and performance. IEEE Transactions on Systems, Man, and Cybernetics **27** (1997) 553–568

20. Feng, J., Brown, D.: Integrate-and-fire models with nonlinear leakage. Bulletin of Mathematical Biology **62** (2000) 467–481

21. Vapnik, V.: Statistical learning theory. Wiley, New York (1998)

22. Sapp, C., Liu, Y.: The Haydn/Mozart string quartet quiz. Center for Computer Assisted Research in the Humanities at Stanford University. Available on `http://qq.themefinder.org/` (2006)

23. Pantev, C., Hoke, M., Lutkenhoner, B., Lehnertz, K.: Tonotopic organization of the auditory cortex: pitch versus frequency representation. Science **246** (1989) 486–488

24. Pantev, C., Hoke, M., Lutkenhoner, B., Lehnertz, K.: Neuromagnetic evidence of functional organization of the auditory cortex in humans. Acta Otolaryngol Supply **491** (1991) 106–115

25. Fernando, C., Sojakka, S.: Pattern recognition in a bucket. In: Proceedings of the 7th European Conference on Artificial Life (2003) 588–597

26. Essid, S., Richard, G., David, B.: Musical instrument recognition based on class pairwise feature selection. In: ISMIR Proceedings 2004 (2004) 560–568

Color Transfer and its Applications

Arvind Nayak[1], Subhasis Chaudhuri[2], and Shilpa Inamdar[2]

[1] ERP Joint Research Institute in Signal and Image Processing, School of
Engineering and Physical Sciences, Heriot-Watt University, Riccarton,
Edinburgh EH14 4AS, UK, amn1@hw.ac.uk
[2] Vision and Image Processing Laboratory, Department of Electrical Engineering,
Indian Institute of Technology – Bombay, Powai, Mumbai 400 076, India,
sc@ee.iitb.ac.in, shilpa@ee.iitb.ac.in

Summary. Varying illumination conditions result in images of a same scene differing widely in color and contrast. Accommodating such images is a problem ubiquitous in machine vision systems. A general approach is to map colors (or features extracted from colors within some pixel neighborhood) from a source image to those in some target image acquired under canonical conditions. This article reports two different methods, one neural network-based and the other multidimensional probability density function matching-based, developed to address the problem. We explain the problem, discuss the issues related to color correction and show the results of such an effort for specific applications.

1 Introduction

Every machine vision system is built with some underlying constraints, and as long as these are satisfied it works beautifully well. Assumptions like, the images have good contrast and the illumination conditions do not change significantly are often implicit. In a controlled environment, for example, automated mounting of electronic components on a printed circuit board, where the direction, intensity and wavelength of the illumination is well known in advance, it is feasible to assume that the illumination conditions do not change. Consider another example, a widely required task in image-based applications of tracking features through image sequences. Human gait analysis, gesture-based telerobotic applications, video surveillance, observing ground activities from aerial images and automatic pointing of space telescopes for studying objects in space are a few examples of the outdoor environment. Images acquired during many such applications suffer from drastic variations in color and contrast due to large variations in illumination conditions. A general approach for correcting colors or improving color contrast is to map colors from a source image to those in some target image acquired

A. Nayak et al.: *Color Transfer and its Applications*, Studies in Computational Intelligence
(SCI) **83**, 217–237 (2008)
www.springerlink.com

under *canonical*[1] conditions. Simply put, this means transforming a source image, taken under unknown illumination conditions, to an image such that it appears as if the original scene is illuminated by the canonical illumination conditions. Sometimes, the illumination conditions in the source and target image could be such that the mapping is nonlinear. The complexity of this problem can be appreciated through the following equation (reproduced here from [6] for brevity), given for a color signal at the sensor. The camera output is affected by surface reflectance and illumination. For the red channel we have:

$$R(x, y) = \int E(\lambda) S(x, y, \lambda) C_R(\lambda) \, d\lambda, \tag{1}$$

where $C_R(\lambda)$ is the spectral sensitivity of camera's red channel (similar equations for the green and blue channels $G(x, y)$ and $B(x, y)$, respectively), $E(\lambda)$ is the spectrum of the incident illumination and $S(x, y, \lambda)$ is the spectral reflectance of the surface. The spectral sensitivity of camera for the three channels (R, G and B) and the spectral reflectance of surface at (x, y) are the properties which remain constant. Even though the equation appears straight forward it is impossible to differentiate between the contributions of S and E without any additional information or scene model [5]. The aim therefore is to correct colors using simple techniques, bypassing the difficult process of estimating S and E.

This article presents two different methods for color correction. First, we show it is possible to use a relatively simple neural network to learn the mapping (build a discrete lookup table) of colors between the image acquired under vision-unfriendly illumination conditions and the image acquired under some *canonical* illumination conditions. Second, we show a multidimensional probability density function (pdf) matching scheme that does the work very well by not only correcting colors but also preserving the statistics within the images. Such a technique can also be used to correct multispectral images obtained by remote sensing satellites. We test our techniques on color and gray level imaging scenarios, and in situations like registered and unregistered image pairs[2].

For the first method (*hereafter referred to as the NN method*), this article discusses an approach based on multilayer feed-forward network using a back-propagation learning rule. The principal application domain is skin color-based hand tracking. The trajectory information obtained through tracking can be used for recognition of dynamic hand gestures [13]. It is well known that any skin color-based tracking system must be robust enough to accommodate varying illumination conditions that may occur while performing gestures. Some robustness may be achieved in cases where the skin color distribution

---

[1] It is an ambient illumination condition under which the machine vision system is known to perform well.

[2] The source and the target images used for training are of dissimilar types, i.e., they do not have a pixel-to-pixel correspondence.

has undergone narrow changes, through the use of luminance invariant color-spaces [14, 20, 22]. But under poor lighting conditions, mere switching to a different color space will not be of any help. One solution is to correct image colors, by appropriately modifying the look up table, before giving it to the tracking algorithm. The neural network learns the mapping between two illumination conditions with the help of one or more pairs of color palettes and then applies this transformation to every pixel in the successive frames of the image sequence. The initial experiments for training the neural network were performed by computer generated standard color palettes (refer Sect. 3) and later, an automatically extracted palm region alone was used as a color palette for training the neural network (refer Sect. 4.1). Finally we use the Condensation algorithm [9] for tracking and verify the results of color transfer. The reason behind using this algorithm is its capability for robust tracking of agile motion in clutter.

We extend the NN method (*hereafter referred to as the NNx method*) and use a cluster-based strategy for contrast restoration[3] in arbitrary gray level images (by arbitrary we mean situations where registration between source and target image pairs is not available or possible). The image is first segmented through a fuzzy c-means clustering technique and correspondences are established for segments in source and target image pairs. The neural network then estimates the color transfer function for pixels using their local histograms in the source and the target images.

The second method (*hereafter referred to as the ND method*) discusses N-dimensional pdf matching which takes into account correlation among various spectral channels. Here N is the number of spectral channels in an image. Although, for comparison with NN method we use $N = 3$, i.e., R, G and B channels of a color image, the method can easily be extended for $N > 3$. This is usually a case with multispectral remote sensing images.

The organization of the article is as follows. In Sect. 2 we very briefly review some of the current approaches to color correction. The proposed NN method for color correction is discussed in Sect. 3 followed by skin color-based hand tracking (principal application domain for the NN method) in Sect. 4. NNx is discussed in Sect. 5 and ND in Sect. 6. Results of various experiments using NN, NNx and ND are shown in respective sections. Finally, Sect. 7 provides a brief summary.

---

[3] The restoration of contrast is one of the main techniques by which the illumination of a scene may appear to have been corrected. True illumination correction in the context of computer vision, however, should take into account the directions of light sources and surface properties. But, usually such techniques require more computations and also prior information about illuminants which are not feasible in many applications.

2 Related Work

Equation 1 showed that the colors apparent to an imaging system depends on three factors: the light source illuminating the object, the physical properties of the object and the spectral response of the imaging sensor. A small change in any of these factors can bring a drastic change in image colors. This can seriously affect the operation of a machine vision system designed to work only under some specific conditions. One solution is to correct colors in the images before passing them to the machine vision system. Barnard et al. [2] define the goal of computational color constancy to account for the effect of illuminant, either by directly mapping the image to a standardized illuminant invariant representation, or by determining a description of the illuminant which can be used for subsequent color correction of the image.

Several algorithms for color constancy have been suggested, the most promising among them fall under the following categories: gray world methods, gamut mapping methods, color by correlation, neural net-based methods and algorithms that are based on the Retinex theory. An extensive comparison of these can be found in [2,3]. Estimating illuminant for color constancy is an underdetermined problem and additional constraints must be added to make the solution feasible. Many papers [4,5,11,12,23] advocate neural network-based methods as they do not make any explicit assumptions and are capable of modeling nonlinearities existing in the mapping function.

Choice of color space is another issue in color correction algorithms. Reinhard et al. [18] argue that the color space, $l\alpha\beta$, developed by Ruderman et al. [19] minimizes correlation among color channels of many natural scenes and little correlation between the axes in $l\alpha\beta$ color space means different operations on different channels are possible without having to worry about the cross-channel artifacts. This algorithm converts the RGB space to $l\alpha\beta$ space, works on that space, and converts back to RGB at the final step. RGB still remains the most popular color space because of its use in display and sensor devices. A comparison of color correction results using different color spaces can be found in [18,24].

Most color transfer techniques mentioned before are sensitive to the size of matching color clusters in the source and the target images. They produce visually unpleasant colors if there are not enough matching regions or a certain region in an image occupies, relatively, a larger fraction than in the other. Pitié et al. [17] develop a nonparametric method that is effective in matching arbitrary distributions. Our work presented in Sect. 6 is an extension to theirs.

Apart from correcting colors for image processing [16,21] or for improving performance of machine vision systems, color transfer techniques have also been employed for region-based artistic recoloring of images [8]. Here the target image is recolored according to the color scheme of a source image. Other applications include color and gray level correction of photos printed in photo printing labs [11] and digital projection environments [23].

3 Color Correction for Similar Images

The first stage in this process involves iteratively estimating the mapping (updating a lookup table) between colors in the source and the target images, and the second stage corrects the image pixel-wise, based on the lookup table. Neural networks are capable of learning nonlinear mappings that approximate any continuous function with any required accuracy. The neural network learns the mapping between two illumination conditions with the help of one or more pairs of color palettes. Each pair of the color palette used for training has one image of the palette observed under unknown illumination conditions, which we refer as *realworld palette*, and the other, image of the same palette observed under canonical lighting conditions, which we refer as *canonical palette*.

We suggest a multilayer feed-forward network with the back-propagation learning rule, for learning the mapping between the realworld palette and the canonical palette. This is illustrated in Fig. 1. The weight adjustment is done with the sigmoid as the activation function. Each layer consists of nodes which receive their inputs from nodes from a layer directly below and send their outputs to nodes in a layer directly above the nodes. The network is thus fully connected but there are no connections within a layer. Empirical studies have shown that two hidden layers with 20 nodes in each of the hidden layers yield acceptable results in learning the mapping between the palettes at a moderate computation. The input nodes are merely *fan-out* nodes; no processing takes place in these nodes. We study the effect of training the neural network by using standard color palettes.

During our initial experimentation we captured images of computer generated standard color palettes in the desired illumination condition, prior to start of a gesture (this comes as an application discussed in Sect. 4). These palettes train the neural network for the mapping between the realworld palette and the canonical palette. Figure 2a shows a computer generated

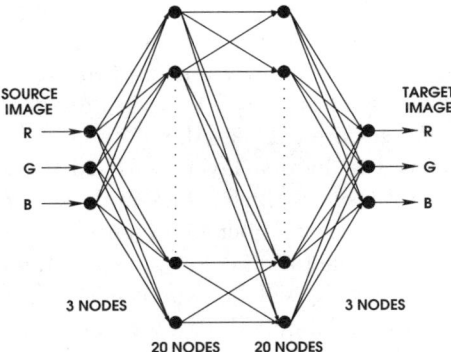

Fig. 1. The multilayer feed-forward neural network used to learn mapping between colors in source and target image pairs

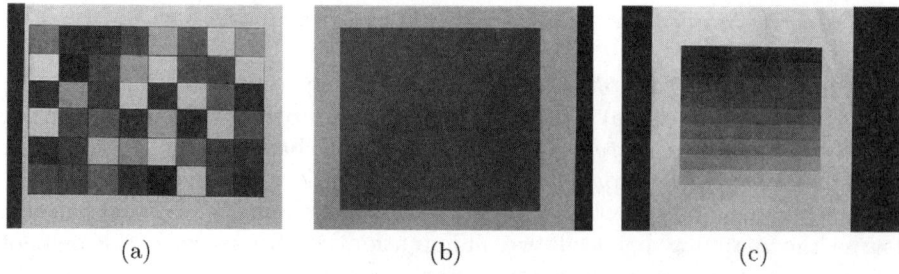

(a)   (b)   (c)

Fig. 2. (a), (b) and (c): Standard color-palettes as observed under canonical illumination conditions

canonical palette for 16 colors (colors are repeated and placed arbitrarily). Figure 2b shows a canonical palette within skin color range (this comes as an application). The palette is generated in such a manner that it covers the entire range of skin color in the $YCbCr$ color space. Similarly, Fig. 2c shows a palette obtained with the whole gray scale range arranged in the ascending order from 0 to 255 for 8-bit levels. Figure 2a–c shows the corresponding realworld palettes observed under an arbitrary illumination condition.

The realworld palette and the canonical palette form a pair for each training set by the neural network. The sizes of the palettes in the image plane are properly scaled to make both the palettes equal in size. The R, G and B values for each pixel in the training palettes are also scaled between 0 and 1. The goal of the neural network is then to estimate the weights for mapping of the scaled RGB values for the pixel at the input nodes to the scaled RGB values for each corresponding pixel in the canonical palette(s).

When a learning process is initiated, in the first phase, the scaled RGB values of the pixels chosen randomly from the realworld palette image are fed into the network. The weights are initially assigned small random values and the activation values are propagated to the output nodes. The actual network output is now compared with the scaled RGB value, i.e., the desired value, of the pixel at the corresponding position in the canonical palette. We usually end up with an error in each of the output nodes. The aim is to reduce these errors. The second phase involves a backward pass through the network during which the error signal is passed to each unit in the network and the appropriate weight changes are calculated.

The network having two hidden layers with 20 nodes in each layer was trained by feeding the matched pixel values multiple times, randomly. The purpose of using a fairly large number of weights is to make the histogram transfer function a smooth one for all color planes. Although there is no guarantee that the learnt mapping will be monotonic in each band, we do not experience any noticeable difficulty during experimentation. The establishment of correspondences between pixels from source and target palettes is quite simple. We detect the top-left and bottom-right corners of the palettes, resize and align them to have pixel wise correspondences. The number of iterations (or equivalently the number of times pixels were fed into the network) is dependent on both the size of the image and the color variations within the image. Empirical studies show that increasing the number of iterations beyond 20 times the number of image pixels does not appreciably reduce the error. The training is now terminated and the weights thus obtained are eventually used to transform successive frames in the image sequence. The expected result at output of the color correction algorithm is a realworld image transformed into an illumination condition had it been illuminated by the canonical lighting conditions.

We present typical results in Figs. 4 and 5. It can be observed that the illumination conditions of Fig. 4a–c are transformed to nearly match the illumination conditions of Fig. 2a–c, respectively. Figure 5a, image from a *Gesture* sequence, was taken under the same illumination conditions as Fig. 3a, and is used as a test image. Figure 5b shows the result when weights obtained by training the neural network with Figs. 3a and 2a as source and target image pair, respectively, were used for color correction.

The previous set of experiments make use of the availability of standard color palettes. In absence of above, one may use any registered pair of images under two different lighting conditions to learn the mapping function which could subsequently be used to transfer the color of any test image accordingly.

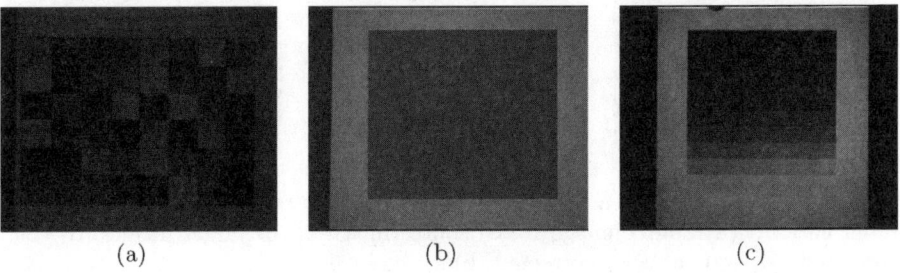

(a)   (b)   (c)

Fig. 3. (a), (b) and (c): Standard color-palettes as observed under arbitrary illumination conditions

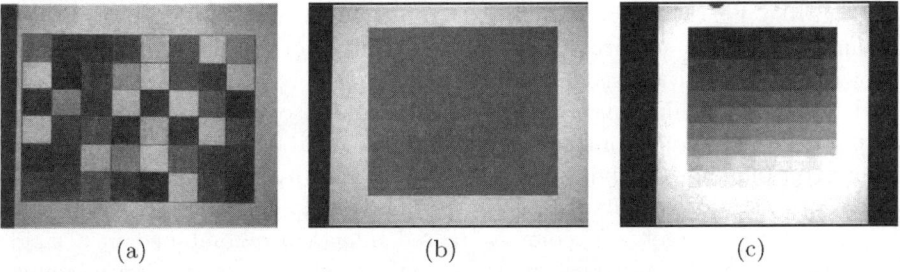

(a)   (b)   (c)

Fig. 4. Results obtained after applying color correction to palettes in Fig. 3a–c, respectively. The target images in this case were Fig. 2a–c, respectively

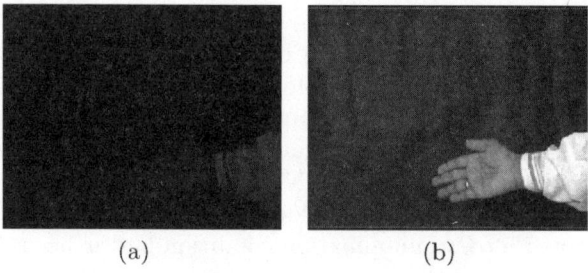

(a)   (b)

Fig. 5. Illustration of color correction for the gesture dataset. (**a**) Input image, (**b**) color corrected (**a**) using weights generated from standard color palette-based training

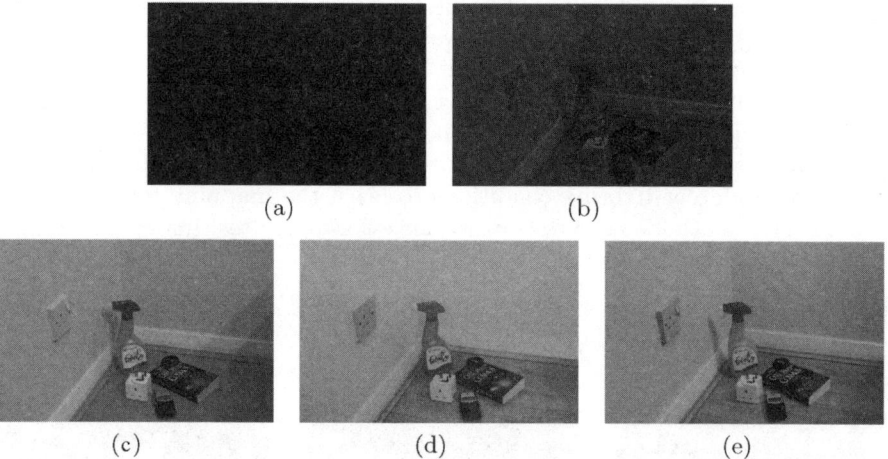

(a)   (b)

(c)   (d)   (e)

Fig. 6. Color correction using registered source and target images in the living room dataset. (**a**) Source image 1, (**b**) source image 2, (**c**) target image, (**d**) color corrected (a), and (**e**) color corrected (b)

However, the learning of the mapping function depends on the richness of the color tones in the training data set.

Figure 6 illustrates the above on a *Living Room* dataset. Two color images of similar contents acquired under two different poor illumination conditions, shown in Fig. 6a,b is used as source images. Figure 6c is the target image with the desired illumination. Figure 6d,e shows color corrected Fig. 6a,b, respectively. The color contrast is very similar to that of Fig. 6c. Figure 7 shows 3D scatter plots for the living room dataset. x-axis, y-axis and z-axis represent R, G and B color triplets, respectively. For visual comparison, scatter plots for the source, target and color corrected images are combined in a single plot. Markers with red color represent the scatter plot for the source image, green for the target image, and blue for color corrected image. Figure 7a,b shows two such combined 3D scatter plots for tests on source images shown in

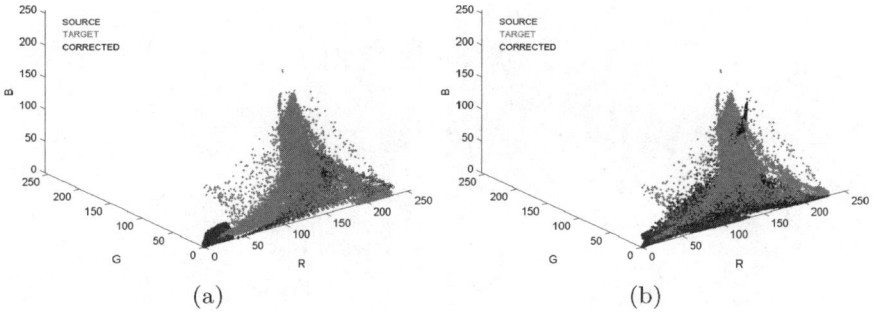

Fig. 7. 3D scatter plots for the living room dataset. x-axis, y-axis and z-axis represent R, G and B color triplets, respectively. Scatter plots for source, target and color corrected images are shown in a single plot. Red color markers represent the source image, green, the target image, and blue, the color corrected image. (**a**) Scatter plots for Fig. 6a,c,d. (**b**) Scatter plots for Fig. 6b,c,e

(a)   (b)   (c)   (d)   (e)

Fig. 8. Color correction in synthetically color distorted images of paintings. (**a**) Source image (synthetically color distorted (b)). (**b**) Target image (original). (**c**) Test image (synthetically color distorted (e) with conditions similar to that used for (a)). (**d**) color corrected (c). (**e**) Original test image before synthetic color distortion, shown here for side-by-side visual comparison with (d). Image source: http://www.renaissance-gallery.net/

Fig. 6a,b. In Fig. 7a, notice the small cluster of red points, this is for the source image. The larger green cluster is for the target image; notice the blue cluster which represents the corrected image. It is stretched to the limits similar to that of the green cluster. This can be seen to overlap very well with the target distribution.

Figure 8 shows another set of experimental results on synthetically color distorted images of paintings. Original images of two paintings downloaded from an Internet source (http://www.renaissance-gallery.net/) are shown in Fig. 8b,e. Both the original images were distorted in color with similar parameters. The color distorted images are shown in Fig. 8a,c, respectively. For training Fig. 8a,b was used as source and target images, respectively. Figure 8c was then used as a test input and the color corrected

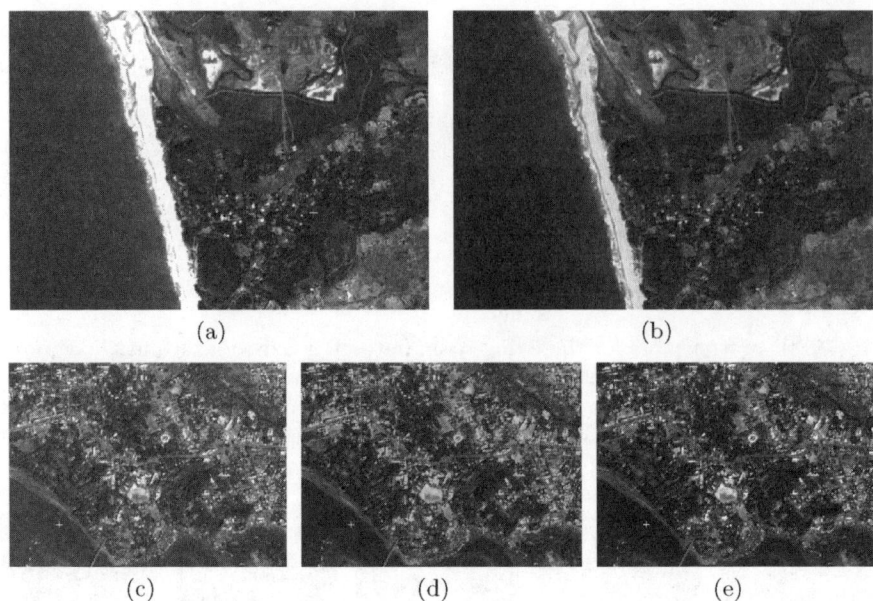

Fig. 9. Color correction in synthetically color distorted satellite images. (a) Source image (synthetically color distorted (b)). (b) Target image (original). (c) Test image (synthetically color distorted (e) with conditions similar to that used for (a)). (d) Color corrected (c). (e) Original test image before synthetic color distortion, shown here for side-by-side visual comparison with (d). Image source: http://www.wikimapia.org/

image is shown in Fig. 8d. Notice the colors in Fig. 8d, which now have characteristics of Fig. 8b. It is interesting to note the color of the yellow fruit lying at the base of the flower-vase. After color correction, it is changed to near white, while the original color was yellow itself. This is due to the fact that the network has failed to learn the look-up table for the yellow region due to its absence during training. Figure 9 shows results for synthetically color distorted satellite images. Satellite images of a portion of Ratnagiri city, in Maharashtra, India, were downloaded from Wikimapia (http://www.wikimapia.org). The figure caption explains further details. Notice the quality of color correction in Fig. 9d, it is almost similar to the original image, Fig. 9e.

It is needless to say that the above technique is applicable to gray scale images also. Figure 10a,b shows two images of similar contents captured using a mobile phone camera, one with a good contrast and other having a very poor contrast. The images are obtained by simply switching on and off some of the light bulbs in the room. The neural network is trained using this pair of images for generating the look up table for contrast transfer. Figure 10c shows another poor quality image captured with a mobile phone camera. The result of contrast transfer is shown in Fig. 10d where the details are visible.

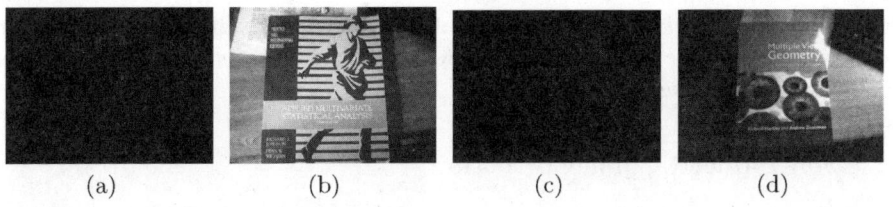

(a)   (b)   (c)   (d)

Fig. 10. Contrast transfer in gray scale images. (**a**) Poor and (**b**) high contrast images obtained using a mobile phone camera (Nokia 3650). (**c**) Test input and (**d**) contrast restored image with NN method

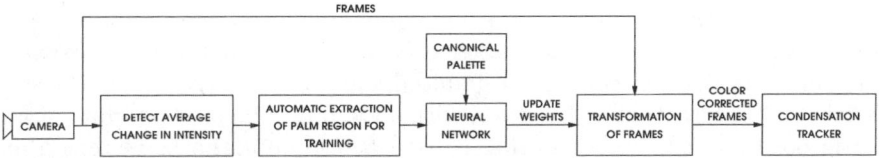

Fig. 11. Block diagram representation of the color correction-based tracking system

4 Application in Skin Color-Based Hand Tracking

In most machine vision-based applications tracking objects in video sequences is a key issue, since it supplies inputs to the recognition engine. Generally, tracking involves following a bounding box, around the desired object, as time progresses. Skin color-based hand tracking is our principal application domain for testing the NN method.

Figure 11 illustrates the entire scheme of the proposed automatic skin color correction and hand tracking algorithm [15]. A digital camera captures gestures under varying lighting conditions. We are required to perform the color correction at the beginning of a sequence to account for illumination changes during each instance of a gesture. Similarly, one may have to update the color correction lookup table during a particular tracking sequence as the ambient illumination may also change while performing a particular gesture. To decide for an initiation of a new training, the average intensity of the palm region from each frame is compared with the average intensity of the palm region from the frame that was used for the last training. Any change in average brightness value above some predetermined threshold triggers a fresh training process. The first few frames from an image sequence are used to automatically extract the palm region of the hand. The neural network then learns the mapping between unknown illumination conditions and the canonical illumination conditions using this palm region and a pre-stored palm palette. The successive frames are then color-corrected using the learnt mapping. The color-corrected frames are then given to the Condensation (particle filter) tracker. By modeling the dynamics of the target and incorporating observations from segmented skin regions, the tracker determines the probability density function of the target's state in each frame with an improved accuracy. In the remaining part of this section we briefly describe our self-induced color correction scheme, the model for the state dynamics and measurements for the condensation tracker.

4.1 Self-Induced Color Correction

Earlier we discussed the use of standard color palettes for training the neural network. Whether it is feasible to display the palette in front of the camera each time we want to capture a gesture sequence, is a question that obviously arises. The answer to which is, definitely no! We, therefore, go ahead in automating the training process and eliminate our previous requirement of displaying the standard color palettes every time. We use the palm region for training which we call here as self-induced training process. Since the gesture vocabulary in our studies [13] starts with a flat out palm moving from a fixed, static position, a motion detector is used to crop out the hand region and eventually the palm region. The only constraint that we put during these first few number of frames is that there is low background clutter so that the palm can be extracted reliably. However, unlike in the case of using standard color palettes, one may not actually have a proper pixel wise correspondence among the palm palettes as the palm shape itself may change. Nonetheless, we use the correspondences in pixels in a similar way as before within the rectangles bounding the palms. We reduce the number of nodes in each hidden layer to five since a smaller number of matching pixels are available for training the network. Figure 12a shows an automatically extracted palm region from the realworld image sequence. Figure 12c shows a palm region from a canonical illumination. Figure 12d shows the result of applying the learnt mapping on Fig. 12a. Since Fig. 12a has very little contrast, the contrast was manually adjusted for visualization and comparison purposes (shown in Fig. 12b). The results indicate that the illumination conditions in the transformed image are similar to those in Fig. 12c. Observe that the test and the target images are of different size and thus does not pose any problem during the training. The neural network will be good only for skin region and the color mapping for the rest of the color space is expected to be quite inferior as it is not be trained.

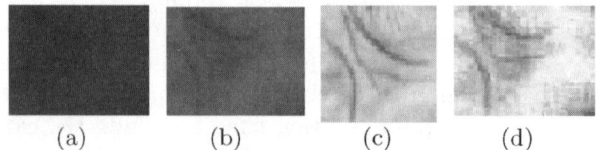

(a)   (b)   (c)   (d)

Fig. 12. Results using the palm itself as a palette. (**a**) Realworld palm as an observed palette; (**b**) Palette (a) enhanced in Linux's XV color editor simply for visualization as (a) has very little contrast; (**c**) canonical palm palette; and (**d**) the palm region after applying color correction on palette in (a)

4.2 Model for State Dynamics

Since our purpose is to track rectangular windows bounding the hands, we select the co-ordinates of the center of each rectangular window $(\bar{x}, \bar{y})$ and its height (h) and width (w) as elements of the four-dimensional state vector $\phi_t = [\bar{x}\ \bar{y}\ h\ w]^T$. We model the state dynamics as a second-order auto-regressive (AR) process.

$$\phi_t = A_2\phi_{t-2} + A_1\phi_{t-1} + v_t, \tag{2}$$

where A_1 and A_2 are the AR-model parameters, v_t is a zero-mean, white Gaussian random vector. This is intuitively satisfying, since the state dynamics may be thought of as a two-dimensional translation and a change in size of the rectangular window surrounding the hand region. We form an augmented state vector X_t for each window as follows:

$$\Phi_t = \begin{pmatrix} \phi_{t-1} \\ \phi_t \end{pmatrix}. \tag{3}$$

Thus we may rewrite (2) as $\Phi_t = A\Phi_{t-1} + V_t$. The Markovian nature of this model is evident in the above equation.

4.3 Observation

In order to differentiate the hand region from the rest of the image we need a strong feature which is specific to the hand. It has been found that irrespective of race, skin color occupies a small portion of the color space [10, 22]. As a result, skin color is a powerful cue in locating the unadorned hand.

Colors are represented as triplets, e.g., RGB values. However, to detect skin color, the effect of luminance needs to be removed. Hence, a suitable color space representation is required which expresses the color independent of intensity or luminance. After an initial survey and experimentation, we chose the $YCbCr$ color representation scheme. The skin pixels lie in a small cluster in the $YCbCr$ space and the RGB to $YCbCr$ conversion is a linear transformation. The intensity information is contained in Y and the position in the color space is given by the Cb and Cr values.

In order to detect pixels with skin-like color, a Bayesian likelihood ratio method is used. A pixel y is classified as skin if the ratio of its probability of being skin to that of it not being skin is greater than some threshold which is found using the likelihood ratio $\ell(y)$.

$$\ell(y) = \frac{P(color|skin)}{P(color|notskin)} > threshold. \tag{4}$$

The likelihood functions $P(color|skin)$ and $P(color|notskin)$ are obtained by learning from a large number of images. Portions of the image containing skin are manually segmented to obtain a histogram. It is interesting to note that, unlike in the work of Sigal [20], there is no need to change the likelihood function for skin color detection, since the pre-processing step of illumination correction keeps the ratio test unchanged.

Once each pixel y is assigned a likelihood ratio $\ell(y)$, a histogram of $\ell(y)$ is obtained. The 99th percentile of $\ell(y)$ is chosen as the upper threshold th_U. We start from pixels having $\ell(y) > th_U$ and form the skin colored blobs around them by including pixels having $\ell(y) > th_L$. We select the 80th percentile of the histogram of $\ell(y)$ as the lower threshold th_L. Thus, we chose pixels which have a very high probability of being skin and starting from them group together other connected pixels having the likelihood ratio above the lower threshold th_L to form a skin-colored blob. In this process, we also make use of the constraint that skin colored pixels will be connected to form a skin colored region. It should be noted that the use of skin color detection will yield regions not only of the hands but also of the face and the neck. Apart from this, even other objects like wooden objects are likely to be classified as skin. Skin color regions are detected only in the predicted window to get better results.

The observation vector at time t is given by $Z_t = [t \ l \ b \ r]^T$ where t, l, b, and r correspond to the top, left, bottom and the right co-ordinates of the bounding box, respectively. In measurements where more than one blob is detected, we select the measurement corresponding to the blob which has the maximum probability of being the skin.

4.4 Tracking Using Color Correction

To test the performance of the proposed algorithm, we capture hand gestures in both uncluttered and cluttered environments with a stationary camera. Each image sequence is of different duration with each frame being of a size 352×288 pixels. We present results for both standard color-palette and self-induced palm-palette (where the palm itself serves as the color reference palette) based training. In case of standard color palette-based training the person displays the color palette in front of the camera before the initiation of the gesture. For the self-induced color correction scheme, there is no such need to flash the palette as the palm itself serves as the palette. For the back-propagation neural network we used two hidden layers, each with five nodes and a sigmoidal activation function. After learning the weights of the neural network, the color look up tables for the subsequent image frames are changed and the condensation tracker is used to track the bounding box for the palm region.

In case of the image sequence shown in Fig. 13 the input to the tracker is a realworld image sequence, without any applied color correction. We observe that the tracker gradually fails. This is due to the poor lighting conditions. The tracker, hence, falsely classifies some part of the background as skin, resulting in a larger bounding rectangle. We now train the neural network using the standard color palettes. Results shown in Fig. 14 depict that now the bounding

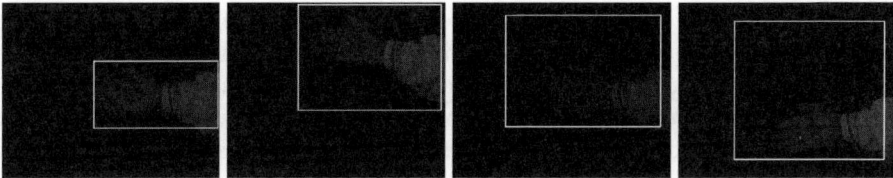

Frames 12, 25, 36 and 48

Fig. 13. Results of the tracker in an uncluttered environment without applying any color correction. All results are given in gray tone images although the original sequence is in color

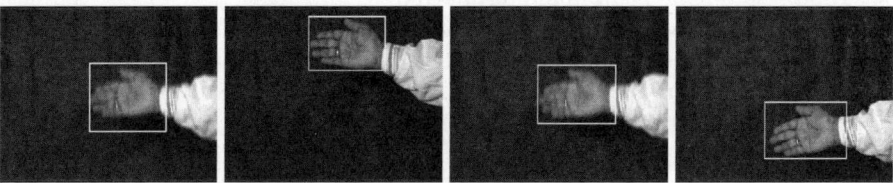

Frames 12, 25, 36 and 48

Fig. 14. Results of the tracker in an uncluttered environment after correcting the color of the corresponding sequence given in Fig. 13. Standard color palettes have been used for training the neural network

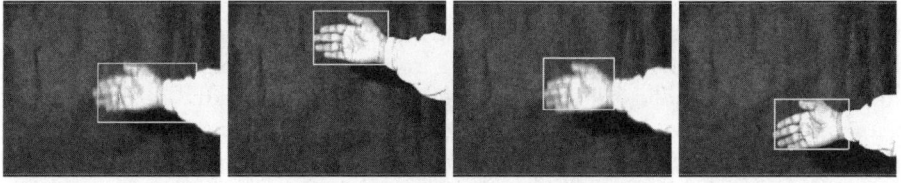

Frames 12, 25, 36 and 48

Fig. 15. Results of the tracker in an uncluttered environment. Here a self-induced skin palette has been used for training the neural network

box is closer to hand for the same video. Though the bounding box is not very tightly fitting the hand region it is definitely much better than the previous results obtained on the direct realworld sequence as shown in Fig. 13 where there was no color correction to the frames. Due to infeasibility of using the standard color palettes for training, now an automatic color correction is done by training the neural network using the self-induced skin-palettes obtained from the initial couple of frames. The corresponding results are shown in Fig. 15. Note that here too the bounding rectangle fits well to the hand and the results are comparable to that obtained after training the neural network with the standard color palettes. We observe that an automatic color correction makes the images a bit more yellowish than what was given in Fig. 14, although the skin color is definitely enhanced. The reason behind this is that since in the later case we are restricting our training by using only the palm region of the hand, the color correction process tries to apply the same transformation for the whole image which disturbs the color of the background and also affects the average intensity as compared to the standard color palette-based training. However, this creates no problem to the tracker, as the tracker is based on the skin color only.

Figures 16 and 17 show the corresponding results for a cluttered environment where a person is moving along with a shadow in the background under extremely poor lighting conditions. Note that the brightness and the contrast of the images in Fig. 16 are changed for display purposes, which otherwise are barely visible due to the poor illumination. The tracker fails within the first 30 frames when no color correction is applied. On training the neural network with the self-induced skin palette and successive color correction of frames, the tracking error reduces substantially. This can be observed by a closely fitting bounding box around the hand region in Fig. 17. Figure 17 shows that the color corrected images appear noisy with a poor quality. This is because of two reasons: the domination of CCD noise in extremely poor lighting conditions and the restricted training using only the palm region. Even though the quality of the transformation for the background region is quite poor, the improvement in the hand (skin) region is quite significant and it helps in tracking the palm.

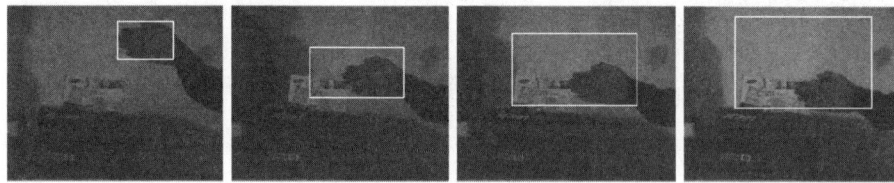

Frames 0, 22, 26 and 30

Fig. 16. Results of the tracker in a cluttered environment, without applying the color correction

Frames 0, 22, 26 and 30

Fig. 17. Results obtained in a cluttered environment using the skin color palette for subsequent color transformation

5 Correction for Dissimilar Images

In Sect. 4 we assumed that the source and the target images are nearly identical so that after scale adjustment between the pair of images, a pixel wise correspondence can be assumed and hence the neural network can be trained accordingly. However, if the canonical observation involves an image which is very different from the source image, the above training procedure is no longer valid. When the source and the target images are of different types, we segment the image into equal number of regions (clusters) and use the correspondence between regions to learn the look-up table for contrast transfer. The learning of the contrast transfer function becomes much easier, and hence we use a simplified network with only a single hidden layer of 10 nodes [7].

We use a fuzzy c-means clustering algorithm to obtain region segmentation in images (both training and test). During training a canonical and realworld palette are segmented into M regions. Histograms of pixels from the segmented regions are used to train a single neural network. The following procedure is adopted to segment gray scale images into regions:

Algorithm to segment images into regions

1. Each image is divided into blocks of 15×15 pixels.
2. Following features are calculated for each block:
   - Average gray level of the block.
   - Entropy of the histogram of the block.
   - Entropy of the co-occurrence matrix of the block.
3. A fuzzy c-means algorithm is applied to the 3D feature vectors obtained from each block of the image to classify it into one of M classes[4].

For color images, one can use just the values of the three channels as the features. For each of the clusters in the source image, we compute the corresponding gray level histograms. The top and bottom 10% population is removed from each histogram to provide robustness against noise perturbations. Using the average gray level of a cluster as the feature, we establish correspondence between clusters in source and target images for all M clusters. It may be noted that each cluster in an image does not necessarily have to represent a contiguous segment. The trimmed histogram for each of the corresponding clusters in source and target images are matched in terms of number of quantization bins and then used to learn the same neural network as discussed in Sect. 4. The training is performed by feeding the input and output gray level relationship taken randomly from any of the segments. Since the segmented regions have typically overlapping histograms, the learnt look-up table is fairly continuous, yielding visually pleasing results. The advantage of this type of color transfer is that each segment can have its own color transfer scheme that smoothly blends across colors of different segments.

---

[4] The number of clusters M is assumed to be known and is same for both source and target images.

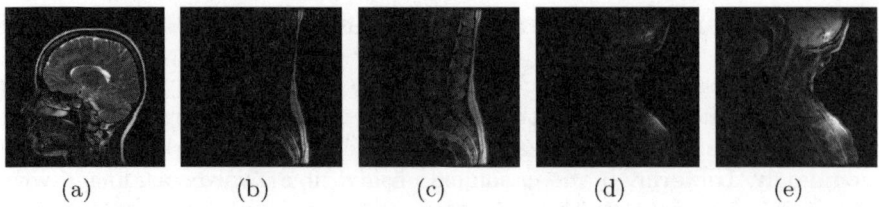

Fig. 18. (a) Target MRI image. (b) and (d) low contrast source images. (c) and (e) contrast restored (b) and (d), respectively for the target contrast of (a)

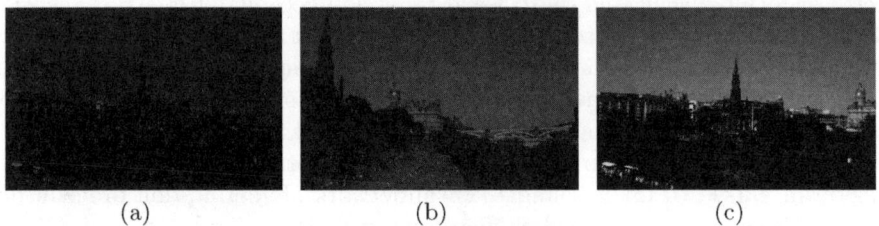

Fig. 19. Contrast transfer in color images. (a) Source image. (b) Target image and (c) result of contrast transfer

In Sect. 4 we used the same type of source and target images for training purposes. Now we show the results of contrast transfer when the canonical image is very different from that of the input image. We use contrast improvement in MRI images as the application domain. Each image is segmented into five clusters and the histograms of corresponding clusters are used for training a simplified neural network as discussed earlier.

Figure 18a shows a good quality MRI scan of brain and this is used as the target contrast. Figure 18b,d shows low contrast MRI images. Figure 18c,e shows the results of contrast enhancement with this. One can clearly see the vertebrae in the contrast restored image shown in Fig. 18c, and the internal details are clear in Fig. 18e. We also demonstrate the correction for dissimilar color images as shown in Fig. 19. Figure 19a,b shows the source and target image pair and Fig. 19c shows the result of contrast transfer. Each RGB image is first converted to $l\alpha\beta$ color space to minimize correlation among color channels [18]. The neural network is trained independently for the corresponding pairs of l, α and β channels. The resultant image obtained after contrast transfer is converted back to RGB. As can be seen from the images, the colors in the result image exhibit more contrast and the finer details in the buildings are more visible than in the source.

6 Multidimensional Pdf Matching

This section discusses a novel multidimensional probability density function (pdf) matching technique for transforming and adapting between them the distributions of two multispectral images. As mentioned earlier, this technique takes into account the correlation among N number of spectral channels.

Let $\mathbf{X}_1$ and $\mathbf{X}_2$ denote the source and the target multispectral images, respectively, made up of N spectral channels. Let $X = (X_1, \ldots, X_N)$ be a multivariate N-dimensional random variable. The bth component $X_b(b = 1, \ldots, N)$ of X represents the random variable associated with the digital number of pixels in the bth spectral band. We denote the N-dimensional pdf of the source image $\mathbf{X}_1$ as $p_1(X)$ and that of the target image $\mathbf{X}_2$ as $p_2(X)$. It is worth noting that in this technique there are no restrictions on the nature and model of the distributions during the pdf matching phase. Our goal is to find a transfer function that can map the function $p_1(X)$ into a new distribution that is as much similar as possible to $p_2(X)$. For a single plane image (single dimension) this problem is much simplified [17] and is known as histogram specification or 1D transfer. The solution is obtained by finding a monotone mapping function $T(X_b)$ such that:

$$T(X_b) = C_2^{-1}[C_1(X_b)], \tag{5}$$

where C_1 and C_2 are the cumulative pdfs of the source and target images, respectively. Next, we include our extension to the algorithm introduced by Pitié et al. [17] for N-dimensional pdf matching.

Algorithm for N-dimensional pdf transfer

1. Select $\mathbf{X}_1$ and $\mathbf{X}_2$, source and target data sets respectively. Both have N components corresponding to the N spectral bands.
2. Pick a randomly generated $N \times N$ rotation matrix R (see Sect. 6.1 on how this is achieved).
3. Rotate the source and target: $\mathbf{X}_1^r \leftarrow R\mathbf{X}_1^{(\rho)}$ and $\mathbf{X}_2^r \leftarrow R\mathbf{X}_2$, where $\mathbf{X}_1^r$ and $\mathbf{X}_2^r$ are the rotated intensity values for the current rotation matrix R; and $\mathbf{X}_1^{(\rho)}$ represents the image derived from the source after ρ rotations iteratively.
4. Find the marginal density functions $p_1(X_b^{(\rho)})$ and $p_2(X_b)$, where $b = 1, 2, \ldots, N$, by projecting $p_1(X^{(\rho)})$ and $p_2(X)$ on each of the axes. Thus, there will be N marginal density functions for each image, one for each spectral band.
5. Find the 1D pdf transfer function for each pair of marginal density functions (corresponding to each spectral band) according to (5).
6. For each pair of marginals, perform histogram matching on rotated bands individually: $p_1(X_1^{(\rho)})$ to $p_2(X_1)$; $p_1(X_2^{(\rho)})$ to $p_2(X_2)$; $\ldots$ $p_1(X_N^{(\rho)})$ to $p_2(X_N)$. At the end, the source image is modified.
7. Rotate back the source image: $\mathbf{X}_1^{(t+1)} \leftarrow R^T\mathbf{X}_1^{(\rho)}$.
8. Move to the next iteration: $t \leftarrow t + 1$
9. Go to Step 2 and compute a new random rotation matrix, until convergence.

The theoretical justification of carrying out the above steps has been given in [17]. In Step 9, a convergence is reached when any further iteration will fail to change the pdf of the modified source image in the N-dimensional space. Step 2 requires that an N-dimensional rotation matrix R be generated randomly at each iteration. The algorithm to generate such a rotation matrix is discussed next.

6.1 Generation of N-Dimensional Rotation Matrix

The generalized approach described in [1] for performing general rotations in a multidimensional Euclidean space is used to obtain $N \times N$ rotation matrix in our case.

An $N \times N$ rotation matrix has degrees of freedom equal to all possible combinations of two chosen from total of N, i.e., $^NC_2$. As an example, for $N = 6$, there are 15 degrees of freedom and hence one needs 15 rotation angles (also known an Euler angles) to represent a 6D rotation. The overall rotation matrix R that depends only on $^NC_2$ independent angular values is obtained by sequentially multiplying each of the corresponding rotation matrices R_i, $i = 1, 2, \ldots, {}^N C_2$. In order to guarantee that each of these $^NC_2$ rotation angles is chosen as independent and uniformly distributed in $[0, 2\pi]$, we propose to adopt the following procedure of generation of the random matrix:

Algorithm for generating a random rotation matrix R

1. Pick $^NC_2$ angles $\theta_1, \theta_2, \ldots, \theta_{NC_2}$ randomly in the uniformly distributed interval of angles $[0, 2\pi]$.
2. Generate $^NC_2$ matrices $R_1, R_2, \ldots, R_{NC_2}$ of size $N \times N$ by considering one angle at a time, each describing a rotation about an $(N\text{-}2)$-dimensional hyperplane. As an example, for N=6, the matrices will be constructed as follows:

$$
R_1 = \begin{bmatrix} cos\theta_1 & sin\theta_1 & 0 & 0 & 0 & 0 \\ -sin\theta_1 & cos\theta_1 & 0 & 0 & 0 & 0 \\ 0 & 0 & 1 & 0 & 0 & 0 \\ 0 & 0 & 0 & 1 & 0 & 0 \\ 0 & 0 & 0 & 0 & 1 & 0 \\ 0 & 0 & 0 & 0 & 0 & 1 \end{bmatrix}, R_2 = \begin{bmatrix} 1 & 0 & 0 & 0 & 0 & 0 \\ 0 & cos\theta_2 & sin\theta_2 & 0 & 0 & 0 \\ 0 & -sin\theta_2 & cos\theta_2 & 0 & 0 & 0 \\ 0 & 0 & 0 & 1 & 0 & 0 \\ 0 & 0 & 0 & 0 & 1 & 0 \\ 0 & 0 & 0 & 0 & 0 & 1 \end{bmatrix}, \ldots,
$$

$$
R_{15} = \begin{bmatrix} cos\theta_{15} & 0 & 0 & 0 & 0 & sin\theta_{15} \\ 0 & 1 & 0 & 0 & 0 & 0 \\ 0 & 0 & 1 & 0 & 0 & 0 \\ 0 & 0 & 0 & 1 & 0 & 0 \\ 0 & 0 & 0 & 0 & 1 & 0 \\ -sin\theta_{15} & 0 & 0 & 0 & 0 & cos\theta_{15} \end{bmatrix}. \tag{6}
$$

3. Generate the final rotation matrix R of size $N \times N$ as the product of all above ${}^N C_2$ matrices[5]:

$$R = R_1 \cdot R_2 \cdot \ldots \cdot R_{N_{C_2}}. \tag{7}$$

For a special case, $N = 3$, the rotation matrix R can be constructed by multiplying three separate rotation matrices as:

$$R = R_1 \cdot R_2 \cdot R_3, \tag{8}$$

where

$$R_1 = \begin{bmatrix} cos\theta_1 & sin\theta_1 & 0 \\ -sin\theta_1 & cos\theta_1 & 0 \\ 0 & 0 & 1 \end{bmatrix}, R_2 = \begin{bmatrix} 1 & 0 & 0 \\ 0 & cos\theta_2 & sin\theta_2 \\ 0 & -sin\theta_2 & cos\theta_2 \end{bmatrix}, \text{ and}$$

$$R_3 = \begin{bmatrix} cos\theta_3 & 0 & sin\theta_3 \\ 0 & 1 & 0 \\ -sin\theta_3 & 0 & cos\theta_3 \end{bmatrix}. \tag{9}$$

Here, the 3 Euler angles θ_1, θ_2 and θ_3 should be generated randomly such that $\theta_i \in [0, 2\pi], i = 1, 2, 3$.

6.2 Pdf Matching with Similar Source and Target Images

Figure 20 shows test on the living room dataset. Two color images of similar contents acquired under two different poor illumination conditions, shown previously in Fig. 6a,b. Figure 20a is the target image. The target image is same as the one shown previously in Fig. 6c but is included here again to facilitate side-by-side visual comparison. Figure 20b,c shows color corrected

(c)   (b)   (c)

Fig. 20. (a) Target image (same as Fig. 6c, included here for side-by-side visual comparison). (b) and (c) color corrected Fig. 6a,b, respectively

---

[5] Matrix multiplications in (7) do not commute. A change in the order of the hyperplanes about which the Euler rotations are carried out will result in a totally different rotation matrix R. But this has no bearing to our application, as we are interested only in generating a set of random rotation matrices.

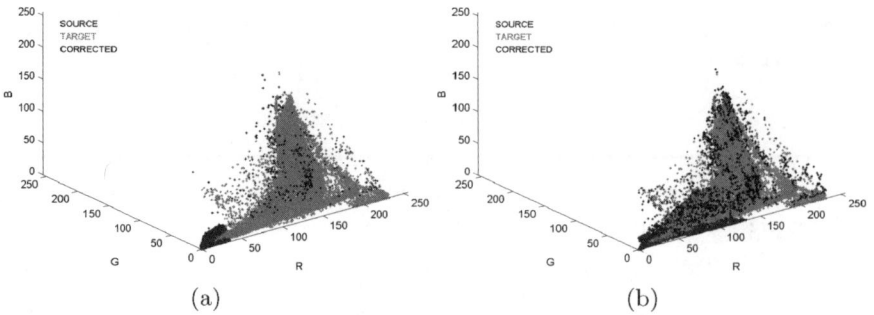

Fig. 21. 3D scatter plots for the living room dataset. Labeling convention in the scatter plots remains the same as followed in Fig. 7. (**a**) Scatter plots for Figs. 6a and 20a,b. (**b**) Scatter plots for Figs. 6b and 20a,c

Fig. 6a,b, respectively. Figure 21 shows 3D scatter plots for the set of results shown here. Labeling convention in the scatter plots remains the same as followed in Fig. 7. Notice the colors in Fig. 20b,c, they are almost similar to those in Fig. 20a.

For the ND method it was observed that for the presented dataset the results converged after around 25 iterations. The proper convergence of the distributions may be observed by the contour plots in the sub-spaces of the global feature domain. It is also worth noting that after de-rotation of the image, it is expected that the result should lie in the same range as the original image. However, during the transformation certain spurious values may be observed due to numerical inaccuracies accruing from (5). To take care of these spurious values we normalized the values accordingly. The normalization logic used was a simple linear mapping of the range of the pixel intensities in the result to that in the target image in each channel separately. This may introduce slight distortions for pixels having values at the limits of the dynamic range of the images. As seen clearly in the results, the ND method effectively corrects the problem due to poor illumination.

6.3 Pdf Matching with Dissimilar Source and Target Image Types

Figure 22a,b shows the source and the target image pair which are dissimilar. It is very important in the ND method that the target is well chosen as the result takes up the color scheme of the target. As illustrated in Fig. 22, the result shows more orangish pixels than there actually are. This can be attributed to the contents of the target image. In the process of matching the shapes of the histograms, the overall occurrence of various colors in the result becomes similar to that in the target. In this case again, we required around 25 iterations for convergence.

Another set of results is shown in Fig. 23. Here, Fig. 23a,b shows the source and the target images for a carefully chosen pair of dissimilar images. The

(a)   (b)   (c)

Fig. 22. (a) Source, (b) target and (c) color correction with ND method

(a)   (b)   (c)

Fig. 23. Color correction in synthetically color distorted images of paintings. (a) Source image (synthetically color distorted, same as Fig. 8a). (b) A well chosen dissimilar target image (notice the contents of the target image are mostly similar to (a)). (c) Color corrected (a)

quality of color correction is obvious in Fig. 23c. It may be noticed that the yellow fruit at the base of the flower-vase has now picked up a very natural hue. The convergence is achieved in relatively lesser iterations; 15, in this particular case. Figure 24a–c shows contour plots in GB subspace for source, target and corrected images, respectively. Note the resultant distribution in GB subspace, it is more similar to the target distribution.

7 Conclusions

We explained two different techniques for color correction. The first uses a neural network to learn mapping between colors in a source image and some target image. This requires that training image pairs have pixel-to-pixel correspondence. To provide a qualitative comparison we show results for images in a typical indoor environment, for paintings and for satellite images. Skin

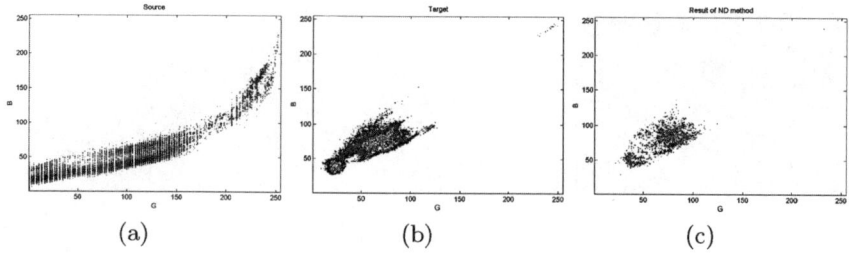

Fig. 24. Contour plots in GB subspace for (**a**) source, (**b**) target and (**c**) result of ND method. Resultant distribution in GB subspace is more similar to the target distribution

color based hand tracking is our principal machine vision-based application for which the color corrected images bring an improved performance. An improved version of the first technique rules out the requirement of registered training image pairs as it uses a cluster-based strategy to match regions between a source and target image pair. For this, results of experiments on contrast restoration in both MRI images as well as colored images are shown. The second technique that provides a further improvement over the first, does the work remarkably well by not only correcting colors but also making sure that the statistical properties in images are maintained even after color correction.

References

1. A. Aguilera and R. Pérez-Aguila. General n-dimensional rotations. In *International Conference in Central Europe on Computer Graphics, Visualization and Computer Vision*, Czech Republic, February 2004.
2. K. Barnard, V. Cardei, and B. Funt. A comparison of computational color constancy algorithms – Part I: methodology and experiments with synthesized data. *IEEE Transactions on Image Processing*, 11(9):972–984, 2002.
3. K. Barnard, L. Martin, A. Coath, and B. Funt. A comparison of computational color constancy algorithms – part II: experiments with image data. *IEEE Transactions on Image Processing*, 11(9):985–996, 2002.
4. B. Bascle, O. Bernier, and V. Lemaire. Illumination-invariant color image correction. In *International Workshop on Intelligent Computing in Pattern Analysis/Synthesis*, Xi'an, China, August 2006.
5. V. Cardei. *A Neural Network Approach to Color Constancy*. PhD thesis, Simon Faser University, Burnaby, BC, Canada, 2000.
6. B. Funt, K. Barnard, and L. Martin. Is machine color constancy good enough? In *Fifth European Conference on Computer Vision*, pages 445–459, 1998.
7. A. Galinde and S. Chaudhuri. A cluster-based target-driven illumination correction scheme for aerial images. In *IEEE National Conference on Image Processing*, Bangalore, India, March 2005.
8. G. Greenfield and D. House. A palette-driven approach to image color transfer. In *Computational Aesthetics in Graphics, Visualization and Imaging*, pages 91–99, 2005.

9. M. Isard and A. Blake. Condensation – conditional density propagation for visual tracking. *International Journal of Computer Vision*, 28(1):5–28, 1998.
10. R. Kjeldsen and J. Kender. Finding skin in color images. In *Proceedings of the Second International Conference on Automatic Face and Gesture Recognition*, pages 312–317, 1996.
11. M. Kocheisen, U. Müller, and G. Tröster. A neural network for grey level and color correction used in photofinishing. In *IEEE International Conference on Neural Networks*, pages 2166–2171, Washington, DC, June 1996.
12. H. Lee and D. Han. Implementation of real time color gamut mapping using neural networks. In *IEEE Mid-Summer Workshop on Soft Computing in Industrial Applications*, pages 138–141, Espoo, Finland, June 2005.
13. J. P. Mammen, S. Chaudhuri, and T. Agrawal. Hierarchical recognition of dynamic hand gestures for telerobotic application. *IETE Journal of Research special issue on Visual Media Processing*, 48(3&4):49–61, 2002.
14. J. B. Martinkauppi, M. N. Soriano, and M. H. Laaksonen. Behaviour of skin color under varying illumination seen by different cameras at different color spaces. In M. A. Hunt, editor, *Proceedings SPIE, Machine Vision in Industrial Inspection IX*, volume 4301, pages 102–113, San Jose, CA, 2001.
15. A. Nayak and S. Chaudhuri. Automatic illumination correction for scene enhancement and object tracking. *Image and Vision Computing*, 24(9):949–959, September 2006.
16. B. Pham and G. Pringle. Color correction for an image sequence. *IEEE Computer Graphics and Applications*, 15(3):38–42, 1995.
17. F. Pitié, A. Kokaram, and R. Dahyot. N-dimensional probability function transfer and its application to colour transfer. In *IEEE International Conference on Computer Vision*, volume 2, pages 1434–1439, 2005.
18. E. Reinhard, M. Ashikhmin, B. Gooch, and P. Shirley. Color transfer between images. *IEEE Computer Graphics and Applications*, 21(5):34–41, 2001.
19. D. Ruderman, T. Cronin, and C. Chiao. Statistics of cone responses to natural images: implications for visual coding. *Journal of Optical Society of America*, 15(8):2036–2045, 1998.
20. L. Sigal, S. Sclaroff, and V. Athitsos. Estimation and prediction of evolving color distributions for skin segmentation under varying illumination. In *Proceedings IEEE Conference on Computer Vision and Pattern Recognition*, volume 2, pages 152–159, June 2000.
21. C. Wang and Y. Huang. A novel color transfer algorithm for image sequences. *Journal of Information Science and Engineering*, 20(6):1039–1056, 2004.
22. J. Yang, W. Lu, and A. Waibel. Skin-color modeling and adaptation. *CMUCS-97-146*, May 1997.
23. J. Yin and J. Cooperstock. Color correction methods with applications to digital projection environments. *Journal of the Winter School of Computer Graphics*, 12(3):499–506, 2004.
24. M. Zhang and N. Georganas. Fast color correction using principal regions mapping in different color spaces. *Journal of Real-Time Imaging*, 10(1), 2004.

A Neural Approach to Unsupervised Change Detection of Remote-Sensing Images

Susmita Ghosh[1], Swarnajyoti Patra[1], and Ashish Ghosh[2]

[1] Department of Computer Science and Engineering, Jadavpur University, Kolkata 700032, India, susmita_de@rediffmail.com, patra_swarna@rediffmail.com
[2] Machine Intelligence Unit and Center for Soft Computing Research, Indian Statistical Institute, 203 B.T. Road, Kolkata 700108, India, ash@isical.ac.in

Summary. Two unsupervised context-sensitive change detection techniques, one based on Hopfield type neural network and the other based on self-organizing feature map neural network, for remote sensing images have been proposed in this chapter. In the presented Hopfield network, each neuron corresponds to a pixel in the difference image and is assumed to be connected to all its neighbors. An energy function is defined to represent the overall status of the network. Each neuron is assigned a status value depending on an initialization threshold and updated iteratively until converges. On the other hand, in the self-organizing feature map model, number of neurons in the output layer is equal to the number of pixels in the difference image and the number of neurons in the input layer is equal to the dimension of the input patterns. The network is updated depending on some threshold. For both the cases, at convergence, the output statuses of neurons represent a change detection map. Experimental results confirm the effectiveness of the proposed approaches.

1 Introduction

In remote sensing applications, change-detection is the process aimed at identifying differences in the state of a land-cover by analyzing a pair of images acquired on the same geographical area at different times [1,2]. Such a problem plays an important role in many different domains, like studies on land-use/land-cover dynamic [3], monitoring shifting cultivations [4], burned areas identification [5], analysis of deforestation processes [6], assessment of vegetation changes [7] and monitoring of urban growth [8]. Since all of these applications usually require an analysis of large areas, development of completely automatic and unsupervised change-detection techniques is of high relevance in order to reduce the time effort required by manual image analysis and to produce objective change-detection maps.

In the literature, several supervised [4, 9–12] and unsupervised [13–17] techniques for detecting changes in remote-sensing images have been proposed. The former requires the availability of a "ground truth" from which

S. Ghosh et al.: *A Neural Approach to Unsupervised Change Detection of Remote-Sensing Images*, Studies in Computational Intelligence (SCI) **83**, 243–264 (2008)
www.springerlink.com

a training set is derived containing information about the spectral signatures of the changes that occurred in the considered area between the two dates. The latter detects changes without any additional information besides the raw images considered. Therefore, from an operational point of view, it is obvious that use of unsupervised techniques is mandatory in many remote-sensing applications, as suitable ground-truth information is not always available.

Most widely used unsupervised change-detection techniques are based on a three-step procedure [1]: (i) preprocessing; (ii) pixel-by-pixel comparison of two multitemporal images; and (iii) image analysis. The aim of the preprocessing step is to make the considered images as comparable as possible with respect to operations like co-registration, radiometric and geometric corrections and noise reduction. The comparison step aims at producing a further image called "difference image", where differences between the two considered acquisitions are highlighted. Different mathematical operators can be adopted to perform image comparison. Once image comparison is performed, the change-detection process can be carried out adopting either context-insensitive or context-sensitive procedures.

The most widely used unsupervised context-insensitive analysis techniques are based on histogram thresholding of the difference image [13, 18]. The thresholding procedures do not take into account the spatial correlation between neighboring pixels in the decision process. To overcome this limitation of neglecting the interpixel class dependency, context-sensitive change-detection procedures based on Markov random field (MRF) have been proposed in [19–21]. These context-sensitive automatic approaches to change-detection require the selection of a proper model for the statistical distributions of changed and unchanged pixels. In order to overcome the limitations imposed by the need for selecting a statistical model for changed and unchanged class distributions, in this chapter we describe two unsupervised context-sensitive change-detection techniques based on artificial neural networks [22]. One method is based on modified Hopfield network [23] and the other on modified self-organizing feature map network [24].

This chapter is organized into seven sections. Section 2 provides a procedure to generate the difference image. Section 3 describes an unsupervised change-detection technique based on Hopfield type neural network. Section 4 provides another unsupervised change-detection technique based on modified self-organizing feature map neural network. The data sets used in the experiments are described in Sect. 5. Experimental results are discussed in Sect. 6. Finally, in Sect. 7, conclusions are drawn.

2 Generation of Difference Image

In the literature concerning unsupervised change-detection, several techniques are used to generate the "difference image". The most widely used operator is the difference. This operator can be applied to: (i) a single spectral band

(Univariate Image Differencing) [1,25,26]; (ii) multiple spectral bands (change vector analysis) [1,13]; (iii) vegetation indices (vegetation index differencing) [1,27] or other linear or nonlinear combinations of spectral bands. Each choice gives rise to a different technique. Among these, the most popular change vector analysis (CVA) technique is used here to generate the difference image. This technique exploits a simple vector subtraction operator to compare two multispectral images, under analysis, pixel-by-pixel. In some cases, depending on the specific type of changes to be identified, the comparison is made on a subset of the spectral channels. The difference image is computed as the magnitude of spectral change vectors obtained for each pair of corresponding pixels.

Let us consider two co-registered and radiometrically corrected γ-spectral band images X_1 and X_2, of size $p \times q$, acquired over the same area at different times T_1 and T_2, and let $D = \{l_{mn}, 1 \leq m \leq p, 1 \leq n \leq q\}$ be the difference image obtained by applying the CVA technique to X_1 and X_2. Then

$$l_{mn} = (int)\sqrt{(l_{mn}^1(X_1) - l_{mn}^1(X_2))^2 + \cdots + (l_{mn}^\gamma(X_1) - l_{mn}^\gamma(X_2))^2}.$$

Here $l_{mn}^\alpha(X_1)$ and $l_{mn}^\alpha(X_2)$ are the gray values of the pixel at the spatial position (m,n) in αth band of images X_1 and X_2, respectively.

3 Change Detection Based on Hopfield Type Neural Network

3.1 Background: Hopfield Neural Network

A Hopfield neural network consists of a set of neurons (or units) arranged in one layer. The output of each neuron is fed back to each of the other units in the network. There is no self-feedback loop and the synaptic weights are symmetric [22]. Hopfield defined the energy function of the network by using the network architecture, i.e., the number of neurons, their output functions, threshold values, connection between neurons and the strength of the connections [28]. Thus the energy function represents the complete status of the network. Hopfield has also shown that at each iteration of the processing of the network the energy value decreases and the network reaches a stable state when its energy value reaches a minimum [29,30]. Since there are interactions among all the units, the collective property inherently reduces the computational complexity.

The input U_i to the generic ith neuron comes from two sources: (i) input V_j from jth units to which it is connected; and (ii) external input bias I_i, which is a fixed bias applied externally to the unit i. Thus the total input to a neuron i is given by

$$U_i = \sum_{j=1, j \neq i}^{n} W_{ij} V_j + I_i, \tag{1}$$

where the weight W_{ij} represents the synaptic interconnection strength from neuron j to neuron i, and n is the total number of units in the network. The connection strengths are assumed to be symmetric, i.e., $W_{ij} = W_{ji}$. Each neuron takes its input at any random time and updates its output independently. The output V_i of neuron i is defined as

$$V_i = g(U_i), \tag{2}$$

where $g(.)$ is an activation function. There are two types of Hopfield models (discrete and continuous), which differ on the output values a neuron can take.

Discrete Model

In the discrete model, neurons are bipolar, i.e., the output V_i of neuron i is either $+1$ or -1. In this model the activation function $g(.)$ is defined according to the following threshold function

$$V_i = g(U_i) = \begin{cases} +1, & if \ U_i \geq \theta_i \\ -1, & if \ U_i < \theta_i \end{cases}, \tag{3}$$

where θ_i is the predefined threshold of neuron i. The energy function E of the discrete model is given by [29]

$$E = -\sum_{i=1}^{n} \sum_{j=1, i\neq j}^{n} W_{ij} V_i V_j - \sum_{i=1}^{n} I_i V_i + \sum_{i=1}^{n} \theta_i V_i. \tag{4}$$

The change of energy $\triangle E$ due to a change of output state of the neuron i equal to $\triangle V_i$ is

$$\triangle E = - \left[\sum_{j=1, i\neq j}^{n} W_{ij} V_j + I_i - \theta_i \right] \triangle V_i = - \left[U_i - \theta_i \right] \triangle V_i. \tag{5}$$

If $\triangle V_i$ is positive (i.e., the state of the neuron i is changed from -1 to $+1$), then from (3) we can see that the bracketed quantity in (5) is also positive, making $\triangle E$ negative. When $\triangle V_i$ is negative (i.e., the state of the neuron i is changed from $+1$ to -1), then from (3) we can see that the bracketed quantity in (5) is also negative. So, any change in E according to (3) is negative. Since E is bounded, the time required by the system to reach convergence is associated to a motion in the state space that seeks out minima of E and stops at such points.

Continuous Model

In this model the output of a neuron is continuous [30] and can assume any real value between $[-1, +1]$. In the continuous model the activation function $g(.)$ must satisfy the following conditions:

- It is a monotonic nondecreasing function;
- $g^{-1}(.)$ exists.

A typical choice of the function $g(.)$ is

$$g(U_i) = \frac{2}{1 + e^{-\phi_i(U_i - \tau)}} - 1, \tag{6}$$

where the parameter τ controls the shifting of the sigmoidal function $g(.)$ along the abscissa and ϕ_i determines the steepness (gain) of neuron i. The value of $g(U_i)$ lies in $[-1, +1]$ and is equal to 0 at $U_i = \tau$. The energy function E of the continuous model is given by [30]

$$E = -\sum_{i=1}^{n} \sum_{j=1, i \neq j}^{n} W_{ij} V_i V_j - \sum_{i=1}^{n} I_i V_i + \sum_{i=1}^{n} \frac{1}{R_i} \int_0^{V_i} g^{-1}(V_i) dV. \tag{7}$$

This function E is a Lyapunov function and R_i is the total input impedance of the amplifier realizing a neuron i. It can be shown that when neurons are updated according to (6), any change in E is negative. The last term in (7) is the energy loss term, which becomes zero at the high gain region. If the gain of the function becomes infinitely large (i.e., the sigmoidal nonlinearity approaches the idealized hard-limiting form), the last term will become negligibly small. In the limiting case, when $\phi_i = \infty$ for all i, the maxima and minima of the continuous model become identical to those of the corresponding discrete Hopfield model. In this case the energy function is simply defined by

$$E = -\sum_{i=1}^{n} \sum_{j=1, i \neq j}^{n} W_{ij} V_i V_j - \sum_{i=1}^{n} I_i V_i, \tag{8}$$

where the output state of each neuron is ± 1. Therefore, the only stable points of the very high-gain, continuous, deterministic Hopfield model correspond to the stable points of the discrete stochastic Hopfield model [22].

3.2 Network Architecture Used for Change Detection

In order to use Hopfield networks for solving the change-detection problem, we assign to each spatial position $(m, n) \in D$ (difference image) a neuron of the network. The spatial correlation between neighboring pixels is modeled by defining the spatial neighborhood systems N of order d, for a given spatial position (m, n) as $N_{mn}^d = \{(m, n) + (i, j), (i, j) \in N^d\}$. The neuron in position (m, n) is connected to its neighboring units included in N^d. According to the value of d, the neighborhood system assumes different configurations. For the first- and second-order neighborhood systems, $N^1 = \{(\pm 1, 0), (0, \pm 1)\}$ and $N^2 = \{(\pm 1, 0), (0, \pm 1), (1, \pm 1), (-1, \pm 1)\}$. Figure 1 depicts a second-order (N^2) topological network. Let $W_{mn,ij}$ be the connection strength between the

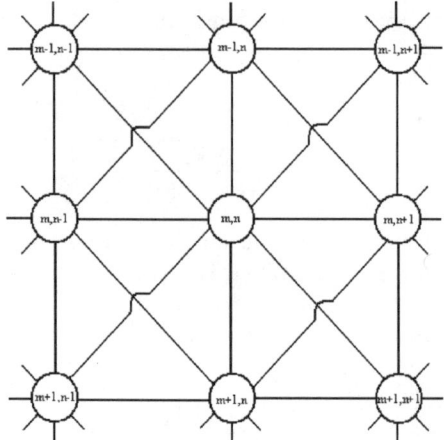

Fig. 1. Second-order topological network: each neuron in the network is connected only to its eight neighbors. The neurons are represented by circles and lines represent connections between neurons

$(m, n)th$ and $(i, j)th$ neuron. We assume that $W_{mn,ij} = 1$ if $(i, j) \in N_{mn}^d$, otherwise $W_{mn,ij} = 0$. Hence, the presented architecture can be seen as a modified version of the Hopfield network [23, 31] in which the connection strength to all neurons outside the neighborhood (N^d) is zero. As each neuron is connected only to its neighboring units, the output of a neuron depends only on its neighboring elements. In this way, the network architecture is intrinsically able to model and properly consider the spatio-contextual information of each pixel.

From (6) we note that the output V_{mn} for the neuron at position (m, n) is given by

$$V_{mn} = \lim_{U_{mn} \to \infty} g(U_{mn}) = +1, \ and \ V_{mn} = \lim_{U_{mn} \to -\infty} g(U_{mn}) = -1.$$

Thus, the domain of U_{mn} is $(-\infty, +\infty)$. In order to simplify the problem, here we use the generalized fuzzy S-function [32] defined over a finite domain as input/output transfer function (activation function). The form of the S-function is the following

$$V_{mn} = g(U_{mn}) = \begin{cases} -1 & U_{mn} \leq a \\ 2^r \left\{ \frac{(U_{mn}-a)}{(c-a)} \right\}^r - 1 & a \leq U_{mn} < b \\ 1 - 2^r \left\{ \frac{(c-U_{mn})}{(c-a)} \right\}^r & b \leq U_{mn} \leq c \\ 1 & U_{mn} \geq c \end{cases}, \quad (9)$$

where $r \geq 2$ and $b = \frac{a+c}{2}$. In this case $g(U_{mn})$ lies in $[-1, +1]$ with $g(U_{mn}) = 0$ at $U_{mn} = b$. The domain of U_{mn} is $[a, c]$. The value of r tunes the sharpness

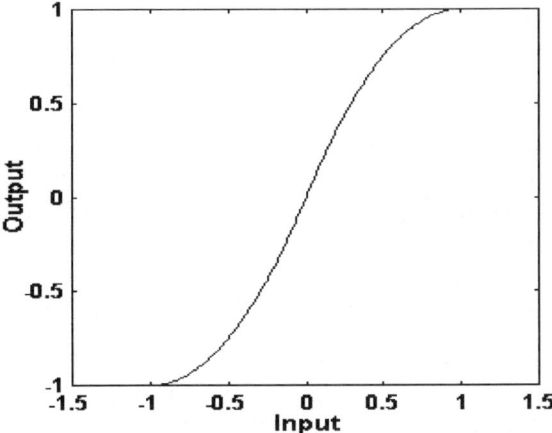

Fig. 2. Behavior of the activation function defined in $[-1, +1]$, assuming $r = 2$

(steepness) of the function. If the number of neighbors is four (N^1 neighborhood), the input value to a neuron lies in $[-5, +5]$, i.e., $a = -5$, $c = 5$. However, for quick convergence one can use the domain of U_{mn} in $[-1, +1]$ and an activation function $g(.)$ (shown in Fig. 2) is thus defined as follows

$$V_{mn} = g(U_{mn}) = \begin{cases} -1 & U_{mn} \leq -1 \\ (U_{mn} + 1)^r - 1 & -1 \leq U_{mn} \leq 0 \\ 1 - (1 - U_{mn})^r & 0 \leq U_{mn} \leq 1 \\ 1 & U_{mn} \geq 1 \end{cases} . \tag{10}$$

3.3 Energy Function for Change Detection

The aim of the presented neural architecture is to separate changed pixels from unchanged ones in D. In order to accomplish this task, we should define the energy function of the network in such a manner that when the network reaches the stable state having minimum energy value the changed regions are clearly separated from the unchanged areas. The basic idea exploited in the proposed approach is inspired by the energy function of the Hopfield model formulated for object background classification in [31].

Let us consider the energy function defined in (4) that has three parts. The first part models the local field (or feedback), while the second and third parts correspond to the input bias (I_i) and the threshold value (θ_i) of each neuron in the network, respectively. In terms of images, the first part can be viewed as the impact of the gray values of the neighboring pixels on the energy function, whereas the second and the third part can be attributed to the gray value of the pixel under consideration. Let us assume, without loss of generality, that the output value of each neuron lies in $[-1, +1]$. Then the energy function can be defined in the following way. If the output of a neuron at position (m, n) is $+1$, then it corresponds to a pixel that belongs to the changed area, while if the output is -1, then it corresponds to a pixel that belongs to the unchanged area. So the threshold between the changed and unchanged pixels can logically be taken as 0 (i.e., $\theta_{mn} = 0, \forall(m, n)$). This helps us to omit the third part of the energy expression (see (4)). Each neuron at position (m, n) has an input bias I_{mn} which can be set proportional to the actual gray value of the corresponding pixel. If the gray value of a pixel is high (low), the corresponding intensity value of the scene is expected to be high (low). The input bias value is taken in the range $[-1, +1]$. If a neuron has a very high positive bias (close to $+1$) or very high negative bias (close to -1), then it is very likely that in the stable state the output will be $+1$ or -1, respectively. So, the product $I_{mn}V_{mn}$ should contribute less towards the total energy value, and the second part of the energy expression may be written as:

$$-\sum_{m=1}^{p}\sum_{n=1}^{q} I_{mn}V_{mn}.$$

Depending on the nonimpulsive autocorrelation function of the difference image D, we can assume that the gray value of a pixel is highly influenced by the gray values of its neighboring pixels. So, if a pixel belongs to a changed region, the probability that its neighboring pixels belong to the same region is very high. This suggests that if a pair of adjacent pixels have similar output values, then the energy contribution of this pair of pixels to the overall energy function should be relatively small. If the gray values of two adjacent pixels (m, n) and (i, j) are given by V_{mn} and V_{ij}, then a reasonable choice for the contribution of each of these pairs to the overall energy function is $-W_{mn,ij}V_{mn}V_{ij}$. Thus, taking into account the contribution of all pixels, the first part of energy can be written as

$$-\sum_{m=1}^{p}\sum_{n=1}^{q}\sum_{(i,j)\in N_{mn}^{d}} W_{mn,ij}V_{mn}V_{ij}$$

$$= -\sum_{m=1}^{p}\sum_{n=1}^{q}\left(\sum_{(i,j)\in N_{mn}^{d}} W_{mn,ij}V_{ij}\right)V_{mn}$$

$$= -\sum_{m=1}^{p}\sum_{n=1}^{q} h_{mn}V_{mn}, \tag{11}$$

where h_{mn} is termed as the local field, and models the neighborhood information of each pixel of D in the energy function. On the basis of the above analysis, the expression of the energy function can be written as

$$E = -\sum_{m=1}^{p}\sum_{n=1}^{q}\sum_{(i,j)\in N_{mn}^{d}} W_{mn,ij}V_{mn}V_{ij} - \sum_{m=1}^{p}\sum_{n=1}^{q} I_{mn}V_{mn}, \tag{12}$$

where $W_{mn,ij} = 1$ if $(i,j) \in N_{mn}^d$, else $W_{mn,ij} = 0$. Minimization of (12) results in a stable state of the network in which changed areas are separated from unchanged ones. The energy function in (12) can be minimized by both the discrete and the continuous model.

In the case of the discrete model, both the initial external input bias I_{mn} and the input U_{mn} to a neuron are taken as $+1$ if the gray value of the corresponding pixel of D is greater than a specific global threshold value t, otherwise they have a value -1. As the threshold t is used to set the initial value of each neuron (note that this initialization threshold t and the threshold θ_i defined in (3) are different). Here the status updating rule is the same as in (3) with the threshold value $\theta_{mn} = 0$, $\forall (m,n)$.

In the case of the continuous model, both the initial external input bias and the input to a neuron at position (m,n) are proportional to $(l_{mn}/t) - 1$ (if $(l_{mn}/t) - 1 > 1$ then the value $+1$ is used for initializing the corresponding neuron). The activation function defined in (10) is used to update the status of the network iteratively, until the network reaches the stable state. When the network reaches the stable state, we consider the value of the gain parameter $r = \infty$ (i.e., the sigmoidal nonlinearity approaches the idealized hard-limiting form), and update the network. This makes the continuous model behave like a discrete model; thus, at the stable state, the energy value of the network can be expressed according to (12).

For both the discrete and the continuous models, at each iteration itr, the external input bias $I_{mn}(itr)$ of the neuron at position (m,n) updates its value by taking the output value $V_{mn}(itr - 1)$ of the previous iteration. When the network begins to update its state, the energy value is gradually reduced until the minimum (stable state) is reached. Convergence is reached when $\frac{dV_{mn}}{dt} = 0$, $\forall (m,n)$, i.e., if for each neuron it holds that $\Delta V_{mn} = V_{mn}(itr - 1) - V_{mn}(itr) = 0$.

3.4 Threshold Selection Techniques

As stated above, the values of the neurons are initialized depending on a threshold value t and the network is updated till convergence. After convergence, the output value V_{mn} of neuron (m,n) is $+1$, else -1 for all (m,n). Thus in the output layer the neurons are divided into two groups G_u (represents *unchanged regions*) and G_c (represents *changed regions*) and a change-detection map corresponding to threshold t is generated. By varying t (threshold values are varied by an amount 1), different change-detection maps are generated. To select a threshold t_1 near the optimal threshold t_0 (corresponding change-detection results provide minimum error), we propose two different criteria which are described below.

Correlation Maximization Criterion

The correlation coefficient [33] between the input sequence (difference image) and the output sequence (change-detection map generated considering threshold t) is maximized to select the near optimal threshold t_1. As the threshold t is varied, the change-detection map also varies, thereby the value of the correlation coefficient is changed. When the threshold corresponds to the boundary between changed and unchanged pixels, the correlation coefficient becomes maximum. Let Y and Z be two random variables that correspond to the input sequences and output sequences, respectively. The correlation coefficient between Y and Z for threshold t (denoted as, $R_{Y,Z}(t)$) is defined as

$$R_{Y,Z}(t) = \frac{cov(Y,Z)}{\sigma_Y . \sigma_Z}.$$

Here $cov(Y, Z)$ is the covariance of Y and Z, σ_Y and σ_Z are the standard deviations of Y and Z, respectively. As l_{mn} and V_{mn} are the values of the difference image (input sequence) and output image (output sequence) at the spatial position (m, n), respectively, the above formula may be rewritten as

$$R_{Y,Z}(t) = \frac{\sum_{m,n} l_{mn}.V_{mn} - \frac{1}{p \times q} \sum_{m,n} l_{mn}.\sum_{m,n} V_{mn}}{\sqrt{\sum_{m,n} l_{mn}^2 - \frac{1}{p \times q}\left(\sum_{m,n} l_{mn}\right)^2} \sqrt{\sum_{m,n} V_{mn}^2 - \frac{1}{p \times q}\left(\sum_{m,n} V_{mn}\right)^2}}.$$

$$(13)$$

We compute the correlation coefficient assuming different thresholds and t_1 is assumed to be a near optimal threshold, if the correlation coefficient $R_{Y,Z}(t_1)$ is maximum. Considering this automatically derived threshold t_1, we initialize the neurons of the network and when the network converges the neurons implicitly generates a change-detection map.

Energy-Based Criterion

From (12) we can see that the energy value is minimum when V_{mn} are either all $+1$ or all -1, $\forall (m, n)$, i.e., the whole output image belongs either to changed or unchanged areas. In the proposed procedure, we first compute the energy at convergence for various threshold values (see Fig. 3). By analyzing the behavior of this graph, it is seen that initially the energy value increases with threshold as the number of regions (changed and unchanged) increases. After a certain threshold value, the energy decreases as the number of regions decreases (some unchanged regions are merged together). After that, the energy does not change significantly with increase of threshold value, i.e., changed and unchanged regions are not significantly altered by the specific range of values considered. We expect that this stable behavior of the energy function is reached around the optimal threshold t_0 (see Fig. 3). If the threshold value increases more, the energy changes slowly and reaches a minimum when the whole output image belongs to the class of unchanged areas. By observing this general behavior, we propose a heuristic technique

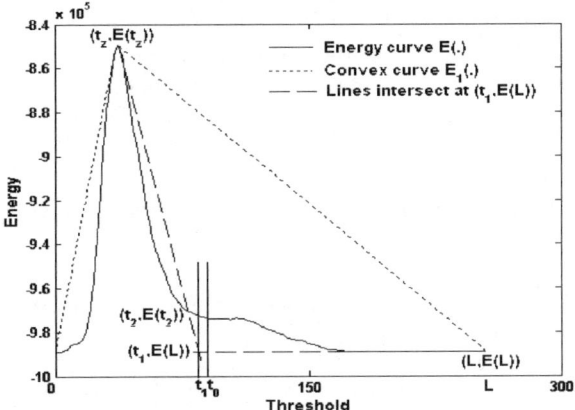

Fig. 3. Behavior of the energy value with threshold. Automatically detected threshold t_1 is close to the optimal threshold t_0

Table 1. Algorithm for automatically deriving a threshold

Phase 1: Generate the smallest convex curve $E_1(.)$ containing the energy curve $E(.)$

Step 1: Initialize $k = 0$;

Step 2: While $k \neq L$ (L is the maximum gray value of D)

Step 3:   For $i = k + 1$ to L

Step 4:   Compute the gradient (slope) of the line passing through the points $(k, E(k))$ and $(i, E(i))$.

Step 5:   End For

Step 6:   Find out a point $(s, E(s))$, $k < s \leq L$, such that the slope of the line passing through the points $(k, E(k))$ and $(s, E(s))$ is maximum.

Step 7:   Join the two points $(k, E(k))$ and $(s, E(s))$ by a straight line. This line is a part of the convex curve $E_1(.)$ from threshold value k to s.

Step 8:   Reset $k = s$.

Step 9: End While

Phase 2: Derive the threshold t_1

Step 10: Select the maximum energy value point $(t_z, E(t_z))$ in energy curve $E(.)$ (see Fig. 3).

Step 11: Select t_2, so that $\{E_1(t_2) - E(t_2)\} = \max_i\{E_1(i) - E(i)\}$, $t_z \leq i \leq L$.

Step 12: Select the threshold t_1, at the intersection between the straight line connecting $(t_z, E(t_z))$ and $(t_2, E(t_2))$ and the straight line parallel to the abscissa and passing through minimum energy value $E(L)$ (see Fig. 3).

Step 13: Stop.

---

that generates the smallest convex curve $E_1(.)$ containing the energy curve $E(.)$ using a concavity analysis algorithm [34] and exploiting these two curves we derive a threshold t_1 which is close to the optimal one. The technique is described in Table 1.

Each neuron in the network is initialized by considering this automatically derived threshold value t_1 and is allowed to update its status with time. When the network reaches a stable state it implicitly generates a change-detection map.

4 Change Detection Based on Self-Organizing Feature Map Type Neural Network

4.1 Background: Kohonen's Model of Self-Organizing Feature Map Neural Network

Kohonen's Self-Organizing Feature Map (SOFM) network [35, 36] consists of an input and an output layer. Each neuron in the output layer is connected to all the neurons in the input layer, i.e., the y-dimensional input signal $\mathbf{U} = [u_1, u_2, \ldots, u_y]$ can be passed to all the output neurons. Let the synaptic weight vector of an output neuron j be denoted by $\mathbf{W}_j = [w_{j1}, w_{j2}, \ldots, w_{jy}]^T, j = 1, 2, \ldots, n$, where n is the total number of neurons in the output layer and w_{jk} is the weight of the jth unit for the kth component of the input. If the synaptic weight vector $\mathbf{W}_i$ of output neuron i is best matched with input vector $\mathbf{U}$, then $\sum_{k=1}^{y} w_{ik}.u_k$ will be maximum among $\sum_{k=1}^{y} w_{jk}.u_k, \forall j$. The neuron i is then called the *wining neuron* for the input vector $\mathbf{U}$. The *wining neuron* is located at the center of a topological neighborhood of cooperating units. Let $h_i(itr)$ denote the topological neighborhood of *wining neuron* i at epoch number itr. There are several ways to define a topological neighborhood [36, 37] such as Gaussian, rectangular, etc. The size of the topological neighborhood shrinks with increase in itr. Figure 4 shows how the size of the topological neighborhood of the *wining neuron* i decreases over itr (in case of rectangular topological neighborhood). For the ith unit and all its neighbors (within a specified radius defined by $h_i(itr)$) the following weight updating rule is applied

$$\mathbf{W}_i(itr + 1) = \frac{\mathbf{W}_i(itr) + \eta(itr)(\mathbf{U} - \mathbf{W}_i(itr))}{\|\mathbf{W}_i(itr) + \eta(itr)(\mathbf{U} - \mathbf{W}_i(itr))\|}. \tag{14}$$

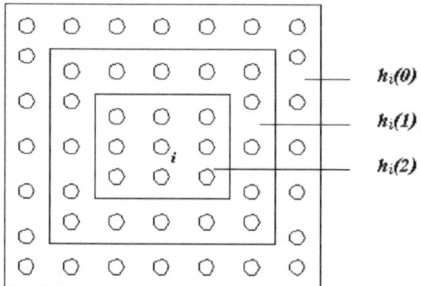

Fig. 4. Rectangular topological neighborhood of neuron i over itr

Here, η $(0 < \eta < 1)$ is the learning rate that determines how rapidly the system adjusts over itr. The learning rate parameter η also decreases as itr increases. The above updating procedure moves the weight vector $\mathbf{W}_i$ towards the input vector $\mathbf{U}$. Note that, the process does not change the length of $\mathbf{W}_i$, rather rotates $\mathbf{W}_i$ towards $\mathbf{U}$. So repeated presentation of training patterns tries to make the synaptic weight vectors tend to follow the distribution of the input vectors.

4.2 Network Architecture Used for Change Detection

In order to use a SOFM network for solving the change-detection problem by exploiting both image radiometric properties and spatial-contextual information, we assign a neuron in the output layer of the network corresponding to each spatial position (m, n) in D. The difference image D is generated by following the technique as described in Sect. 2. The spatial correlation between neighboring pixels is modeled by generating the input patterns corresponding to each pixel in the difference image D, considering its spatial neighborhood systems N of order d. In the present study we used a SOFM network where the number of neurons in the output layer is equal to the number of pixels in the difference image D and the number of neurons in the input layer is equal to the dimension of the input patterns.

Let $\mathbf{U}_{mn}$ and $\mathbf{W}_{mn}$, respectively be the y-dimensional input and weight vectors corresponding to the neuron (m, n) located at mth row and nth column of the output layer, i.e., $\mathbf{U}_{mn} = [u_{mn,1}, u_{mn,2}, \ldots, u_{mn,y}]$ and $\mathbf{W}_{mn} = [w_{mn,1}, w_{mn,2}, \ldots, w_{mn,y}]^T$. Note that in the feature mapping algorithm, the maximum length of the input and weight vectors are fixed. Let the maximum length for each component of the input vector be unity. To keep the value of each component of the input less than or equal to 1, let us apply a mapping function f where

$$f : [c_{min}, c_{max}] \rightarrow [0, 1].$$

Here c_{min} and c_{max} are the global lowest and highest component (feature) values present in the input vectors. The initial component of the weight vectors are chosen randomly in [0,1].

4.3 Learning of the Weights

The dot product $x(m, n)$ of $\mathbf{U}_{mn}$ and $\mathbf{W}_{mn}$ is obtained as

$$x(m, n) = \mathbf{U}_{mn}.\mathbf{W}_{mn} = \sum_{k=1}^{y} u_{mn,k}.w_{mn,k}. \qquad (15)$$

Now only those neurons for which $x(m, n) \geq t$ (here t is assumed to be a predefined threshold) are allowed to modify (update) their weights along with their neighbors (specified by $h_{mn}(.)$). Consideration of a set of neighbors enables one to grow the region by including those which might have been dropped out because of the initial randomness of weights (i.e., if the assigned weights are such that $x(m, n) \geq t$ for a few (m, n), then the weights of these neurons and their neighboring neurons will also be updated and subsequently categorized into changed region, if they originally were from changed region). The weight updating procedure is performed using (14). The value of learning rate parameter η and the size of the topological neighborhood $h_{mn}(.)$ decreases over itr. To check the convergence, total output $O(itr)$ for each epoch number itr is computed as follows:

$$O(itr) = \sum_{x(m,n) \geq t} x(m, n). \tag{16}$$

The updating of the weights continues until $|O(itr) - O(itr-1)| < \delta$, where δ is a preassigned small positive quantity. After the network is converged, the pixel at spatial position (m, n) in D is assigned to *changed region* if $x(m, n) \geq t$ else to *unchanged region*. The network converges for any value of t (proof is available in [38]). The above-mentioned learning technique is described below algorithmically (Table 2).

Note that, unlike the conventional SOFM, in the present network instead of giving the same input to all output neurons and finding out the *wining neuron*, here different input is given to different output neurons and weight updating is performed based on the assumed threshold t.

4.4 Threshold Selection Techniques

As stated in the above algorithm, the updating of weights depends on the threshold value t. Initially for a particular threshold t, the network is updated till convergence. After convergence, if $x(m, n) \geq t$, then make the output value

Table 2. Learning algorithm of the presented network

Step 1: Initialize each component of $\mathbf{W}_{mn}$ randomly (in [0,1]) for each output neuron (m, n).
Step 2: Set $itr = 0$. Initialize η and $h_{mn}(0)$.
Step 3: Set $O(itr) = 0$.
Step 4: Select an input vector $\mathbf{U}_{mn}$ and corresponding weight vector $\mathbf{W}_{mn}$;
Step 5: Compute their dot product $x(m, n)$ using (15).
Step 6: If $x(m, n) \geq t$ then update the weight vector $\mathbf{W}_{mn}$ along with the weight vectors of the neighboring neurons of the $(m, n)th$ neuron using (14) and set $O(itr) = O(itr) + x(m, n)$.
Step 7: Repeat Steps 4-6 for all input patterns, completing one epoch.
Step 8: If $(|O(itr) - O(itr - 1)| < \delta)$ then goto Step 11;
Step 9: $itr = itr + 1$.
Step 10: Decrease the value of η and $h_{mn}(itr)$ and goto Step 3.
Step 11: Stop.

V_{mn} of $+1$, else make it -1. Thus as in Hopfield model, the neurons of the output layer are also divided into two groups G_u and G_c and a change-detection map corresponding to threshold t is generated. By varying t (threshold values are varied by an amount $1/L$, where L is the maximum gray value of D), different change-detection maps are generated. The threshold selection technique for this network is the same as used in the Hopfield model (see Sect. 3.4). Note that in the case of Hopfield model the threshold value is varied by an amount of 1 to generate change-detection maps and here the threshold values are varied by an amount $1/L$. In the present network the energy value with threshold t is computed as

$$E(t) = -\sum_{m=1}^{p}\sum_{n=1}^{q}\sum_{(i,j)\varepsilon N_{mn}^2} V_{mn}.V_{ij} - \sum_{m=1}^{p}\sum_{n=1}^{q} V_{mn}^2.$$

5 Description of the Data Sets

In order to carry out an experimental analysis aimed at assessing the effectiveness of the proposed approach, we considered two multitemporal data sets corresponding to geographical areas of Mexico and Island of Sardinia, Italy. A detailed description of each data set is given below.

5.1 Data Set of Mexico Area

The first data set used for the experiment is made up of two multispectral images acquired by the Landsat Enhanced Thematic Mapper Plus (ETM+) sensor of the Landsat-7 satellite over an area of Mexico on 18th April 2000 and 20th May 2002. From the entire available Landsat scene, a section of 512×512 pixels has been selected as test site. Between the two aforementioned acquisition dates a fire destroyed a large portion of the vegetation in the considered region. Figure 5a,b shows channel 4 of the 2000 and 2002 images, respectively. In order to be able to make a quantitative evaluation of the effectiveness of the proposed approach, a reference map was manually defined (see Fig. 5d) according to a detailed visual analysis of both the available multitemporal images and the difference image (see Fig. 5c). Different color composites of the above-mentioned images were used to highlight all the portions of the changed area in the best possible way. Experiments were carried out to produce, in an automatic way, a change-detection map as similar as possible to reference map that represents the best result obtainable with a time consuming procedure.

Analysis of the behavior of the histograms of multitemporal images did not reveal any significant difference due to light and atmospheric conditions at the acquisition dates. Therefore, no radiometric correction algorithm was applied. The 2002 image was registered with the 2000 one using 12 ground control points. The procedure led to a residual average misregistration error on ground control points of about 0.3 pixels.

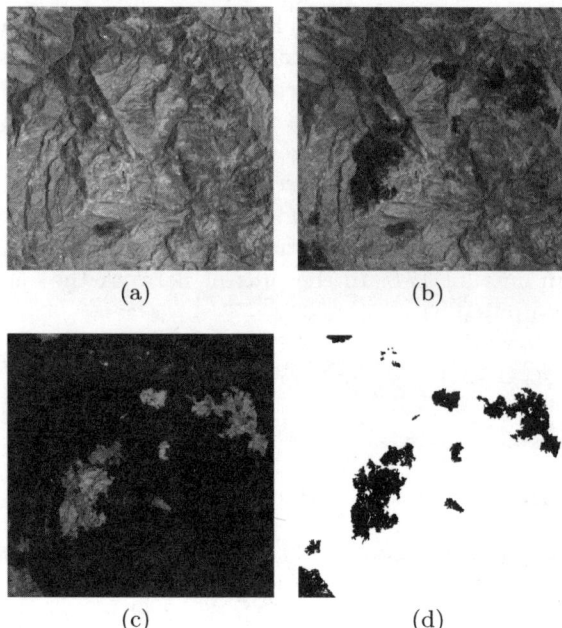

(a)   (b)

(c)   (d)

Fig. 5. Image of Mexico area. (**a**) Band 4 of the Landsat ETM+ image acquired in April 2000, (**b**) band 4 of the Landsat ETM+ image acquired in May 2002, (**c**) corresponding difference image generated by CVA technique, and (**d**) reference map of the changed area

5.2 Data Set of Sardinia Island, Italy

The second data set used in the experiment is composed of two multispectral images acquired by the Landsat Thematic Mapper (TM) sensor of the Landsat-5 satellite in September 1995 and July 1996. The test site is a section of 412 × 300 pixels of a scene including lake Mulargia on the Island of Sardinia (Italy). Between the two aforementioned acquisition dates the water level in the lake increased (see the lower central part of the image). Figure 6a,b shows channel 4 of the 1995 and 1996 images. As done for the Mexico data set, in this case also a reference map was manually defined (see Fig. 6d) according to a detailed visual analysis of both the available multitemporal images and the difference image (see Fig. 6c). As histograms did not show any significant difference, no radiometric correction algorithms were applied on the multitemporal images. The images were co-registered with 12 ground control points resulting in an average residual misregistration error of about 0.2 pixels on the ground control points.

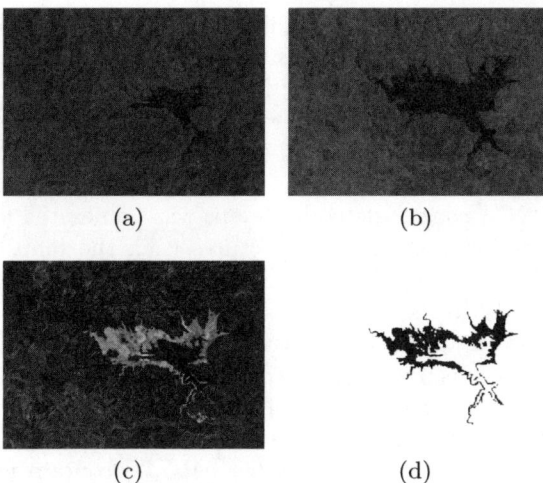

Fig. 6. Image of Sardinia island, Italy. (**a**) Band 4 of the Landsat TM image acquired in September 1995, (**b**) band 4 of the Landsat TM image acquired in July 1996, (**c**) difference image generated by CVA technique using bands 1, 2, 4, & 5; and (**d**) reference map of the changed area

6 Description of the Experiments

Two different experiments were carried out to test the validity of both the proposed neural network based techniques. The first experiment aims at assessing the validity of the proposed threshold selection criteria (as described in Sect. 3.4). To this end, the optimal threshold t_0 is chosen by a trial-and-error procedure where the change-detection maps (at convergence of the network) are generated by varying threshold t and computing the change-detection error corresponding to each threshold with the help of the reference map (please note that the reference map is not available in real situation). The threshold t_0 corresponds to the minimum change-detection error. The change-detection results obtained considering the threshold detected by the proposed criteria were compared with the change-detection results produced by assuming the optimal threshold t_0.

In order to establish the effectiveness of the proposed techniques, the second experiment compares the results (change-detection maps) provided by our method with a context-insensitive manual trial and error thresholding (MTET) technique [21] and a context-sensitive technique presented in [13] based on the combined use of the EM algorithm and MRFs (we refer to it as EM+MRF technique). The MTET technique generates a minimum error change-detection map under the hypothesis of spatial independence among pixels by finding a minimum error decision threshold for the difference image. The minimum error decision threshold is obtained by computing change-detection errors (with the help of the reference map) for all values of the decision threshold. Note that, this minimum error context-insensitive threshold is different from the context-sensitive optimal threshold t_0 as obtained in the first experiment. Comparisons were carried out in terms of both overall change-detection error and number of false alarms (i.e., unchanged pixels identified as changed ones) and missed alarms (i.e., changed pixels categorized as unchanged ones).

In case of HTNN, change-detection maps were generated for each data set by using two different network architectures (i.e., the maps were produced by considering both the discrete and continuous Hopfield networks with first-order neighborhood). In the continuous model, the activation function defined in (10) was used for updating the status of the network setting $r = 2$ in order to have a moderate steepness. It is worth noting that the change-detection process is less sensitive to variations of r.

In case of self-organizing feature map type neural network (SOFMTNN), the input vectors $\mathbf{U}_{mn}$ (for each pixel $(m, n) \in D$) contain nine components considering the gray value of the pixel (m, n) and the gray values of its eight neighboring pixels (here second-order neighborhood is considered). To map the value of each component of the input vectors in $[0, 1]$, the following formula is used

$$\frac{u_{mn,k} - c_{min}}{c_{max} - c_{min}}, \tag{17}$$

where $u_{mn,k}$ is the kth component value of input vector $\mathbf{U}_{mn}$ and c_{min} and c_{max} are the global lowest and highest component (feature) values present in the input vectors. The weight vector also have nine components. Initial weights are assigned randomly in $[0, 1]$. The learning rate parameter η is chosen as $\eta(itr) = \frac{1}{itr}$, i.e., the value of η at epoch itr is taken as $\frac{1}{itr}$. This ensures $0 < \eta < 1$ and it decreases with itr. Initial size of the topological neighborhood $h_{mn}(itr), \forall (m, n)$ was taken as a 11×11 rectangular window, and gradually reduced to 3×3 after five epochs and kept constant for the remaining epochs (until converges). The value of the convergence parameter δ is considered as 0.01.

6.1 Result Analysis: Mexico Data Set

First of all we performed some trials in order to determine the most effective spectral bands for detecting the burned area in the considered data set. On the basis of the results of these trials, we found that band 4 is more effective to locate the burned area. Hence we generated difference images by considering only spectral band 4.

In order to assess the validity of the proposed threshold selection criteria, in the first experiment a comparison is made between the optimal threshold t_0 (which was derived manually) and the thresholds t_1 detected by the criteria described in Sect. 3.4. As mentioned, in case of HTNN-based technique we considered two different network architectures (discrete and continuous with first-order neighborhood) and the initialization of each architecture was

Table 3. Change-detection results obtained by the proposed HTNN and SOFMTNN based techniques (Band 4, Mexico data set)

Proposed techniques		Initialization thresholds			Missed alarms	False alarms	Overall error
		Optimal (t_0)	Using correlation (t_1)	Using energy (t_1)			
HTNN	First order – discrete	34	–	–	1,252	1,640	2,892
		–	41	–	2,553	836	3,389
		–	–	33	1,102	1,802	2,904
	First order – continuous	34	–	–	1,035	1,554	2,589
		–	42	–	2,527	781	3,308
		–	–	31	660	2,157	2,817
SOFMTNN		0.216	–	–	1,406	1,573	2,979
		–	0.232	–	2,039	1,178	3,217
		–	–	0.183	583	2,929	3,512

carried out by considering the above-mentioned threshold values, t_0 and t_1. Table 3 reports the obtained change-detection results for both the proposed HTNN and SOFMTNN based techniques. From an analysis of the table, one can deduce that the proposed criteria detected proper threshold values as the automatically detected threshold t_1 is always near to the manually derived optimal thresholds t_0. Since energy-based criterion detected a threshold t_1 (considering both discrete and continuous Hopfield networks) which is closer to the optimal threshold t_0 ($t_0 = 34$), the overall error produced by the proposed HTNN technique based on energy criterion are close to the optimal results. Figure 7a shows the behavior of energy value with threshold for continuous HTNN considering first-order neighborhood. In case of SOFMTNN, as the correlation-based criterion selects higher threshold value ($t_1 = 0.232$), it generates higher missed alarms and lower false alarms as compared to energy based one ($t_1 = 0.183$). Since correlation-based criterion is able to detect a threshold t_1 which is closer to the optimal threshold t_0 ($t_0 = 0.216$), the overall error produced by the network using this criterion (3,217 pixels) is close to the optimal one (2,979 pixels). Figure 7b shows the behavior of correlation coefficient with threshold for SOFMTNN.

In the second experiment, the change-detection maps produced by the proposed techniques based on HTNN and SOFMTNN were compared with those obtained by the context-insensitive MTET procedure and the context-sensitive EM+MRF technique. Table 4 shows that the overall error obtained by the proposed techniques based on HTNN (considering the first-order continuous architecture with energy-based criterion) and SOFMTNN (considering correlation criterion) are much smaller than the overall error incurred

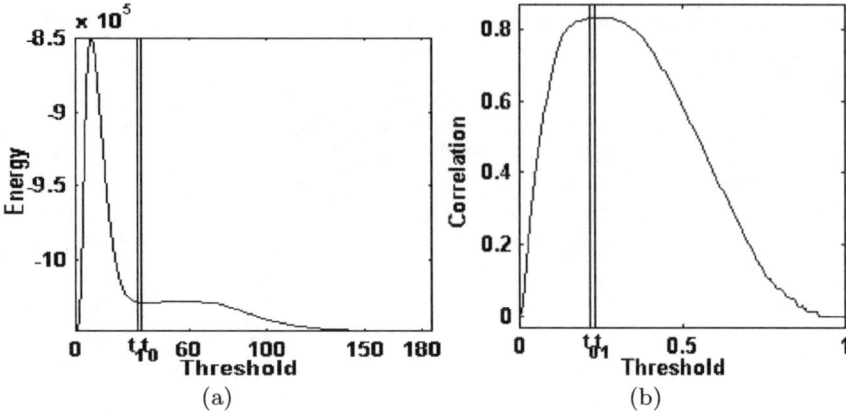

Fig. 7. Variation of (**a**) energy value with threshold considering continuous HTNN with first-order neighborhood, and (**b**) correlation coefficient with thresholds considering SOFMTNN (Band 4, Mexico data)

Table 4. Overall error, missed alarms and false alarms resulting by MTET technique, EM+MRF technique, proposed HTNN-based technique, and proposed SOFMTNN-based technique (Band 4, Mexico data set)

Techniques	Missed alarms	False alarms	Overall error
MTET	2,404	2,187	4,591
EM+MRF ($\beta = 1.5$)	946	2,257	3,203
HTNN (continuous with energy criterion)	660	2,157	2,817
SOFMTNN (with correlation criterion)	2,039	1,178	3,217

by the context-insensitive MTET technique. Table 4 also presents the best change-detection results obtained by the context-sensitive EM+MRF technique, when the parameter β of MRF [13] was set to 1.5 (this value was defined manually and corresponds to the minimum possible error). The overall change-detection error produced by the HTNN-based technique (2,817 pixels) is smaller than the minimum error yielded by EM+MRF technique (3,203 pixels) whereas the overall error produced by the SOFMTNN-based technique (3,217 pixels) is close to the error yielded by EM+MRF technique. It is also seen from the Tables 3 and 4 that the change-detection results obtained by the proposed techniques (considering both criteria) are comparable with the results obtained by the existing EM+MRF technique. Figure 8 shows the change-detection maps obtained by the two proposed context-sensitive neural network based techniques.

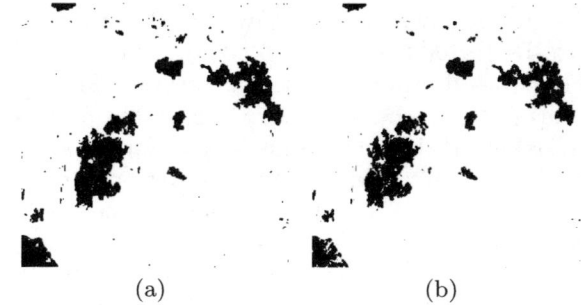

(a)   (b)

Fig. 8. Change-detection maps obtained for Mexico data by the proposed technique based on (**a**) HTNN (considering first-order continuous model with energy criterion), and (**b**) SOFMTNN (with correlation criterion)

Table 5. Change-detection results obtained by the proposed HTNN- and SOFMTNN-based techniques (Sardinia Island data set)

Proposed techniques		Initialization thresholds			Missed alarms	False alarms	Overall error
		Optimal (t_0)	Using correlation (t_1)	Using energy (t_1)			
HTNN	First order – discrete	90	–	–	862	614	1,476
		–	87	–	754	821	1,575
		–	–	86	667	1,065	1,732
	First order – continuous	90	–	–	1,060	455	1,515
		–	82	–	779	1,056	1,835
		–	–	90	1,060	455	1,515
SOFMTNN		0.368	–	–	1,090	558	1,648
		–	0.337	–	721	1,164	1,885
		–	–	0.356	935	729	1,664

6.2 Result Analysis: Sardinia Island Data Set

We applied the CVA technique to spectral bands 1, 2, 4, and 5 of the two multispectral images, as preliminary experiments show that the above channels contain useful information on the changes in water level.

In the first experiment, as in Mexico data set, here also we considered two HTNN models and the obtained change-detection results are reported in Table 5. By analyzing the results displayed in the table, one can deduce that, the proposed threshold selection criteria detect thresholds t_1 which are close to the optimal thresholds t_0. As discrete HTNN with correlation-based criterion detected a threshold t_1 ($t_1 = 87$) which is close to the optimal threshold

t_0 ($t_0 = 90$), the overall error produced by this HTNN-based model (1,575 pixels) is close to the optimal error (1,476 pixels). Further, as the continuous HTNN with energy-based criterion detected a threshold t_1 ($t_1 = 90$) same as the optimal threshold t_0, the overall error produced by this HTNN-based model (1,515 pixels) is equal to the optimal error. Figure 9a shows the variation of energy value with thresholds considering first-order continuous HTNN. Table 5 also shows the change-detection results obtained using SOFMTNN-based technique considering different threshold values. From the table it is shown that, the automatically detected thresholds t_1 (using both the criteria) are close to the optimal threshold t_0. As the energy-based criterion detects a threshold t_1 ($t_1 = 0.356$) which is more close to the optimal threshold t_0 ($t_0 = 0.368$), the produced overall error (1,664 pixels) is more close to the optimal one (1,648 pixels). Figure 9b shows the variation of energy value with thresholds considering SOFMTNN-based technique.

Concerning the second experiment, the change-detection results obtained using different techniques are shown in Table 6. From the table we can conclude that the overall change-detection error obtained by the proposed context-sensitive techniques based on HTNN and SOFMTNN are better than

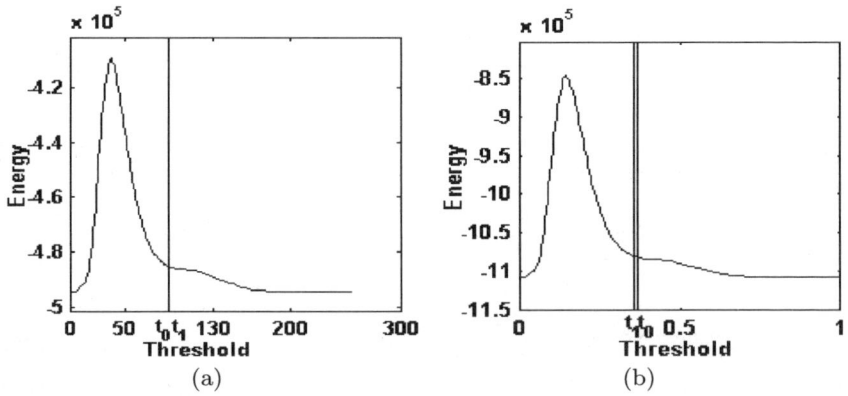

Fig. 9. Variation of energy value with threshold considering (a) continuous HTNN with first-order neighborhood, and (b) SOFMTNN (Sardinia Island data)

Table 6. Overall error, missed alarms and false alarms resulting by MTET technique, EM+MRF technique, proposed HTNN-based technique, and proposed SOFMTNN-based technique (Sardinia island data set)

Techniques	Missed alarms	False alarms	Overall error
MTET	1,015	875	1,890
EM+MRF ($\beta = 2.2$)	592	1,108	1,700
HTNN (continuous with energy criterion)	1,060	455	1,515
SOFMTNN (with energy criterion)	935	729	1,664

(a)   (b)

Fig. 10. Change-detection maps obtained for the data set related to the Island of Sardinia, Italy by the proposed technique based on (**a**) HTNN (first-order continuous model with energy criterion), and (**b**) SOFMTNN (with energy criterion)

the overall error produced by the MTET procedure. It is also seen that the proposed context-sensitive neural network based techniques provide better accuracy than the best result yielded by the context-sensitive EM+MRF technique. The proposed context-sensitive techniques based on HTNN (first-order continuous model) and SOFMTNN with energy criterion incurred an overall error of 1,515 pixels (1,060 missed alarms and 455 false alarms) and 1,664 pixels (935 missed alarms and 729 false alarms), respectively whereas the overall error for the context-sensitive EM+MRF approach was equal to 1,700 pixels (592 missed alarms and 1,108 false alarms). By analyzing the results shown in Tables 5 and 6 we can say that the change-detection results obtained by the proposed neural network based techniques (using any of the threshold selection criterion) are comparable with the best results obtained by existing EM+MRF based technique (with $\beta = 2.2$). Figure 10 shows the change-detection maps produced by both the proposed techniques.

7 Discussion and Conclusion

In this chapter two unsupervised context-sensitive techniques for change-detection in multitemporal remote-sensing images have been proposed. The first one is based on HTNN and the second one is based on SOFMTNN. Different network architectures are used to incorporate the spatial-contextual information of the difference image. In case of HTNN, the network architecture represents the structure of the difference image by associating a neuron to each pixel; and each neuron is connected only to its neighboring neurons. This allows us to incorporate the spatial-contextual information of the difference image. In case of SOFMTNN, the number of neurons in the input layer is equal to the dimension of the input patterns. In order to take into account the spatial-contextual information from neighborhood, the input patterns are generated considering each pixel in the difference image along with its neighboring pixels. Both the techniques require a proper threshold value to generate the change-detection map. For detecting this threshold, two robust threshold selection criteria are presented here.

The proposed techniques show the following advantages with respect to the context-sensitive change-detection method based on the EM+MRF [13]: (i) distribution free, i.e., does not require any explicit assumption on the statistical model of the distributions of classes of changed and unchanged pixels, (ii) completely automatic, i.e., the proposed techniques does not require the setting of any input parameters manually. It is to be noted that the EM+MRF technique requires the definition of the regularization parameter β that tunes the effect of the spatial-context information in the energy function to be optimized. Although both HTNN and SOFMTNN themselves have some parameters, but there is no need to set them manually.

Experimental results obtained on different real multitemporal data sets confirm the effectiveness of the proposed approaches. The neural network based approaches significantly outperform the standard optimal manual context-insensitive technique and provide an overall change-detection error comparable (sometimes better) to the one achieved with the context-sensitive EM+MRF technique.

Acknowledgements

The authors would like to thank the Department of Science and Technology, Government of India for sponsoring a project titled "Advanced Techniques for Remote Sensing Image Processing" and Prof. L. Bruzzone of University of Trento, Italy, for his comments on this chapter and providing the data.

References

1. A. Singh. Digital change detection techniques using remotely sensed data. *Int. J. Remote Sensing*, 10(6):989–1003, 1989.
2. J. A. Richards and X. Jia. *Remote Sensing Digital Image Analysis*. 4th ed. Berlin: Springer-Verlag, 2006.
3. J. Cihlar, T. J. Pultz and A. L. Gray. Change detection with synthetic aperture radar. *Int. J. Remote Sensing*, 13(3):401–414, 1992.
4. L. Bruzzone and S. B. Serpico. An iterative technique for the detection of land-cover transitions in multitemporal remote-sensing images. *IEEE Trans. Geosci. Remote Sensing*, 35:858–867, 1997.
5. L. Bruzzone and D. F. Prieto. An adaptive parcel-based technique for unsupervised change detection. *Int. J. Remote Sensing*, 21(4):817–822, 2000.
6. T. Hame, I. Heiler and J. S. Miguel-Ayanz. An unsupervised change detection and recognition system for forestry. *Int. J. Remote Sensing*, 19(6):1079–1099, 1998.
7. P. S. Chavez Jr. and D. J. MacKinnon. Automatic detection of vegetation changes in the southwestern United States using remotely sensed images. *Photogram. Eng. Remote Sensing*, 60(5):1285–1294, 1994.

8. K. R. Merril and L. Jiajun. A comparison of four algorithms for change detection in an urban environment. *Remote Sensing Environ.*, 63:95–100, 1998.
9. S. Gopal and C. Woodcock. Remote sensing of forest change using artificial neural networks. *IEEE Trans. Geosci. Remote Sensing*, 34(2):398–404, 1996.
10. F. Yuan, K. E. Sawaya, B. C. Loeffelholz and M. E. Bauer. Land cover classification and change analysis of the Twin cities (Minnesota) metropolitan area by multitemporal Landsat remote sensing. *Remote Sensing Environ.*, 98:317–328, 2005.
11. M. Alberti, R. Weeks and S. Coe. Urban land cover change analysis in Central Puget Sound. *Photogram. Eng. Remote Sensing*, 70(9):1043–1052, 2004.
12. S. Ghosh, S. Patra, M. Kothari and A. Ghosh. Supervised change detection in multi-temporal remote-sensing images using multi-layer perceptron. *ANVESA: J. Fakir Mohan Univ.*, 1(2):48–60, 2005.
13. L. Bruzzone and D. F. Prieto. Automatic analysis of the difference image for unsupervised change detection. *IEEE Trans. Geosci. Remote Sensing*, 38(3):1171–1182, 2000.
14. R. Wiemker. An iterative spectral-spatial Bayesian labeling approach for unsupervised robust change detection on remotely sensed multispectral imagery. In *Proceedings of CAIP*, pages 263–270, 1997.
15. M. J. Canty and A. A. Nielsen. Visualization and unsupervised classification of changes in multispectral satellite imagery. *Int. J. Remote Sensing*, 27(18):3961–3975, 2006.
16. M. Kothari, S. Ghosh and A. Ghosh. Aggregation pheromone density based change detection in remotely sensed images. In *Sixth International Conference on Advances in Pattern Recognition (ICAPR-2007), Kolkata, India.* World Scientific Publishers, pages 193–197, 2007.
17. S. Patra, S. Ghosh and A. Ghosh. Unsupervised change detection in remote-sensing images using one-dimensional self-organizing feature map neural network. In *Ninth International Conference Conf. on Information Technology (ICIT-2006), Bhubaneswar, India*, pages 141–142. IEEE Computer Society Press, 2006.
18. F. Melgani, G. Moser and S. B. Serpico. Unsupervised change-detection methods for remote-sensing data. *Opt. Eng.*, 41:3288–3297, 2002.
19. T. Kasetkasem and P. K. Varshney. An image change-detection algorithm based on Markov random field models. *IEEE Trans. Geosci. Remote Sensing*, 40(8):1815–1823, 2002.
20. L. Bruzzone and D. F. Prieto. An adaptive semiparametric and context-based approach to unsupervised change detection in multitemporal remote-sensing images. *IEEE Trans. Image Process.*, 11(4):452–466, 2002.
21. Y. Bazi, L. Bruzzone and F. Melgani. An unsupervised approach based on the generalized Gaussian model to automatic change detection in multitemporal SAR images. *IEEE Trans. Geosci. Remote Sensing*, 43(4):874–887, 2005.
22. S. Haykin. *Neural Networks: A Comprehensive Foundation.* Pearson Education, Fourth Indian Reprint, 2003.
23. S. Ghosh, L. Bruzzone, S. Patra, F. Bovolo and A. Ghosh. A context-aensitive technique for unsupervised change detection based on Hopfield type neural networks. *IEEE Trans. Geosci. Remote Sensing*, 45(3):778–789, 2007.
24. S. Patra, S. Ghosh and A. Ghosh. Unsupervised change detection in remote-sensing images using modified self-organizing feature map neural

network. In *Internationa Conference on Computing: Theory and Applications (ICCTA-2007), Kolkata, India.* IEEE Computer Society Press, pages 716–720, 2007.

25. T. Fung. An assessment of TM imagery for land-cover change detection. *IEEE Trans. Geosci. Remote Sensing,* 28:681–684, 1990.

26. D. M. Muchoney and B. N. Haack. Change detection for monitoring forest defoliation. *Photogram. Eng. Remote Sensing,* 60:1243–1251, 1994.

27. J. R. G. Townshend and C. O. Justice. Spatial variability of images and the monitoring of changes in the normalized difference vegetation index. *Int. J. Remote Sensing,* 16(12):2187–2195, 1995.

28. S. V. B. Aiyer, M. Niranjan and F. Fallside. A theoretical investigation into the performance of the Hopfield model. *IEEE Trans. Neural Netw.,* 1(2):204–215, 1990.

29. J. J. Hopfield. Neural networks and physical systems with emergent collective computational abilities. *Proc. Natl Acad. Sci. USA,* 79:2554–2558, 1982.

30. J. J. Hopfield. Neurons with graded response have collective computational properties like those of two state neurons. *Proc. Natl Acad. Sci. USA,* 81:3088–3092, 1984.

31. A. Ghosh, N. R. Pal and S. K. Pal. Object background classification using Hopfield type Neural Network, *Int. J. Pattern Recognit. Artif. Intell.,* 6(5):989–1008, 1992.

32. L. A. Zadeh. Fuzzy sets. *Information Control,* 8:338–353, 1965.

33. S. M. Ross. *Introduction to Probability and Statistics for Engineers and Scientists.* New York: Wiley, 1987.

34. A. Rosenfeld and P. De La Torre. Histogram concavity analysis as an aid in threshold selection. *IEEE Trans. SMC,* 13(3):231–235, 1983.

35. T. Kohonen. Self-organized formation of topologically correct feature maps. *Biol. Cybernet.,* 43:59–69, 1982.

36. T. Kohonen. *Self-Organizing Maps.* 2nd edn. Berlin: Springer-Verlag, 1997.

37. Z.-P. Lo, Y. Yu and B. Bavarian. Analysis of the convergence properties of topology preserving neural networks. *IEEE Trans. Neural Netw.,* 4:207–220, 1993.

38. A. Ghosh and S. K. Pal. Neural network, self-organization and object extraction. *Pattern Recognit. Lett.,* 13:387–397, 1992.

Fisher Linear Discriminant Analysis and Connectionist Model for Efficient Image Recognition

Manjunath Aradhya V N, Hemantha Kumar G, and Noushath S

Department of Studies in Computer Science, University of Mysore, India,
mukesh_mysore@rediffmail.com

Summary. Subspace analysis is an effective technique for dimensionality reduction, which aims at finding a low-dimensional space of high-dimensional data. Fisher linear discriminant analysis (FLD) and Neural Networks are commonly used techniques of image processing and recognition. In this paper, a new scheme of image feature extraction namely, the FLD and Generalized Regression Neural Networks (GRNN) approach for image recognition is presented. Linear discriminant analysis is usually performed to investigate differences among multivariate classes, to determine which attributes discriminate between the classes, and to determine the most parsimonious way to distinguish among classes. Two-dimensional linear discriminative features and GRNN are used to perform the classification. Experiments on the image database (Face, Object and Character) demonstrate the effectiveness and feasibility of the proposed method.

1 Introduction

The problem in object/image recognition is to determine which, if any, of a given set of objects appear in a given image or image sequence. Thus object recognition is a problem of matching models from a database with representations of those models extracted from the image luminance data. Early work involved the extraction of three-dimensional models from stereo data, but more recent work has concentrated on recognizing objects from geometric invariants extracted from the two-dimensional luminance data. Object recognition finds applications in the areas of machine/computer vision, industrial inspection, robotic manipulation, character recognition, face recognition, fingerprint identification, medical diagnosis, signature verification, etc. Feature selection for any representation is one of central solutions to any recognition systems. Among various solutions to the problem, the most successful seems to be those appearance-based approaches, which generally operate directly on images or appearances of objects and process the images as two-dimensional (2-D) holistic patterns. Appearance based methods avoid

Manjunath Aradhya V N et al.: *Fisher Linear Discriminant Analysis and Connectionist Model for Efficient Image Recognition*, Studies in Computational Intelligence (SCI) **83**, 269–278 (2008)
www.springerlink.com   © Springer-Verlag Berlin Heidelberg 2008

difficulties in modeling 3D structures of objects by considering possible object appearances under various conditions. When using appearance-based methods, we usually represent an image of size n-by-m pixels by a vector in an $n \times m$ dimensional space. However, these spaces are too large to allow robust and fast object recognition. A common war to resolve this problem is to use dimensionality reduction techniques. Feature extraction from data or a pattern is a necessary step in pattern recognition and can raise generalization of subsequent classification and avoid notorious curse of dimensionality. Among numerous feature extraction methods, Principal Component Analysis [1] and Fisher Linear Discriminant Analysis [2] are the most popular, relatively effective and simple methods and widely used in recognition and computer vision. It is generally believed that, when it comes to solving problems of pattern classification, LDA-based algorithms outperforms PCA-based ones, since the former optimizes the low-dimensional representation of the objects with focus on the most discriminant feature extraction while the latter achieves simply object reconstruction.

Many computer vision systems reported in the literature now employ the appearance-based paradigm for image recognition. One primary advantage of appearance-based methods is that it is not necessary to create representation or models for objects since, for a given object, its model is now implicitly defined by the selection of the sample images of the object. To further exploit the potentiality of PCA and FLD methods many contemporary methods have been proposed for the face/object recognition purpose and some of the related work on neural networks for character recognition are also proposed, which are described in Sect. 2. The rest of the paper is as follows: Some of the related work is reported in Sect. 2. In Sect. 3, we briefly introduce FLD technique. In Sect. 4, a brief GRNN is described. Experimental results and Comparative study are reported in Sect. 5. Finally conclusions are drawn at the end.

2 Related Work

A detection system based on principal component analysis (PCA) subspace or eigenface representation is described in [1]. Whereas only likelihood in the PCA subspace is considered in the basic PCA method, [3] also consider the likelihood in the orthogonal complement subspace; using that system, the likelihood in the image space is modeled as the product of the two likelihood estimates, which provide a more accurate likelihood estimate for the detection. Method by [4] presented a image feature extraction and recognition method coined two-dimensional linear discriminant analysis (2D-LDA). 2D-LDA provides a sequentially optimal compression mechanism, making the discriminant information compact into the up-left corner of the image. The idea of 2D-LDA is to perform image matrix LDA twice: the first one is in horizontal direction and the second is in vertical direction. After the two sequential image matrix LDA transforms, the discriminant information is compacted into the up-left corner of the image. A feature selection mechanism is followed to select the most discriminative features from the corner. Although these methods proved to be efficient in terms of both computational time and accuracy, a vital unresolved problem is that these methods require huge feature matrix for the representation. To alleviate this problem, the $2D^2LDA$ [5] method was proposed which gave the same or even higher accuracy than the 2DLDA method. The main idea behind the 2DLDA is that it is based on 2D matrices and essentially working in row direction of images. By simultaneously considering the row and column information, they described 2-Directional 2DLD, i.e. $2D^2LDA$ for representation and recognition.

Method by [6] proposed a system to use the traditional 3D Gabor filter and the 3D spherical Gabor filter for face recognition based on range and gray-level facial images. Gabor wavelets and general discriminant analysis for face identification and verification is proposed in [7]. They first discussed how to design Gabor filters empirically for facial feature extraction and demonstrate that the proposed novel Gabor + GDA framework is robust and uniform for both identification and verification. Second, they showed GDA outperforms other subspace projection techniques such as PCA, LDA, and KPCA.

A unified framework for subspace face recognition is described in [8]. They showed that PCA, LDA and Bayesian analysis are the most representative subspace approaches and unified under the same framework. A class-information-incorporated PCA (CIPCA) is presented in [9]. The objective of this is to sufficiently utilize a given class label in feature extraction and the other is to still follow the same simple mathematical formulation as PCA. They also discussed batch methods for CIPCA and PCA computation in there experiments, in practice the idea from the Hebbian-based neural networks to more effectively implement CIPCA with aiming at overcoming the batch methods.

Over the last two decades, Neural Networks have been widely used to solve complex classification problems [13]. On the other hand, there is a consensus in machine learning community that Support Vector Machines (SVM) is most promising classifiers due to their excellent generalization performance [14]. However, SVMs for multi-classes classification problems are relatively slow and their training on a large data set is still a bottle-neck. In [15], they proposed an improved method of handwritten digit recognition based on neural classifier Limited Receptive Area (LIRA) for recognition purpose. The classifier LIRA contains three neuron layers: sensor, associative and output layers. An efficient three-stage classifier for handwritten digit recognition is proposed in [10]. They propose efficient handwritten digit recognition based on Neural Networks and Support Vector Machine classifiers. Combining multiple classifiers based on third-order dependency for handwritten numeral recognition is presented in [16]. New approximation scheme is proposed to optimally approximate the probability distribution by the third–order dependency, and then multiple classifiers are combined using such approximation scheme. Eigen-deformations for elastic matching based on handwritten character recognition are presented in [17].

The objective of the FLD is to find the optimal projection so that the ratio of the discriminant of the between class and the within class scatter matrices of the projected samples reaches its maximum. Artificial neural networks, which are also referred to as neural computation, network computation, connectionist models, and parallel distributed processing (PDP), are massively parallel computing systems consisting of an extremely large number of simple processors with many interconnection between them. The success of FLD based method is one of the reasons that instigated us to further explore their potentiality by combining them with neural network for subsequent classification purpose. Hence, in this work we adopted generalized regression neural network (GRNN) for efficient and subsequent classification purpose.

3 The FLD Method

Feature extraction is the identification of appropriate measures to characterize the component images distinctly. There are many popular methods to extract features. Amongst which PCA and FLD is the state-of-the art methods for feature extraction and data representation technique widely used in the area of computer vision and pattern recognition. Using these techniques, an image can be efficiently represented as a feature vector of low dimensionality. The features in such subspace provide more salient and richer information for recognition than the raw image. Linear discriminant analysis [18] is usually performed to investigate differences among multivariate classes, to determine which attributes discriminate between the classes, and to determine the most parsimonious way to distinguish among classes. Similar to analysis of variance for single attribute, we could compute the within-class variance to evaluate the dispersion within class, and between-class variance to examine the differences between the classes. In this section we briefly explain FLD method for the sake of continuity and completeness. It is this success, which instigated us to investigate FLD for feature extraction and subsequent image classification. Steps involved in the feature extraction using FLD for a set of images are as follows.

Suppose that there are M training samples $A_k (k = 1, 2 \cdots, M)$, denoted by m by n matrices, which contain C classes, and the ith class C_i has n_i samples. For each training image, define the corresponding image as follows: Calculate the within class scatter matrix S_w for the ith class, a scatter matrix S_i is calculated as the sum of the covariance matrices of the centered images in that class.

$$S_i = \sum_{x \in x_i} (x - m_i)(x - m_i)^T \tag{1}$$

where m_i is the mean of the images in the class. The within class scatter matrix S_w is the sum of all the scatter matrices.

$$S_w = \sum_{i=1}^{c} S_i \tag{2}$$

Calculate the between class scatter matrix S_b : It is calculated as the sum of the covariance matrices of the differences between the total mean and mean of each class.

$$S_b = \sum_{i=1}^{c} n_i(m_i - m)(m_i - m)^T \tag{3}$$

Solve the generalized eigenvalue problem: Solve for the generalized eigenvectors (v) and eigenvalues (λ) of the within class and between class scatter matrices:

$$S_B V = S_W V \tag{4}$$

Keep first C-1 eigenvectors: Sort the eigenvectors by their associate's eigenvalues from high to low and keep the first C-1 eigenvectors W. For each sample Y in training set, extract the feature $Z = Y^T * W$ Then, use the generalized regression neural network for classification.

4 Generalized Regression Neural Network (GRNN)

Work on artificial neural networks, commonly referred to as "neural networks", has been motivated right from its inception by the recognition that the human brain computes in an entirely different way from the conventional digital computer [19]. Statistical prediction methods are widely used for many purposes: to predict tomorrow's weather, the sales of newly introduced product and various financial indices. Regression analysis is used to fit a smooth curve to a number of sample data points which represents some continuously varying phenomena. The generalized regression neural network was introduced to perform general (linear and non-linear) regression. As in the case of the Probabilistic Neural Network (PNN), it was also introduced by Specht [20]. Generalized regression neural networks are paradigms of the Radial Basis Function (RBF) used to functional approximation. To apply GRNN to classification, an input vector x (FLD projection feature matrix F_i) is formed and weight vectors W are calculated using (5). The output y_x is the weighted average of the target values T_k of training cases x_i close to a given input case x as given by:

$$y_x = \frac{\sum_{i=1}^{n} T_i W_i}{\sum_{i=1}^{n} W_i} \tag{5}$$

The architecture of the GRNN is shown in Fig. 1. It consists of an input-layer followed by three computational layers: pattern, summation and output layers.

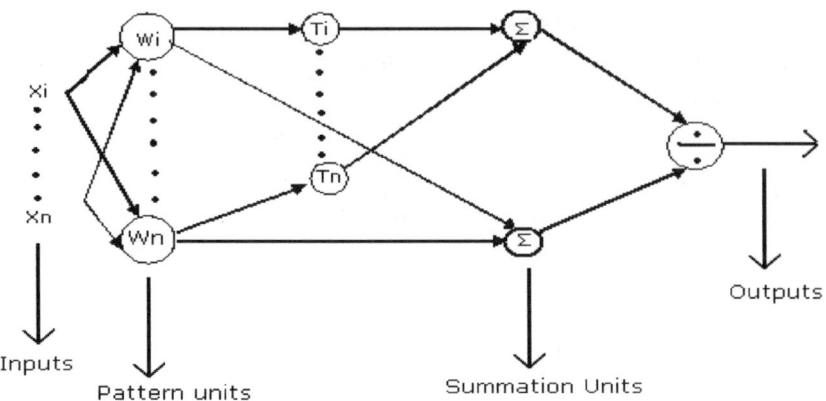

Fig. 1. Scheme of GRNN

The input layer is used to distribute the input patterns to the next layer, the pattern layer. The pattern unit layer has K units, one for each exemplar or one for each pattern cluster. The adjusted weights are wet equal to the exemplar patterns or to the cluster center values (centriod values). A detailed review of this method is described in [19].

Brief algorithm of the proposed method is shown below:
Algorithm: FLD Training
Input: A set of training sample images
Output: A knowledge Base (KB):
Method:

1. Acquire the training samples A_is
2. Apply FLD technique for feature extraction on set of training samples.
3. Store the feature matrix in KB

Algorithm FLD Training Ends
The corresponding testing algorithm is as follows:
Algorithm: FLD Testing
Input:
(i) The Knowledge Base:
(ii) Test Image, I.
Output: Class label / Identity of I.
Method

1. Obtain the feature matrix F of the pattern I
2. Apply GRNN method for subsequent classification
3. Label the class/Identity using step 2.

Algorithm FLD Testing Ends

5 Experimental Results

Each experiment is repeated 28 times by varying number of projection vectors t (where $t = 1...20, 25, 30, 35, 40, 45, 50, 100$, and 200). Since t, has a considerable impact on recognition accuracy, we chose the value that corresponds to best classification result on the image set. All of our experiments are carried out on a PC machine with P4 3GHz CPU and 512 MB RAM memory under Matlab 7.0 platform.

5.1 Image Databases

In order to test the proposed algorithm, experiments are made on face, object and character image databases. We carry out the experiments on ORL face database. In ORL database, there are 400 images of 40 adults, 10 images per person. Besides these variations images in ORL database also vary in facial details (with or without glasses) and head pose. The COIL-20 database contains 1,440 views of 20 objects where each object is represented by 72 views obtained by rotating each object through 3,600 in steps of 50. Finally, we conduct an experiment on printed and handwritten (English and Kannada) character database.

5.2 Results on ORL Database

The ORL database contains 112×92 sized 400 frontal faces: 10 tightly, cropped images of 40 individuals with variation in pose, illumination, facial expression (open/closed eyes, smiling/not smiling) and facial details (glasses/ no glasses). Some of the samples images of ORL face database are shown in Fig. 2. In our experiments, we split the whole database into two parts evenly.

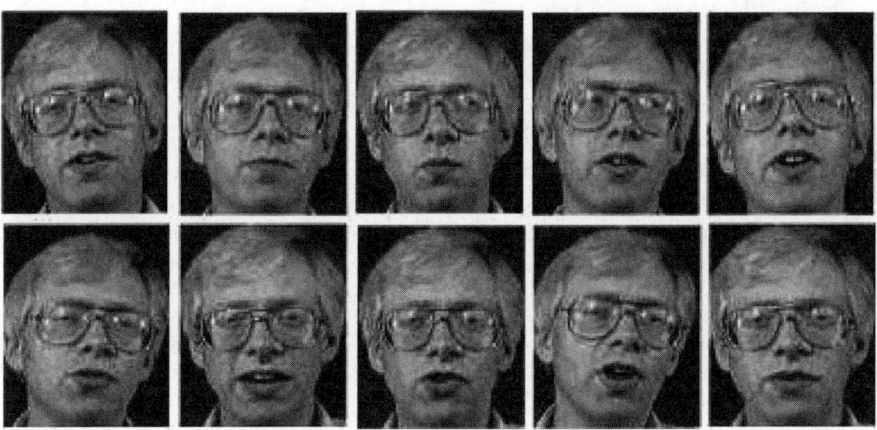

Fig. 2. Some Sample Images of the ORL Face Database

Table 1. Comparison of six methods on ORL

Methods	Accuracy (%)
PCA [1]	95.50
FLD [2]	94.50
2DPCA [11]	97.00
2DFLD [12]	98.00
A2DLDA [5]	98.00
$2D^2LDA$ [5]	98.00
FLD-GRNN	98.50

Table 2. Comparison of different approaches for varying number of training samples

Methods	Number of training samples (p)			
	2 (%)	4 (%)	6 (%)	8 (%)
PCA [1]	85.5	93.25	98.0	99.0
FLD [2]	86.0	94.25	98.0	99.0
2DPCA [11]	89.75	95.75	98.5	99.25
2DFLD [12]	90.25	95.25	98.5	99.50
A2DLDA [5]	88.75	95.25	98.25	99.75
$2D^2LDA$ [5]	90.25	96.00	98.75	99.75
FLD-GRNN	91.00	95.50	98.50	99.75

First five images of each class are used for training and the rest of the five images are used for testing. This experiment is repeated 28 times by varying projection vectors d (where $d = 1 \ldots 20, 25, 30, 35, 40, 45, 50, 100$, and 200). Since d, the number of projection vectors, has a considerable impact on different algorithms, we chose the value that corresponds to the best classification result on the image set. Table 1 gives the comparisons of six methods on top recognition accuracy. It can be found from Table 1 that the top recognition accuracy of proposed FLD-GRNN method is comparable with other methods.

To make full use of the available data, we randomly select p images from each subject to construct the training data set, the remaining images being used as the test images. To ensure sufficient training a value of at least 2 is used for p. Table 2 shows the top recognition accuracy achieved by all the methods for varying number of training samples. The superiority of the proposed methods in terms of accuracy used over their counterparts and other discrimination methods are quite evident from Table 2.

5.3 Results on the COIL-20 Object Database

In this section, we experimentally evaluate the performance of proposed method with PCA and FLD based methods on COIL-20 database, a standard

Fig. 3. Objects considered in COIL-20 database

Table 3. Comparison of different methods on COIL-20 database

Methods	Best Accuracy (%)
PCA [1]	93.89
FLD [2]	91.25
FLD-GRNN	94.25

object database commonly used by most researchers. The database contains 1,440 grayscale images of 20 objects. The objects were placed on a motorized turntable against a black background. The turntable was rotated through 360° to vary object pose with respect to fixed camera. Images of the objects were taken at pose intervals of 5°, which corresponds to 72 images per object. Figure 3 shows the sample objects considered from COIL-20 database. We have conducted a series of experiments by varying the number of training views (p) of objects. For comparative analysis of various approaches, we consider first 36 views of each object for training and remaining 36 views for testing. So, size of both the training and the testing database in this experiment is 720. The superiority in terms of accuracy of proposed method over other methods is quite evident from Table 3.

5.4 Results on Character Database

In this section, we empirically evaluated the performance of proposed method with character database consisting of around 30,000 samples. In order to conduct the experimentation, we considered both printed and handwritten characters. The following section shows the results obtained from the character data set.

Experimentation on Printed Kannada and English Characters

We evaluated the performance of the proposed system of the recognition process on printed Kannada and English character set. Subsets from this dataset were used for training and testing. The size of the input character is normalized to 50×50. Totally we conducted three different types of experimentation (1) Clear and degraded characters (2) font independence (3) noisy characters.

In the first experimentation, we considered the font –specific performance of Kannada and English, considering two fonts for each of these languages. Results are reported in Table 4. From Table 4 it is noticed that for clear and degraded characters the recognition rate achieved is 99.35%.

A font is a set of printable or displayable text characters in a specific style and size. Recognizing different font when the character class is huge is really interesting and challenging. In this experimentation we considered different font independent characters of Kannada and English and totally around 10,000 samples for font independence and achieved around 95.62%.

Noise plays a very important role in the filed of document image processing. To show the proposed method is robust to noise we conducted series of experimentation on noisy characters varying from noise density 0.01 to 0.5. For this, we randomly selected one character image from each class and generated 50 corresponding noisy images (here "salt and pepper" noise is considered). We noticed that the proposed system achieves 97.98% recognition rate and remained robust to noise by withstanding noise density upto 0.3. We also compared our proposed method with Eigencharacter (PCA) and Fishercharacter (FLD) based technique [1, 2]. Table 4 shows the performance accuracy of the PCA, FLD based technique with clear and degraded characters, Font independence, and Noisy characters. From Table 4 it is clear that the proposed technique perform well for all the conditions considered.

Table 4. Recognition rate of the proposed system and comparative study with PCA and FLD method

Experimental details	Language	Fonts	FLD-GRNN (%)	PCA (%)	FLD (%)
Clear And Degraded	Kannada	Kailasam			
	Kannada	Kailasam			
	English	New Roman	99.35	97.4	98.2
	English	Arial			
Font Independence	Kannada	N.A.	95.62	91.75	93.0
	Kannada	N.A.			
Noisy Data	Kannada	Kailasam			
	Kannada	Kailasam			
	English	New Roman	97.98	94.00	96.3
	English	Arial			

Experimentation on English Handwritten Characters

Handwritten recognition is an active topic in OCR application and pattern classification/learning research. In OCR applications, English alphabets recognition is dealt with postal mail sorting, bank cheque processing, form data entry, etc. For these applications, the performance of handwritten English alphabets recognition is crucial to the overall performance. In this work, we also extended our experiment to handwritten characters of English alphabets. For experimentation, we considered samples from 200 individual writers and total of 12,400 character set are considered. Some of the sample images of handwritten characters are shown in Fig. 4. We train the system by varying the number of training samples by 50, 75, 125, 175 and remaining samples of each character class are used during testing. Table 5 shows the best recognition accuracy obtained from the proposed system, PCA and FLD based techniques for varying number of samples. From Table 5 it is clear that by varying the number of training samples by 50, the highest recognition rate achieved is 58.26%. For 75 training samples with remaining 125 samples for testing, the highest recognition rate achieved is around 78.69%. Similarly for 125 samples of training and remaining 75 samples for testing, the highest recognition rate achieved is 90.15%. Finally for 175 training samples, the recognition rate achieved is 94.54%.

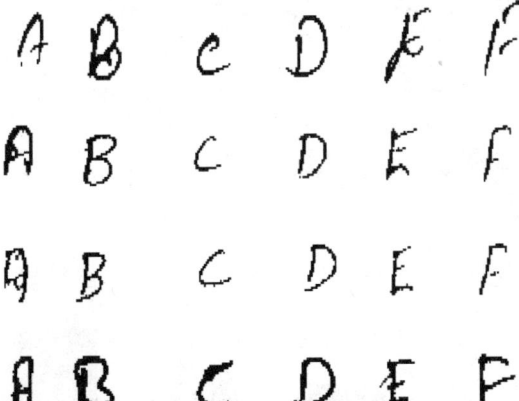

Fig. 4. Some of the sample images of English handwritten characters

Table 5. Best recognition accuracy for handwritten English characters

No. of training samples	FLD-GRNN (%)	PCA (%)	FLD (%)
50	58.26	57.5	57.5
75	78.69	76.7	76.9
125	90.15	86.0	86.1
175	94.54	90.8	91.1

Experimentation on Kannada Numeral Handwritten Characters

We also extended our experiment to handwritten Kannada numerals. For experimentation, we considered samples from 200 individual writers and total of 2,000 character set are considered. Some of the sample images of handwritten Kannada numerals are shown in Fig. 5. We train the system by 100 samples and remaining samples are used during testing. Figure 6 shows the best recognition accuracy obtained from the proposed system, PCA and FLD based technique. From Fig. 6 it is clear that the proposed method is performed well and achieved better accuracy compared to PCA and FLD based technique.

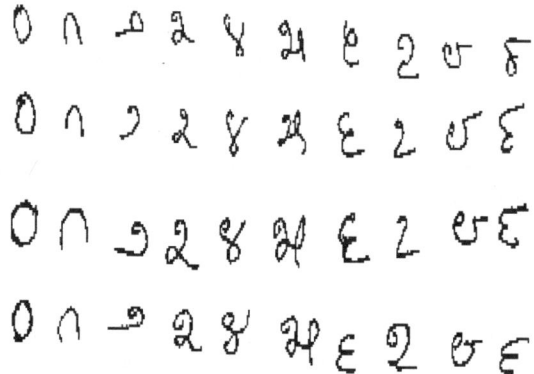

Fig. 5. Some of the sample images of Kannada handwritten characters

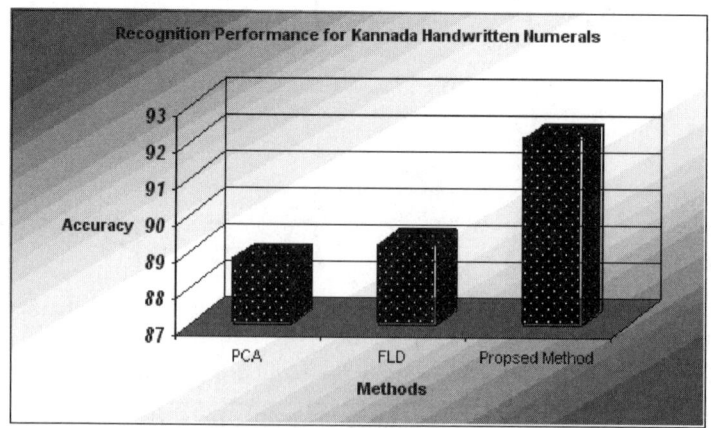

Fig. 6. Best recognition accuracy comparisons for various approaches

6 Discussion and Conclusion

FLD has recently emerged as a more efficient approach for extracting features or many pattern classification problems as compared to traditional PCA. FLD and PCA on the recognition problem were reported independently by numerous authors, in which FLD outperformed PCA significantly. These successful applications of FLD have drawn a lot of attention on this subject and the ensuing years have witnessed a burst of research activities to pattern recognition problems. In the same way, many researchers from various scientific disciplines are designing artificial neural networks (ANN's) to solve a variety of problems in decision making, optimization, prediction, and control. The GRNN was applied to solve variety of problems like predication, control and mapping problems and GRNN are paradigms of the Radial Basis Function (RBF) used to functional approximation. In the classification of genetic data, GRNN has performed better than Support Vector Machines (SVM). This success made us to further explore the potentiality of FLD and GRNN based methods to classification of different image domains. We tested our method on three different image domains namely, Face, Objects, and Character data set. Experiments on face and objects database, reveled that the proposed method outperforms other contemporary linear discriminant methods in all the aspects. Series of experiments are also conducted on printed and handwritten character dataset. Three types of experiments are conducted for printed characters and the proposed method worked well for all the condition. Test on English and Kannada handwritten characters are also showed and performed better compared to other methods. In future, there is a need to further explore the propose method for different face and object database. However, with respect to handwritten characters, there is still need to test on large standard database to further explore the proposed method.

References

1. Turk M, Pentland A (1991) Journal of Cognitive Neuroscience 3:71–86
2. Belhumeur P N, Hespanha J P, Kriegman D J (1997) IEEE Transactions on Pattern Analysis and Machine Intelligence 19:711–720
3. Moghaddam B, Pentland A (1997) IEEE Transactions on Pattern Analysis and Machine Intelligence 7:696–710
4. Yang J, Zhang D, Yong X, Yang J.-Y (2005) Pattern Recognition 38:1125–1129
5. Noushath S, Hemantha Kumar G, Shivakumara P (2006) Pattern Recognition 39:1396–1400
6. Wang Y, Chua C.-S (2005) Image and Vision Computing 23:1018–1028
7. Shen L, Bai L, Fairhurst M (2006) Image and Vision Computing 23:30–50
8. Wang X, Tang X (2004) IEEE Transaction on PAMI 26:1222–1227
9. Chen S, Sun T (2005) Neurocomputing 69:216–223
10. Gorgevik D, Cakmakov D (2004) International Conference on Pattern Recognition (ICPR) 45–48

11. Xiong H, Swamy M N S, Ahmad M O (2005) Pattern Recognition 38:1121–1124
12. Yang J, Zhang D, Frangi A F, Yang J (2004) IEEE Transaction on PAMI 26:131–137
13. Bishop C M (1995) Neural Networks for Pattern Recognition. In: Clarendon, Oxford
14. Burges C (1998) Knowledge Discovery and Data Mining 2:1–47
15. Kussul E, Baidyk T (2004) Image and Vision Computing 22:971–981
16. Kang H J (2003) Pattern Recognition Letters 24:3027–3036
17. Uchida S, Sakoe H (2003) Pattern Recognition 36:2031–2040
18. Fisher R A (1936) Annals of Eugenics 7:179–188
19. Patterson D W (1995) Artificial Neural Networks. In: Prentice Hall, Englewood Cliffs, NJ
20. Specht A (1991) IEEE Transactions on Neural Networks 2:568–576

Detection and Recognition of Human Faces and Facial Features

Miloš Oravec, Gregor Rozinaj, and Marian Beszédeš

Department of Telecommunications, FEI STU Bratislava, Ilkovičova 3, 812 19 Bratislava, Slovakia, oravec@ktl.elf.stuba.sk, gregor@ktl.elf.stuba.sk, beszedes@ktl.elf.stuba.sk

Summary. The area of face recognition, face detection, and facial features detection is considered. We discuss both linear and nonlinear feature extraction methods from face images. Second-order feature extraction methods are presented. Higher-order and nonlinear methods of feature extraction as well as feature extraction by kernel methods are also considered. Relevant types of classifiers are introduced. We illustrate presented methods by results of several face recognition systems. Next, we propose description of preprocessing techniques that can improve the performance of face recognition methods. Normalization of face position, size and rotation are crucial tasks during the preprocessing phase. Due to these tasks precise face detection techniques that use neural networks are proposed. First, skin color (SOM, MLP) and face detection (MLP) systems descriptions are given. Then, comparison with other detection systems is given as well. The last part is devoted to several methods for facial feature detection.

1 Introduction

In this chapter, we consider the area of face recognition, face detection, and facial features detection. We show the importance of neural networks in this area. At the same time, we discuss and compare neural and nonneural (standard) approaches and methods.

The area of face recognition is closely related to biometrics. Biometrics deals with automated procedures of recognizing a person that is based on physiological or behavioral characteristic [1]. Biometrics is important in order to achieve security in an open society. Face recognition is used both in unimodal and multimodal biometric systems (systems using single or multiple biometric).

Neural networks offer several advantageous properties [4]. They are characterized by massive parallel and distributed structure and by ability to learn and generalize. Neural networks can implement nonlinearity easily, they can implement input–output mappings, neural network are adaptive systems and they are fault tolerant.

M. Oravec et al.: *Detection and Recognition of Human Faces and Facial Features*, Studies in Computational Intelligence (SCI) **83**, 283–301 (2008)
www.springerlink.com

Output of a face recognition system is the identity of the input subject. While it is possible to recognize face images directly, face recognition systems usually consist of more stages: image preprocessing, feature extraction, and classification. From general point of view, image preprocessing deals with digital image processing procedures such as image resampling, histogram equalization, color balance, etc. Other important procedures are connected to localization of a face in an image, normalization of face to scale, light condition, rotation, etc. When we do not count the image preprocessing, we can divide pattern recognition systems into two classes – one and two-stage pattern recognition systems. One-stage pattern recognition system classifies input data directly. Two-stage pattern recognition system consists of feature extractor, followed by some form of classifier. The process of feature extraction transforms original data (of dimensionality p) to a feature space, where it is possible to reduce dimensionality of transformed data ($m < p$).

2 Feature Extraction

Feature extraction transforms original space into a feature space. Dimensionality reduction is one of the required results of this process. Retained features contain most important information from original image and they are useful for image recognition and also for image compression. At first, we present linear second-order feature extraction methods. Their goal is to find exact, faithful representation of data from the point of view of mean-square error of image reconstruction [2]. Principal component analysis (PCA), linear discriminant analysis (LDA), linear PCA neural networks, and linear multilayer perceptron (MLP) are examples of such methods [3, 4].

2.1 Principal Component Analysis

Principal component analysis PCA [3–6] is a standard statistical method used for feature extraction. It transforms input data represented by a random vector $\mathbf{x} = [x_0, x_1, x_2, \ldots, x_{p-1}]^T$, $E[x] = 0$, with correlation matrix $\mathbf{R_x} = E[\mathbf{xx}^T]$ to a set of coefficients (principal components) represented by the vector $\mathbf{a} = [a_0, a_1, a_2, \ldots, a_{p-1}]^T$. Transform matrix $\mathbf{U} = [\mathbf{u}_0, \mathbf{u}_1, \mathbf{u}_2, \ldots, \mathbf{u}_{p-1}]^T$ consists of unit vectors $\mathbf{u}_j = [u_{j0}, u_{j1}, u_{j2}, \ldots, u_{jp-1}]^T$ that are eigenvectors of the correlation matrix $\mathbf{R_x}$ associated with the eigenvalues $\lambda_0, \lambda_1, \lambda_2, \ldots, \lambda_{p-1}$, where $\lambda_0 > \lambda_1 > \cdots > \lambda_{p-1}$ and $\lambda_0 = \lambda_{max}$. The most important eigenvectors are those corresponding to largest eigenvalues of $\mathbf{R_x}$.

The representation of input data (analysis, forward transform) is defined by

$$\mathbf{a} = \mathbf{x}^T \mathbf{U} = \mathbf{U}^T \mathbf{x} \tag{1}$$

and synthesis (inverse transform) is represented by

$$\mathbf{x} = \mathbf{U}\mathbf{a} = \sum_{j=0}^{p-1} a_j \mathbf{u}_j. \tag{2}$$

It is possible to represent input data by a reduced number of principal components (dimensionality reduction). The transform uses eigenvectors corresponding to largest eigenvalues of $\mathbf{R_x}$, and those corresponding to small eigenvalues are discarded

$$\mathbf{x}' = \sum_{j=0}^{m-1} a_j \mathbf{u}_j, m < p. \tag{3}$$

Then vector $\mathbf{x}'$ is an approximation of $\mathbf{x}$, while $\lambda_0 > \lambda_1 > \cdots > \lambda_{m-1} > \lambda_m > \cdots > \lambda_{p-1}$. Many neural models performing PCA were published, they are summarized in [5]. They can be divided into two categories: Algorithms based on Hebbian learning rule and algorithms using least squares learning rules, for instance backpropagation. Generalized Hebbian algorithm (GHA) based on linear neural network [10] and adaptive principal component extraction (APEX), containing also lateral connections [11] are examples of former. Multilayer perceptron trained by backpropagation algorithm working in autoassociative mode is the example of latter.

2.2 Feature Extraction by Neural Networks and Nonlinear PCA

Multilayer perceptron (MLP) consists of artificial neurons that compute the weighted sum of the inputs plus the threshold weight and pass this sum through the activation function [4]. Often a special form of sigmoidal (nonconstant, bounded, and monotone-increasing) activation function – logistic function is used. In a multilayer perceptron, the outputs of the units in one layer form the inputs to the next layer. The weights of the network are usually computed by the backpropagation (BP) algorithm.

Now we consider MLP from feature extraction point of view. Figure 1 (left) shows multilayer perceptron with one hidden layer [12, 13]. It contains p input, p output neurons and m hidden neurons, where $m < p$. It works in autoassociative mode, i.e. input pattern is associated to itself. The goal of such network is to reproduce the input. Since hidden layer contains less neurons than input or output layers, output of the network will be approximation of input rather than its exact copy. Hidden layer neurons represent features that output layer uses for reconstruction of input data. MLP working in such mode reduces dimension of original data by means of its hidden layer.

Operation of the MLP that contains linear neurons only is closely related to linear PCA. Input data are projected onto linear combinations of first m eigenvectors of the correlation matrix. In contrast to PCA, variances of hidden layer outputs are approximately equal and contribution of each hidden neuron to reconstruction error is distributed uniformly.

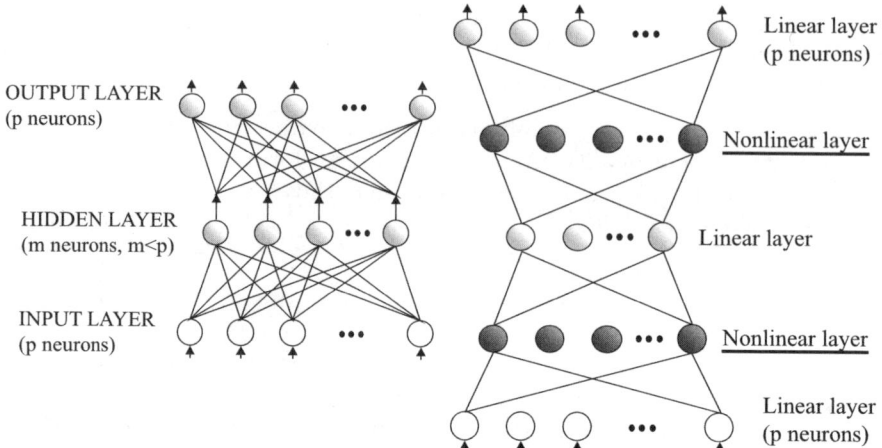

Fig. 1. MLP in autoassociative mode for dimensionality reduction (*left*) and nonlinear MLP performing nonlinear PCA (*right*)

As a next step, we are concerned with nonlinear methods of feature extraction, where nonlinear PCA using nonlinear MLP is probably the best-known example. When an MLP contains nonlinear neurons (i.e. the activation function is nonlinear) it is not further possible to speak about linear PCA. In [14], it was shown that nonlinear PCA in such form does not have an advantage compared to linear networks and reconstruction of input data by such network is comparable to linear network.

But this does not mean that nonlinear PCA is useless. MLP of configuration shown in Fig. 1 (right) working in associative mode is presented in [15]. It was used in first stage of face recognition system (feature extraction stage). Its first hidden layer and third hidden layer are nonlinear, while remaining layers are linear. Such network is capable of nonlinear PCA computation and it performs better compared to linear network.

2.3 Data Representation and Classification and Linear Discriminant Analysis

From the pattern classification point of view PCA belongs to methods based on unsupervised techniques. They take no account of the target data, and thus can give results which are not optimal [3]. Dimensionality reduction involves a loss of information and sometimes this information could be important for classification, although it is not very important for representation of input data.

Fisher's linear discriminant (FLD) [3, 7, 8] tries to shape the scatter of all projected samples in order to make it more reliable for classification. It is based on the principal generalized eigenvector of symmetric matrices $\mathbf{S}_B$

(between-class scatter matrix) and $\mathbf{S}_W$ (within-class scatter matrix). They are defined by

$$\mathbf{S}_B = \sum_{i=1}^{c} N_i (\mu_i - \mu)(\mu_i - \mu)^T, \tag{4}$$

$$\mathbf{S}_W = \sum_{i=1}^{c} \sum_{\mathbf{x}_k \in X_i} (\mathbf{x}_k - \mu_i)(\mathbf{x}_k - \mu)^T, \tag{5}$$

respectively, where c is the number of classes, N_i is the number of samples in class X_i and μ_i is the mean image of class X_i. The transform matrix $\mathbf{U}$ is chosen to maximize the ratio of the determinant of the between-class scatter matrix of the projected samples to the determinant of the within-class scatter matrix of the projected samples. In [8], the columns of $\mathbf{U}$ are the eigenvectors of $\mathbf{S}_W^{-1}\mathbf{S}_B$. The singularity of $\mathbf{S}_W$ is the problem in face recognition, since the number of sample images N is much smaller than the number of pixels n in each image (so called small-sample size problem). One possible solution is called Fisherfaces, it firstly uses PCA to reduce the dimension of the feature space, and then standard FLD is applied to reduce the dimension further.

It is generally believed that algorithms based on linear discriminant analysis (LDA) are superior to those based on PCA. LDA is insensitive to large variation in lighting direction [7, 8] and facial expression [7]. In [9], however, authors show that this is not always true. Their conclusion is that when the training data set is small, PCA can outperform LDA, and also that PCA is less sensitive to different training data sets.

2.4 Higher-Order Feature Extraction Methods

Suitable data representation is a key problem in data analysis, recognition, compression, visualization, etc. For simplicity, such representation is done by linear transforms. Second-order methods are very popular – they use information contained in data correlation matrix. This is appropriate for Gaussian signals. Example of such methods is PCA.

The second-order methods try to find a faithful representation of the data (from the mean-square error point of view). Higher-order methods try to find a meaningful representation [2].

Higher-order methods use information that is not contained in covariance matrix. Such information is contained in non-Gaussian signals. An example of higher-order methods is projection pursuit [16], which can be used also for dimensionality reduction. Independent component analysis is another example. Independent component analysis ICA [17,18] is an extension of standard PCA, it can be applied also for feature extraction purposes. Estimating ICA basis vectors and their use in signal processing is more complicated task, since they are not orthogonal. Complete ICA can be done using multilayer feedforward neural networks [18].

2.5 Kernel-Based Learning Methods

At present, so called kernel methods are intensively studied and became very popular [19–23]. Kernel-based principal component analysis KPCA, kernel-based linear discriminant analysis KLDA (or kernel Fisher discriminant analysis KFDA), kernel radial basis function networks KRBF or support vector machines SVM are examples of kernel methods. KPCA and KLDA show better results in face recognition than linear subspace methods [20].

Kernel algorithms map data $\mathbf{x}_1, \mathbf{x}_2, \ldots, \mathbf{x}_n \in R^p$ from original space $\mathbf{x}$ into higher dimensional feature space F using nonlinear mapping Φ [19]

$$\Phi : R^p \to F, \mathbf{x} \to \Phi(\mathbf{x}). \tag{6}$$

For the learning in feature space, the original algorithm from original space is used. But high-dimensional spaces increase also the difficulty of a solved problem. However, learning in high-dimensional feature space can be simpler, if one uses simple class of linear classifiers. For certain feature space and corresponding mappings, a trick for computing scalar products in feature space exists - it consists in kernel functions. Computation of a scalar product between two feature space vectors can be done using kernel function k

$$\Phi(\mathbf{x}) \cdot \Phi(\mathbf{y}) = k(\mathbf{x}, \mathbf{y}). \tag{7}$$

It means that the feature space does not need to be computed explicitly. Only inner products in the kernel feature space have to be taken into account [20]. Most common kernel functions are Gaussian radial basis function, polynomial, sigmoidal, and inverse multiquadrics function. Every (linear) algorithm that only uses scalar products can implicitly be executed in high-dimensional feature space by using kernels. In such way, a nonlinear version of a linear algorithm can be constructed [19].

Considerable amount of papers using kernel methods for face recognition appeared, e.g. [20–23]. The computational complexity of kernel algorithms is high, but kernel methods outperform linear methods. KPCA and KLDA are often used for input images of dimensions 28×23 pixels [20, 21] or 80×80 pixels [23].

3 Classification

Result of classification in face recognition system is the identity of an input subject. Many different classifiers are used in face recognition systems, e.g. simple metrics of feature space as well as neural network classifiers. Multilayer perceptron (MLP) and radial-basis function (RBF) network [3,4] are popular neural network classifiers.

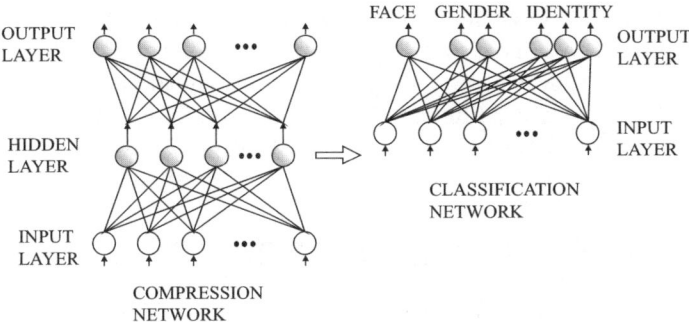

Fig. 2. Two-stage MLP face recognition system

3.1 Classification by Multilayer Perceptron

Frequently presented example of MLP in the role of both feature extractor and classifier (two-stage MLP system) is shown in Fig. 2 [24, 25]. Feature extraction stage is based on compression network shown in Fig. 1, classification stage takes hidden layer outputs of compression MLP and can classify them not only for identity purposes, but also gender, face emotions, etc. can be classified ("face" output neuron can discriminate between face and nonface images).

3.2 Classification by Radial-Basis Function Network

Radial basis function (RBF) network [4, 26, 27] is another example of feed-forward neural network used for classification purposes. It is based on a multivariable interpolation: Given a set of N distinct vectors $\{\mathbf{x}_i \in R^p \mid i = 1, \ldots, N\}$ and N real numbers $\{d_i \in R \mid i = 1, \ldots, N\}$, the aim is to find a function $f : R^p \rightarrow R$ satisfying the condition $f(\mathbf{x}_i = d_i, \forall i = 1, \ldots, N)$. RBF approach works with N radial basis functions (RBF) ϕ_i, where $\phi_i :$ $R^p \rightarrow R, i = 1, \ldots, N$ and $\phi_i = \phi(\|\mathbf{x} - \mathbf{c}_i\|)$, where $\phi : R_0^+ \rightarrow R$, $\mathbf{x} \in R^p$, $\mathbf{c}_i \in R^p$ are centers of RBFs. Centers are set to $\mathbf{c}_i = \mathbf{x}_i \in R^p, i = 1, \ldots, N$. Functions $\phi_i, i = 1, \ldots, N$ form the basis of a liner space and interpolation function f is their linear combination

$$f(\mathbf{x}) = \sum_{j=1}^{N} w_j \phi(\|\mathbf{x} - \mathbf{c}_j\|). \tag{8}$$

Interpolation problem is simple to solve, in contrast to approximation problem (there is N given points and n_0 functions ϕ, where $n_0 < N$), which is more complicated. Then it is a problem to set centers $\mathbf{c}_i, i = 1, \ldots, n_0$, also parameter σ of each RBF can be not the same for all RBFs. One possible solution for RBF approximation problem is a neural network solution. RBF network is a feedforward network consisting of input, one hidden, and output layer. Input layer distributes input vectors into the network, hidden layer represents RBFs. Linear output neurons compute linear combinations of their inputs. RBF network learning consists of more different steps (description of RBF network learning can be found in [26, 27]).

4 Examples of Face Recognition Systems

In order to illustrate described approaches, we present several face recognition systems [28] applied to the concrete face database of order of hundreds face images. We use the face database from Massachusetts Institute of Technology (MIT) [29], which consists of face images of 16 people (shown in Fig. 3), 27 of each person under various conditions of illumination, scale, and head orientation. It means, total number of face images is 432. Each image is 64 (width) by 60 (height) pixels, 8-bit grayscale. An example of different face images (patterns) belonging to the same class is shown at bottom of Fig. 3.

We use several different methods, they are shown in Fig. 4.

(a) One-stage systems: Direct classification of input face images by multilayer perceptron (MLP) and radial-basis function network (RBF), Fig. 4a. The configuration of MLP is $64 \times 60\text{-}16$ (i.e. 3,840 input neurons and 16 output neurons), MLP was trained on training face set containing 48 faces (those 16 shown on top of Fig. 3 plus other 32 images – two different scales of top of Fig. 3). MLP correctly classifies 78.12% of test faces, (300 succesfully recognized faces from total 384 test faces). The configuration of RBF network was $64 \times 60\text{-}48\text{-}16$ (48 training faces for RBF network classifier, which gives best results with 48 RBF neurons in hidden layer). RBF behavior is comparable to MLP – this network correctly classifies 78.12% of test faces. These results are shown as methods No. 2 and 3 in Fig. 5.

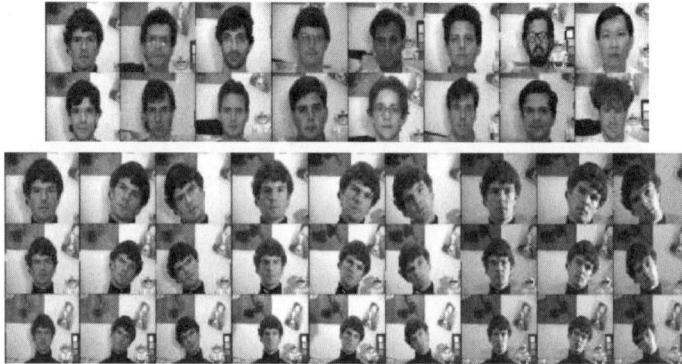

Fig. 3. Subjects in the face database (*top*) and examples of one subject (*bottom*)

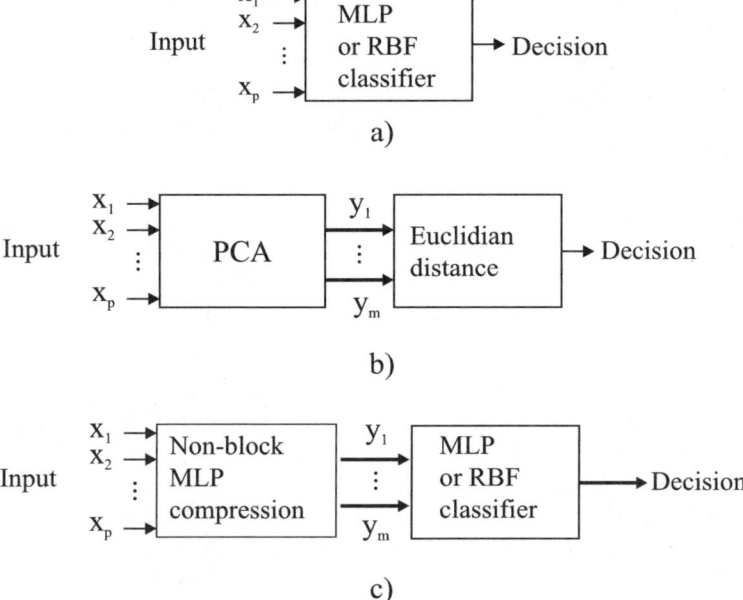

Fig. 4. Presented face recognition methods

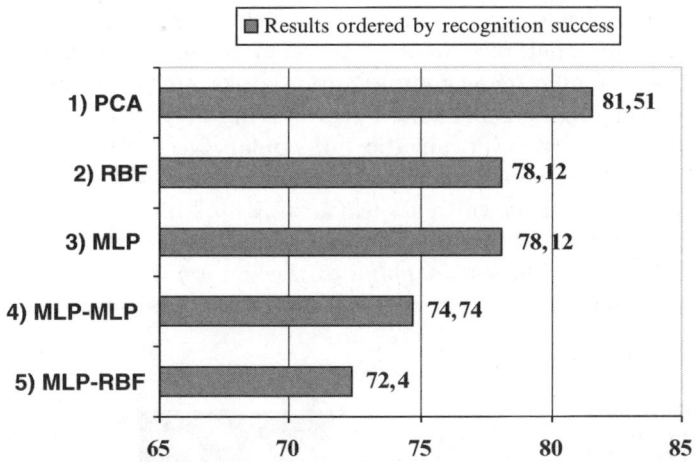

Fig. 5. Comparison of presented methods

(b) Two-stage systems: Feature extraction by PCA applied directly to face images where Euclidian distance is used as a classification measure, Fig. 4b. The correlation matrix was computed from 48 training faces (the same as previous method) and for classification first 48 eigenvectors of the correlation matrix are used. 81.51% of test faces was recognized successfully (313 from total 384). This result corresponds to method No. 1 in Fig. 5.

(c) In order to compare results of recognition using compression networks for feature extraction, we present also method shown in Fig. 4c. It consists of nonblock compression MLP working in autoassociative mode [12, 13] according to Fig. 1 (left) followed by MLP or RBF network classifier (principle is shown in Fig. 2). The configuration of compression MLP was 64×60-48-64×60. Hidden layer outputs serve as input to classification networks. Best classification results were obtained by MLP 48-16 (74.74%, i.e. 287 of 384 faces were recognized successfully) and RBF network 48-32-16 (72.40%, i.e. 278 of 384). These results correspond to methods No. 4 and 5 in Fig. 5.

In [28], two new methods using MLP and RBF networks are proposed, recognition success is 83.07 and 82.29%, respectively. Presented face recognition methods cover one- and two-stage recognition systems and they include feedforward neural networks both in the role of feature extractor and classifier. Image preprocessing for normalization of pose, illumination, distance from camera, etc. could improve recognition results.

5 Face Detection

As can be seen from Sect. 4 the task of face recognition is very complex and sophisticated methods have to be used to achieve sufficient successfulness. Even though these complex approaches are used, the recognition results are still often insufficient. Preprocessing steps can usually be used for improvement. Precise localization of human face and subsequent background elimination, scene light conditions normalization, face orientation and expression normalization, or facial features highlighting are just some of the normalization procedures that can be mentioned. Neural network approaches can be applied with success in many of these normalization procedures. In this section, we will describe normalization scheme given on Fig. 6.

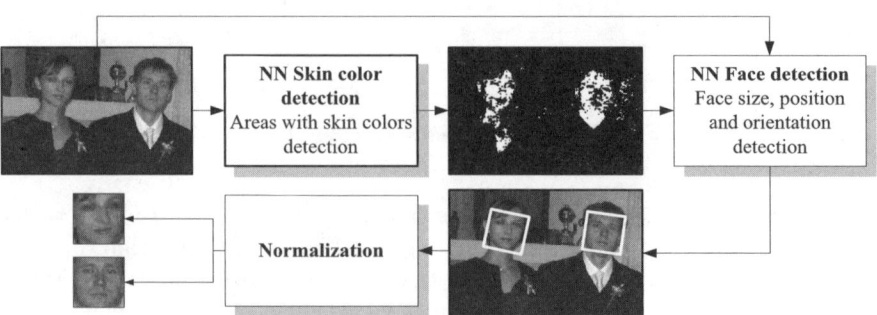

Fig. 6. Face recognition normalization scheme

5.1 Neural Network Skin Color Detection

Color detection is often used for its simplicity and speed as an important step in image or video processing methods like video indexing, object detection, object tracking, intelligent segmentation and compression. Over the past decade it has been proved in that color of human skin is distinctive enough from other object colors and thus can be powerful indicator for detecting people in unconstrained color images. Skin color detection modules are frequently used as the first stage of more sophisticated human detectors because they can mark out possible object locations accurately and very fast and hence eliminate the need for cumbersome scanning of the whole image area.

The problem of skin color detection can be defined as technique that classifies each pixel of input image independently on its neighbors as skin or nonskin using color information. This is often a demanding task because the color of the skin pixels in digital photo can vary with ambient light, properties of camera, or human race. The distribution of skin colors in RGB color space is given in Fig. 7a.

Skin color classification methods have two central tasks. The first task is to construct a (statistical) color model in suitable color space according to training set which represents skin/nonskin colors. When the size of training sets is not sufficient, sophisticated learning algorithms can be used to interpolate between samples. The second task is to use the constructed model for the decision making whether a pixel extracted from an image has skin color or not.

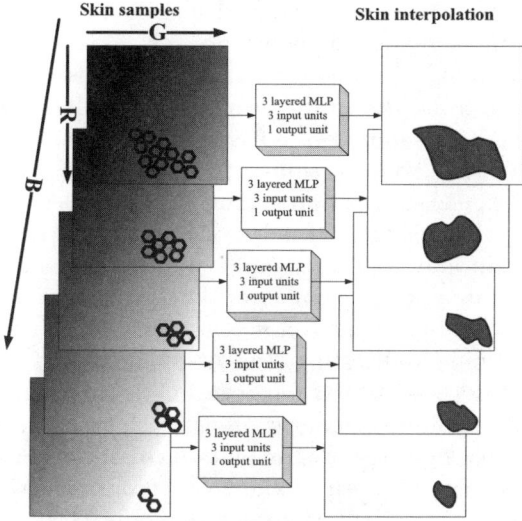

Fig. 7. (a) Skin color distribution in RGB color space. The bigger the dot the higher the frequency of RGB color in training set. (b) Neural network skin color RGB distribution interpolation

There are numerous color classification methods. Explicit defined, histogram, Bayesian, or Gaussian classifiers are probably the best known. More elaborate are multiple Gaussian clusters, dynamic and adaptive skin color distribution models and neural network based approaches (the color classification methods survey can be found in [32,40]). Next sections will aim reader's attention to self-organizing map (SOM) and multilayered perceptron (MLP) based on skin color detection methods.

SOM Skin Color Detection

Self-organizing map (SOM) has been devised by T. Kohonen [49] and has become one of the most popular and used unsupervised artificial neural network. It has been used to find patterns in unknown data and classify data with no restrictions on its dimensionality.

Brown, Crow and Lewthwaite use in [41] SOM for skin color detection in two ways. In the first case, a set of 30,000 skin pixels from landmarked face images is collected, subsequently randomly ordered list is formed. Finally, the set is sequentially presented to the SOM. After this training, given samples are classified by measuring their quantization error, which can be computed as Euclidean distance between the codebook vector of the winning neuron and the given sample. The fact that skin samples usually have smaller quantization error than nonskin samples is successfully applied here. The correct threshold for the classification is chosen as a final step of training procedure. Consequently, in the process of classification, pixel's color is declared to be skin if its quantization error is lower than chosen threshold. While the first SOM based skin color detector is trained just on skin samples, the training of second one requires in addition nonskin pixels. Skin and nonskin sets, both containing 15,000 samples, are used for training. Nonskin set is collected from randomly chosen Internet images, which were carefully checked to filter out images with parts of human bodies. After training on randomly ordered list of skin and nonskin pixels, which is similar to the first type of classifier, the calibration phase takes place. Labeled training data is presented to the SOM and the data label (skin/nonskin) and index of the winning neuron are recorded. The number of skin and nonskin samples won by the neuron is counted for each of them separately. If the number of skin/nonskin samples prevails, then each new sample won by the neuron is classified as skin/nonskin pixel.

During the training and testing four dissimilar color spaces (HSV, Cartesian SH, TSL, normalized RGB) for both SOMs are analyzed. All of these color spaces have an ability to remove intensity component which can improve the invariantness of a detector to illumination conditions in scenes. Moreover, hexagonal lattices using between 16 and 256 neurons are tested for both types of SOM and different threshold values for skin-only SOM.

Authors compared 300 configurations of SOM classifiers to various Gaussian-mixture model configurations. They conclude that in general, there is just a little difference in overall performance of detectors. This is due to the fact that over 94% of tested samples were classified correctly by the best configurations of all detector types. Although there is little variation between top performances, an important difference occurs when we look at the top 20–30 results for each detector. Skin-only SOM detector is observed to be the most consistent because it shows just 1% performance loss over its top 30 configurations while mixture model and skin/nonskin SOM shows 5% performance loss over their top 20 configurations. This suggests that skin-only SOM is less critical to parameter settings. Although skin/nonskin SOM is generally less consistent than skin-only SOM, it achieves the best results of all experiments (95.5%). In addition, it is threshold-free, it has small memory requirements, it is easily adaptive and even suitable for hardware implementation (AXEON learning processor). Both SOM detectors perform well with a variety of different SOM sizes (63–196 neurons), and the experiments show that they are invariant to the color space selection, unlike Gaussian-mixture models.

Multilayer Perceptron Skin Color Detection

Seow, Valaparla and Asari propose in [42] another neural network based approach for skin color detection. They describe a training process which has three stages: skin color collection, skin color interpolation, and skin color classification.

During the skin color collection stage they created data set containing 41,000 samples collected from various races. This data set is used to approximately describe the distribution of skin colors in RGB color space, the so-called "RGB skin color cube". While collecting images that contain skin colors is uncomplicated, it is much harder to get a representative sample of those, which do not, since many objects can have color similar to the skin. For this reason, an interpolation strategy is used in further steps.

To estimate the boundaries of real skin color cluster in RGB color space a MLP neural network with backpropagation learning algorithm is used. "RGB skin color cube" is first split into 256 RG slices where each slice belongs to specific B value. Labeled (skin/nonskin) RGB values from specific slice are then used to train neural network to classify the slice's RGB color values as skin/nonskin. 256 three-layered neural networks are trained in this manner to extract skin regions from the slices. The final step of the interpolation process is to get skin and nonskin labels for all colors in RGB color space using the 256 trained networks with trained optimum decision boundaries. The whole interpolation process can be seen in Fig. 7b. Using this labeled "RGB skin color cube" in subsequent learning steps eliminates the difficulties of finding the nonskin part of training samples. This is due to the fact that the interpolated data is considered as skin and rest of the color cube as nonskin.

Skin/nonskin labels for RGB colors retrieved from labeled "RGB skin color cube" are used in final stage of skin color classification training. Again, three-layered perceptron network using backpropagation learning algorithm is used.

This configuration is chosen because it is capable to span complex skin cluster boundaries.

Unfortunately, the authors do not give any statistical evaluation of the performance speed or classification accuracy. They just mention that the final classifier is used to segment still images and track faces in real-time video sequences in nonuniform lighting conditions and that computation time is reduced considerably compared to conventional techniques.

5.2 Neural Network Face Detection

The task of face detection can be defined as technique that determines the locations, rotation and sizes of all human faces in arbitrary digital images. In the last decade lots of approaches dealing with this nontrivial problem were published (exhaustive survey can be found in [33]). Since it can be treated as a two class pattern recognition problem, many different types of neural networks architectures have been proposed. The benefit of using neural networks in this area compared with other computer vision approaches, is the possibility of training a system to capture all variations of various face patterns. However, to get an acceptable performance the basic parameters related to neural networks (network type and topology, learning method, learning parameters) must be adjusted very accurately.

Probably the most famous neural network based approach for face detection is work of Rowley et al. [34]. Their system directly analyzes image intensities using multilayer perceptron, parameters of which are learned automatically from training examples. Trained neural network is then capable to distinct the face pattern and the nonface. This detection scheme is then invariant to position, size, partial rotation (up to $\pm 10°$), differences between individuals and number of faces in the scene. In the following years authors proposed an extension to this method using a neural network capable of detecting rotations of faces in scene (from $0°$ to $360°$) [43]. In further sections we will discuss this face detection system in details.

The most important parts of the proposed system, given in Fig. 8, are two neural networks processing intensity images of constant size (20×20 pixels). First one, termed "router", is used to estimate the angle of face given in input image. The image can be then "derotated" to make the face upright. If the input image does not contain a face pattern, then meaningless rotation angle is returned. The second one termed "detector", is used as a filter that produces an output of $+1.0$ if a face is present in input "derotated" image and -1.0 otherwise. To detect rotated faces of any size and anywhere in the image, the input image is repeatedly subsampled to produce a pyramid of images. The "router" and the "detector" networks then process subimages extracted from every location in the image pyramid. To improve the classification results of the networks, several image preprocessing steps are applied to extracted windows. Histogram equalization, rotation classification and derotation, and lighting correction steps are used before the face detection. The

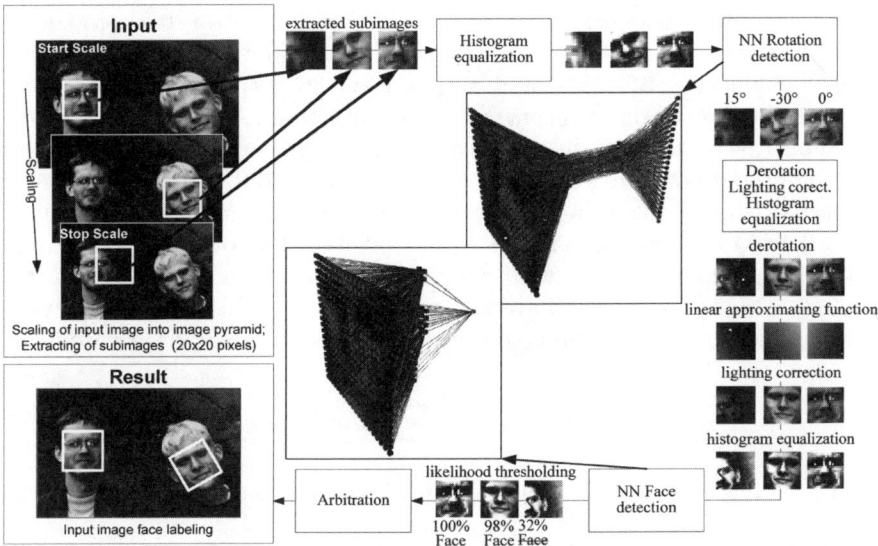

Fig. 8. Face detection system based on [43]

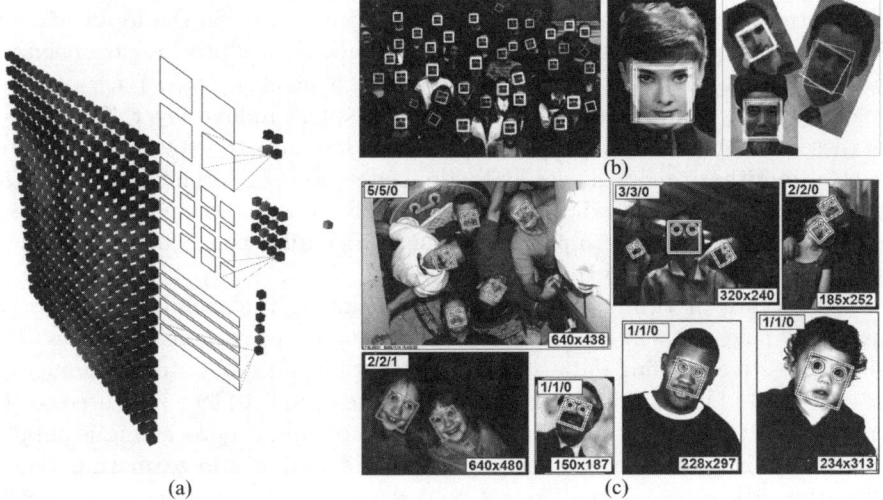

Fig. 9. Neural network for face detection architecture (**a**); Face detection results without arbitration post-processing [35] (**b**); Face detection results proposed in [43]. The numbers in the top left corner describes the number of faces in scene/the number of correct classifications/the number of misclassifications. The number in bottom right corner describes the size of image in pixels

system described so far can detect faces with different rotation angles at adjacent pixel locations in the image (Fig. 9b). An arbitration heuristic technique is used to counter these anomalies and reinforce correct detections as the last processing step of the system.

The "router" network is trained using standard error backpropagation algorithm and consists of three fully connected layers. Input layer has 400 units, hidden layer 15 units, and output layer 36 output units. Each of the units uses a hyperbolic tangent activation function. A basic training set is based on 1,047 images of different upright faces with normalized position and size, preprocessed with histogram equalization. 15,720 training samples are generated from the training basic set, using scaling (random factor between $1/\sqrt{1.2}$ to $\sqrt{1.2}$) translation (up to 0.5 pixels) and rotation ($0°$ to $360°$). If the image contains face with rotation angle α, then the output units i, $i = 0, \ldots, 35$, are trained to have values $cos(\alpha - i * 10°)$. The resulting values of the output units are evaluated during the classification phase as follows:

$$(\sum_{i=0}^{35} output_i * cos(i * 10°), \sum_{i=0}^{35} output_i * sin(i * 10°)), \tag{9}$$

where the direction of this average vector is interpreted as the predicted face angle.

The "detector" network is trained using standard error backpropagation algorithm and has input layer with 400 units, 26 units in one (or more) hidden layer and a single, real-valued output. The value of output indicates whether or not the window contains a face. As it is given in Fig. 9a the input window is broken down into smaller pieces, namely four 10×10 pixel regions, sixteen 5×5 pixel regions, and six overlapping 20×5 pixel regions. Each of these regions is fully connected to one unit in subsequent hidden layer. The shapes of these subregions are chosen to allow the hidden units to detect local features such as mouths or pairs of eyes, individual eyes, nose, or corners of mouth. The authors claim that the exact shapes of these regions do not matter. However, it is important that the input is broken into smaller pieces instead of using complete connections to the entire input.

Two types of training images are used during the training process: (1) images of faces and (2) images of all other visual patterns called nonfaces. The size of positive training data set is similar to that of the "router" training set. It is generated in a similar way, but the rotation angle of face is limited to the range of $\pm 10°$. Furthermore, histogram equalization preprocessing is supplemented with lighting correction using linear function approximating overall brightness. It is very difficult to get representative sample of negative samples because almost any pattern present in scenes can be classified as nonfacial. Instead of collecting many proper nonfacial samples before the training process, the nonfacial samples are collected during training using the so called bootstrap algorithm. The first step of bootstrap algorithm is a training of randomly initialized neural network on initial face and nonface training set. Trained network is then tested on subimages extracted from images containing no faces. A specified portion of incorrectly classified subimages is then added to the nonface part of the training set. The network trained in previous iteration is trained further, but the training set with extended nonface part is used instead. This iteration process is repeated until sufficient small amount of incorrectly classified nonface subimages is achieved.

The arbitration part of the system stores each detection result in four-dimensional result space (dimensions: the x and y coordinate of face center, the level in the image pyramid and angle of the face). The number of all results within 4 units (4 pixels, 4 pyramid level, 40°) along each result space dimension is counted then for each detection result. This number is interpreted as a confidence measure. Only the detections with confidence measure passing a threshold are taken as correct detections. Moreover, all detections that overlap the correct detections in result space are discarded. To further reduce the number of false detections and improve detection rate, the results of two independently trained face detection systems are ANDed. Only those image patterns which are detected by both of them are taken as correct result.

We analyzed the training properties of the "router" and the "detector" networks using similar system architectures, and training and testing data set structures [30, 35]. We have found out that the final classification results of both types of neural networks are very sensitive to initialization of network. Therefore, we suggest to use number of different initializations of network weights and to choose just the network with the best classification results. Moreover we suggest to set learning rate parameter to value 0.2, because our training experiments proved it as optimal for both of networks trainings. Furthermore, the classification results on testing set of "detector" network with two hidden layers are slightly better (97.93%) compared to the results of the network with just one hidden layer (97.86%).

The classification results of described face detection system in comparison with other well known face detection approaches can be seen in Tables 1 and 2. Two different testing sets are used for the testing purposes. CMU data set [50] was collected by authors of [34] (130 images, 511 faces, 469 upright ±10°). In [43] a new [50] data set with in image plane rotated faces is proposed (50 images, 223 faces, rotations from 0 to 360°).

As can be seen from the tables, the neural network based approach to face detection achieves classification accuracy comparable or in some cases even

Table 1. Comparison of detection rate vs. number of false detections for various algorithms applied to the CMU data set (expanded version of table appearing in [48])

Method	Number of false detections							
	10 (%)	31 (%)	65 (%)	78 (%)	95 (%)	120 (%)	167 (%)	422 (%)
Boosted cascade detector [37]	78.3	85.2	89.8	90.1	90.8	–	91.8	93.7
Router and detector MLP [34]	83.2	86.0	–	–	89.2	–	90.1	89.9
Naive Bayes classifier [45]	–	–	94.4	–	–	–	–	–
SNoW detector [38]	–	–	–	94.8	–	–	–	
Convolutional face finder [39]	90.5	91.5	92.3	–	–	–	93.1	–

Table 2. Comparison of detection rate vs. number of false detections for various algorithms applied to the CMU in-plane rotations data set

Method	Number of false detections			
	45 (%)	221 (%)	400 (%)	600 (%)
Router and detector MLP [43]	–	89.2	–	–
Pose estimation, boosted cascade detector [46]	87.0	89.7	90.6	91.6
Orientation map detector [47]	89.7	–	–	–

better then other well known face detection methods. Some of the results can be seen in Fig. 9c. The greatest disadvantage of described method is the performance speed. Even though the neural networks can perform the classification rather fast the complex preprocessing steps preceding the classification are very time-consuming. Different normalization methods proposed in [39] or [37] can be possibly utilized to reduce the computation time.

6 Facial Features Detection

In this part we present some conventional or nonneural algorithms for facial features detection localization. A combination of the neural network approach with some other algorithms basically improves the quality of the detection.

6.1 Eyes Localization

To determine the vertical position of the eyes in the face, we use the fact that eye regions correspond to regions of high intensity gradients. We create a vertical gradient-map, from the estimated region of the face using edge-detection.

We then use morphological processing tools to separate the confined area of high intensity gradients from isolated pixels, the next step assumes a horizontal projection by summing the pixels along the horizontal lines. Both eyes are likely to be positioned at the same line; the sum will have a strong peak on that line. We finally obtained a selection of the best representative peaks for eye's latitude. We proceed with the same histogram method by summing this time pixels vertically to determine the potentional eye longitudes on the previous vertical gradient map. By the morphological generalization we could determine which latitude is the most likely to be the latitude of eyes.

6.2 Facial Symmetrical Axis

Our previous eyes localization is the first and principal step of the feature localization, to help the localization of the other facial features. The facial symmetrical axis is a useful tool. The determination of this axis is realized thanks to the computation of the normalized cross-correlation function between the two eye areas. The generality of the function is based on the fact that two eye centers have independent coordinates, so we can easily avoid the head inclination problem.

We considered an image and a mask which size is variable and similar. When the maximum of similarity is reached, the normalized cross-correlation function is maximized. The formula is given by:

$$\frac{\sum(x_{image} + x_{mask})}{\sqrt{\sum(x_{image}^2) + \sum(x_{mask}^2)}}, \tag{10}$$

where

$$x_{image} = (Ri + Gi + Bi)/3, \tag{11}$$

and

$$x_{mask} = (Rm + Gm + Bm)/3 \tag{12}$$

and Ri, Gi, Bi, Rm, Gm, Bm are the red, green and blue value for each pixels of the image and the mask, respectively.

6.3 Mouth Localization

Statistically, eyes are always at the same distance from mouth, and considering the distance between the two eyes a specific potential mouth area could be defined thanks to this geometrical consideration. If the face does not show any special emotion, this feature is easily localizable by the horizontal edge detection method crossed with colorimetric considerations. We eliminate eventual false detection using the morphological treatment of the potential area. The histogram method from previous detection is reused (Fig. 10).

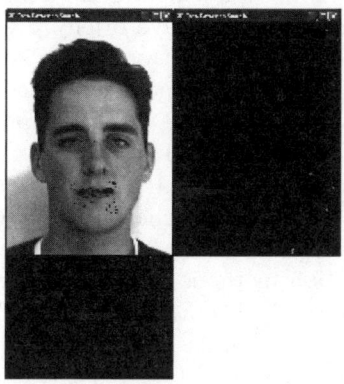

Fig. 10. Histogram method for mouth localization

The horizontal sum allows determining the maximum peak and the width of this peak standing for the mouth high. Considering this more accurate mouth area, the vertical sum is realized and permitted to obtain the mouth corners thanks to the pattern obtained by vertical sum. The corners are then searched more precisely around the previous points standing for it. Thanks to this last step, the two corners are detected independently, and the inclination risk is avoided.

6.4 Nose Localization

Considering all the reference points obtained previously and the facial symmetrical axis, we determine a potential nose area and find in this area the nose itself, this facial feature presents a high horizontal gradient on its base and dark areas standing for the nostrils. These observations and specific treatments allowed us to localize the required points. The width of the nose was determined by the same way as the localization of mouth, using the method of histogram.

First we localize the maximum peak and its width thanks to the sum of horizontal pixels. Then the vertical sum provides the width of the nose. However, nostrils are independently localized thanks to the mass center computation of the two underlined darkest areas on each side of the facial symmetrical axe. After that the tip of the nose can be easily localize between the two nostrils and due to a maximum of picture saturation to the upper direction thanks to the fact that nose always presents higher saturation due to its shape and closer distance from the camera.

7 Conclusions

In this chapter, we have presented face detection and recognition methods. At the beginning, we have discussed neural and nonneural linear and nonlinear feature extraction methods. We have briefly presented also higher-order methods and possibilities of feature extraction by kernel methods, which are intensively studied and became very popular. Then, neural network classification has been taken into account. Described approaches were illustrated by several examples of face recognition systems. A normalization scheme for face recognition has been proposed. MLP and SOM based skin color detection and MLP based face detection approaches have shown that even so complex problems can be successfully solved by neural network methods. Comparison of well known face detection methods to neural network based one has been given. Finally, we have described some methods for facial features detection. The main aim of this chapter is to describe several up-to-date approaches for face detection, recognition and facial features extraction. The combination of neural and conventional methods and algorithms lead to better results in the area of processing of face images.

Acknowledgement

Research described in this chapter was done within the Slovak Grant Agency VEGA under grants No. 1/3117/06 and 1/3110/06.

References

1. Jain A K, Ross A, Prabhakar S (2004) An introduction to biometric recognition, IEEE Trans. Circ. Syst. Video Technol., 14(1):4–20
2. Hyvärinen A (1999) Survey on independent component analysis, Neural Comput. Surveys, 2:94–128
3. Bishop C M (1995) Neural Networks for Pattern Recognition. Oxford University Press, Oxford
4. Haykin S (1994) Neural Networks – A Comprehensive Foundation. Macmillan College Publishing Company, New York
5. Diamantaras K I, Kung S Y (1996) Principal Component Neural Networks (Theory and Applications). Wiley, New York
6. Turk M, Pentland A (1991) Eigenfaces for recognition, J. Cogn. Neurosci., 3(1):71–86
7. Belhumeur P N, Hespanha J P, Kriegman D J (1997) Eigenfaces versus fisherfaces: recognition using class specific linear projection, IEEE Trans. Pattern Anal. Machine Intell., 19(7):711–720
8. Marcialis G L, Roli F (2002) Fusion of LDA and PCA for face recognition, Proceedings of the Workshop on Machine Vision and Perception, Eighth Meeting of the Italian Association of Artificial Intelligence (AI*IA), Siena, Italy, September 10–13
9. Martinez A M, Kak A C (2001) PCA versus LDA, IEEE Trans. Pattern Anal. Machine Intell., 23(2):228–233
10. Sanger T D (1998) Optimal unsupervised learning in a single-layer linear feedforward network, Neural Netw., 2:459–473
11. Kung S Y, Diamantaras K I (1990) A neural network learning algorithm for adaptive principal component extraction (APEX), International Conference on Acoustics, Speech, and Signal Processing ICASSP 90, Albuquerque, New Mexico, USA:861–864
12. Cottrell G W, Munro P (1988) Principal components analysis of images via back propagation, SPIE Vol. 1001, Visual Communications and Image Processing:1070–1077
13. Cottrell G W, Munro P, Zipser D (1988) Image compression by back propagation: an example of extensional programming, in Sharkey N E (Ed.): Models of Cognition: A Review of Cognition Science, Norwood, NJ
14. Bourlard H, Kamp Y (1988) Auto-association by multilayer perceptrons and singular value decomposition, Biol. Cybern., 59:291–294
15. DeMers D, Cottrell G W (1993) Non-linear dimensionality reduction, Adv Neural Inf. Process. Syst., 5:580–587
16. Huber P J (1985) Projection pursuit, Ann. Stat., 13(2):435–475
17. Common P (1994) Independent component analysis: a new concept? Signal Process., 36:287–314

18. Karhunen J, Oja E, Wang L, Vigário R, Joutsensalo J (1997) A class of neu-
    ral networks for independent component analysis, IEEE Trans. Neural Netw.,
    8(3):486–503
19. Muller K, Mika S, Ratsch G, Tsuda K, Scholkopf B (2001) An introduction to
    Kernel-based learning algorithms, IEEE Trans. Neural Netw., 12(2):181–201
20. Wang Y, Jiar Y, Hu C, Turk M (2004) Face recognition based on Kernel radial
    basis function networks, Asian Conference on Computer Vision, Korea, January
    27–30, pages 174–179
21. Gupta H, Agrawal A K, Pruthi T et al. (2002) An experimental eval-
    uation of linear and Kernel-based methods for face recognition, Proceed-
    ings of the Sixth Workshop on Applications of Computer Vision WACV'02,
    http://csdl.computer.org/dl/proceedings/wacv/2002/1858/00/18580013.pdf
22. Yang M (2002) Kernel eigenfaces vs. Kernel fisherfaces: face recognition using
    Kernel methods. IEEE Int. Conf. Automatic Face ang Gesture Recognit.,
    Mountain View, CA, pages 215–220
23. Yang M, Frangi A F, Yang J Y, Zhang D, Jin Z (2005) KPCA Plus LDA:
    a complete Kernel fisher discriminant framework for feature extraction and
    recognition, IEEE Trans. Pattern Anal. Machine Intell., 27(2):230–244
24. Cottrell G W, Fleming M K (1990) Face recognition using unsupervised feature
    extraction, Proceedings of the International Conference on Neural Networks,
    Paris, pages 322–325
25. Fleming M K, Cottrell G W (1990) Categorization of faces using unsupervised
    feature extraction, Proceedings of the International Joint Conference on Neural
    Networks, San Diego, Vol. 2, pages 65–70
26. Hlaváčková K, Neruda R (1993) Radial basis function networks, Neural Netw.
    World, 1:93–102
27. Poggio T, Girosi F (1990) Networks for approximation and learning, Proc. IEEE,
    78(9):1481–1497
28. Oravec M, Pavlovičová J (2004) Face recognition methods based on principal
    component analysis and feedforward neural networks, Proceedings of the Inter-
    national Joint Conference on Neural Networks IJCNN 2004, Budapest, Hungary,
    July 25–29, Vol. 1, pages 437–442
29. MIT face database, ftp://whitechapel.media.mit.edu/pub/images/
30. Beszedes M (2002) Face detection using neural networks. MA Thesis, FEI STU
    Bratislava, Slovak Republic
31. Cristinacce D (2004) Automatic detection of facial features in grey scale images,
    PhD Thesis, University of Manchester, United Kingdom
32. Phung SL, Bouzerdoum A, Chai D (2005) Skin segmentation using color pixel
    classification: analysis and comparison. IEEE Trans. Pattern Anal. Machine
    Intell., 27(1):148–154
33. Yang M, Kriegman D, Ahuja N (2002) Detecting faces in images: a survey. IEEE
    Trans. Pattern Anal. Machine Intell., 24(1):34–58
34. Rowley H A, Baluja S, Kanade T (1998) Neural network-based face detection.
    IEEE Trans. Pattern Anal. Machine Intell., 20(1):23–38
35. Beszedes M, Oravec M (2005) A system for localization of human faces in image
    using neural networks. J. Electrical Eng., 56:195–199
36. Sung K, Poggio T (1998) Example-based learning for view-based face detection.
    IEEE Trans. Pattern Anal. Machine Intell., 20(1):39–51
37. Viola P, Jones M (2002) Fast and robust classification using asymmetric
    adaboost and a detector cascade. Adv. Neural Inf. Process. Syst., 2:1311–1318

38. Roth D, Yang M, Ahuja N (2000) A SNoW-based face detector. In Adv Neural Inf Process Syst 12 (NIPS 12), MIT Press, Cambridge, MA, pages 855–861.
39. Garcia C, Delakis M (2004) Convolutional face finder: a neural architecture for fast and robust face detection. IEEE Trans. Pattern Recogn. Machine Intell., 26(11):1408–1423
40. Vezhnevets V, Sazonov V, Andreeva A (2003) A survey on pixel-based skin color detection techniques. In: Proceedings of Graphicon-2003, Moscow, Russia, September 2003, pages 85–92
41. Brown D, Crawl I, Lewthwaite J (2001) A SOM based approach to skin detection with application in real time systems. In: Proceedings of the British Machine Vision Conference
42. Seow M J, Valaparla D, Asari V (2003) Neural network based skin color model for face detection. In: 32nd Applied Imagery Pattern Recognition Workshop, Washington DC, October
43. Rowley H A, Baluja S, Kanade T (1998) Rotation invariant neural network based face detection. In Computer Vision and Pattern Recognition Conference 1998:38–44
44. Osuna E, Freund R, Girosi F (1997) Training support vector machines: an application to face detection, In Computer Vision and Pattern Recognition Conference
45. Schneiderman H, Kanade T (1998) Probablistic modelling of local appearance and spatial relationships for object recognition. In Computer Vision and Pattern Recognition Conference
46. Jones M, Viola P (2004) Fast multi-view face detection. Technical Report TR2003-96, Mitsubishi Electric Research Labs
47. Froba B, Ernst A (2003) Fast frontal-view face detection using a multi-path decision tree. In Fourth International Conference on Audio- and Video-Based Bio-metric Person Authentication 2003, pages 921–928
48. Viola P, Jones M (2001) Robust real-time object detection. In ICCV Workshop on Statistical and Computation Theories of Vision, Vancouver, Canada
49. Kohonen T (2000) Self-Organizing Maps. Springer Verlag, New York
50. CMU face database, http://vasc.ri.cmu.edu/idb/html/face/frontal_images/

Classification of Satellite Images with Regularized AdaBoosting of RBF Neural Networks

Gustavo Camps-Valls and Antonio Rodrigo-González

Grup de Processament Digital de Senyals, Universitat de València, Spain
C/Dr. Moliner, 50. Burjassot (València), Spain, gustavo.camps@uv.es,
http://www.uv.es/gcamps

Summary. This chapter presents the soft margin regularized AdaBoost method for the classification of images acquired by satellite sensors. The regularized AdaBoost algorithm combines a number of RBF neural networks as base learners efficiently. The method is theoretically motivated by noting the need of regularization and soft margin in the standard AdaBoost method, particularly in the context of high-dimensional possibly noisy problems, such as those posed in the classification of hyperspectral images. Several experiments are conducted to assess method's behaviour in the classification of crop fields using satellite images of different dimensionality. A synthetic multispectral image and a real hyperspectral image acquired with the AVIRIS sensor are considered in our experiments. Performance is analyzed in terms of accuracy, robustness to different levels and nature of additive noise and ill-posing situations. In all cases, the method is compared to state-of-the-art support vector machines (SVM) and the linear discriminant analysis (LDA) as a baseline method. Experimental results suggest that this method can be more useful than SVMs or LDA in low signal-to-noise ratio environments and when input dimension is relatively high.

1 Introduction

Materials in a scene reflect, absorb, and emit electromagnetic radiation in a different way depending of their molecular composition and shape. Remote sensing exploits this physical fact and deals with the acquisition of information about a scene (or specific object) at a short, medium or long distance. The radiation acquired by an (airborne or satellite) sensor is measured at different wavelengths and the resulting spectral signature (or *spectrum*) is used to identify a given material. The field of *spectroscopy* is concerned with the measurement, analysis, and interpretation of such spectra [1]. Figure 1 shows the application of imaging spectroscopy to perform satellite remote sensing.

The earlier sensor developments considered a few number of bands, which readily demonstrated to be a limitation for detecting similar materials. Thus,

G. Camps-Valls and A. Rodrigo-González: *Classification of Satellite Images with Regularized AdaBoosting of RBF Neural Networks*, Studies in Computational Intelligence (SCI) **83**, 307–322 (2008)

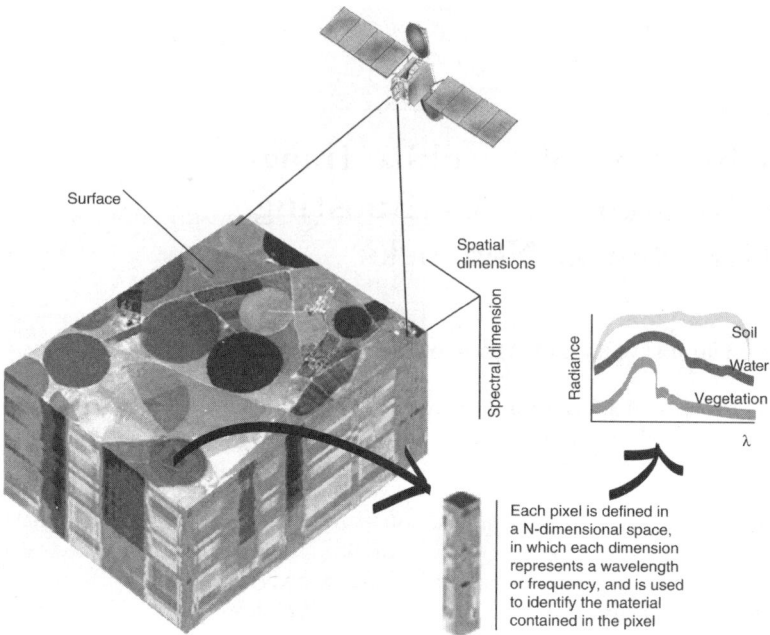

Fig. 1. Principle of imaging spectroscopy

a new class of imaging spectroscopy sensors, which acquire hundreds of contiguous narrow bands or channels appeared, and are called hyperspectral (imaging) sensors. Hyperspectral sensors sample the reflective portion of the electromagnetic spectrum ranging from the visible region (0.4–0.7 μm) through the near-infrared (about 2.4 μm) in hundreds of N narrow contiguous bands about 10 nm wide.[1] Hyperspectral sensors represent an evolution in technology from earlier multispectral sensors, which typically collect spectral information in only a few discrete, non-contiguous bands.

The high spectral resolution characteristic of hyperspectral sensors preserves important aspects of the spectrum (e.g., shape of narrow absorption bands), and makes differentiation of different materials on the ground possible. The spatially and spectrally sampled information can be described as a data cube (colloquially referred to as "the hypercube"), which includes two spatial coordinates and the spectral one (or wavelength). As a consequence, each image pixel is defined in a (potentially very high) dimensional space where each dimension corresponds to a given wavelength interval in the spectrum.

The use of hyperspectral images for Earth Observation is a consolidated technology since the (increasing) high number of spectral bands contained in

---

[1] Other types of hyperspectral sensors exploit the emissive properties of objects by collecting data in the mid-wave and long-wave infrared (MWIR and LWIR) regions of the spectrum.

the acquired image allows excellent characterization, identification, and classification of the land-covers [1]. However, many classifiers are affected by input
sample dimension, tend to overfit data in the presence of noise, or the computational cost poorly scales with the number of samples. All these problems are
related theoretically through the well-known *curse of dimensionality* by which
developing a classifier with low number of high-dimensional samples runs the
risk of overfitting the training data, and then showing poor generalization
performance [2, 3].

There are some techniques to alleviate this problem: (1) perform a feature
selection/extraction before developing the classifier, such as principal component analysis (PCA) [4]; (2) sample replication, such as bootstrap resampling
techniques [5]; (3) regularization of the classifier equations, such as adding
a (small) controlled quantity to covariance matrices in linear discriminant
analysis [3]; and (4) working under the structural risk minimization (SRM)
principle rather than the empirical risk minimization (ERM) principle, such
as kernel methods in general, and support vector machines (SVM) in particular [6–8]. In the last few years, the use of SVMs for hyperspectral image
classification has been paid attention basically because the method integrates
in the same classification procedure: (1) a *feature extraction* step, as samples
are mapped to a higher dimensional space where a simpler (linear) classification is performed, becoming non-linear in the input space; (2) a *regularization*
procedure by which model's complexity is efficiently controlled; and (3) the
minimization of an upper bound of the generalization error, thus following the
SRM principle. These theoretical properties make the SVM very attractive in
the context of hyperspectral image classification where the classifier commonly
deals with a low number of high dimensional training samples [8, 9, 11].

However, SVMs have some problems when working with high levels of
noise, specially when this is (thermal) Gaussian noise. The latter is basically
due to the use of the linear cost function which is more appropriate to uniform
noise from a maximum likelihood perspective [10]. Also, SVMs have the problem of scaling badly (from the point of view of the computational cost) with
the number of training samples. Also, it is claimed that SVMs provide *sparse*
solutions, i.e., the final decision solution is expressed using a reduced number
of weights or model parameters. This, however, though theoretically well-
motivated, is not commonly observed in practical applications, in which the
non-linear nature of the problem is complex enough to preclude the (desirable)
sparsity property, and thus many support vectors are ultimately obtained [9].
Last but not least, the problem of free parameter tuning is an important one
in the SVM framework. Specifically, the choice of the kernel and its associated
free parameter is a crucial (and so far unsolved) problem, making the training
process very heuristic and computationally demanding [8].

In the last years, an active research area in machine learning is that concerned with *boosting* methodologies [12]. Boosting is a general method for
improving the performance of a given algorithm by combining an ensemble
of simpler (or *weak*) classifiers, such as linear discriminants or classification

trees [13]. The reader interested in boosting methods can visit `http://www.bosting.org` for a number of applications, introductory tutorials, publications, and software resources. A recently proposed (and very promising) boosting algorithm is the Adaptive Boosting (AdaBoost) [14], which has demonstrated good performance in many applications such as optical character recognition [15], image retrieval [16], speech recognition [17], etc. The use of AdaBoost has extended rapidly in the remote sensing literature due to the good results offered, model simplicity, and the intuitive nature of the free parameters [18–22]. Roughly speaking, the AdaBoost algorithm combines classifiers, such as neural networks, so it concentrates its efforts on the most difficult samples to be classified, resembling the support vectors in the SVM framework. This method is also related to the SVMs from a theoretically point of view [23], while keeping good characteristics as the sparsity (or compactness) and stability of the free parameters.

However, the formulations of AdaBoost used so far are not regularized, which can be of dramatic consequences in the presence of moderate-to-high levels of noise, particularly when medium-to-high dimension of samples are used. In fact, it is well-known that AdaBoost performs worse on noisy tasks [24, 25]. This poor behaviour suggests that *regularization* of AdaBoost equations becomes necessary, and constitutes a fundamental motivation of this work. As a consequence, in this work, we will concentrate on studying this issue from a theoretical and experimental point of view.

In this chapter, we will introduce the regularized soft margin AdaBoost [26] in the context of hyperspectral image classification and extend previous works in different ways, such as testing the algorithm in different images inducing ill-posed scenarios and robustness on the presence of different levels and nature of additive noise. Also, several issues are analyzed, such as working with different dimension of the training samples and number of classes, and also analyzing the stability of the free parameters selection.

The rest of this chapter is outlined as follows. Section 2 presents an introduction to boosting and the standard AdaBoost formulation. In Sect. 3 we motivate the need for regularization and soft margin in the context of AdaBoost, and thus present the regularized AdaBoost. Also, a discussion on the relationships among the proposed method and SVMs is given. Extensive experimental results are presented in Sect. 4. In Sect. 5, we conclude the chapter with a discussion and some final remarks.

2 Boosting and AdaBoost Methods

Boosting is a general method for improving the performance of any learning algorithm [12, 27]. It consists of generating an ensemble of *weak* classifiers (which need to perform only slightly better than random guessing) that are combined according to an arbitrarily strong learning algorithm. The good performance of boosting algorithms was explained by the PAC theory [28].

Table 1. List of symbols

Symbol	Description
i, n	Counter and number of samples
N	Dimension of samples (spectral channels)
j, J	Counter and number of hypotheses
t, T	Counter and number of iterations
$\mathbf{x}_i$	Training sample i
ε	Accuracy
$\hat{\boldsymbol{\alpha}}$	Hypotheses vector
$\mathcal{F}$	Set of functions
$f_{\hat{\alpha}}$	A set of $\mathcal{F}$ using the weights $\hat{\alpha}$
$\mathbf{d}$	Weighting vector of the training set
C	The regularization parameter (cost)

It was shown that AdaBoost has the so-called PAC boosting property, i.e., if the learner generating the base hypotheses is just slightly better than random guessing, AdaBoost is able to find a combined hypothesis with arbitrarily high accuracy (if enough training examples are available) [28].

In the following sections, we will introduce the standard boosting and AdaBoosting algorithms, and then we will present its regularized version. Table 1 shows a list of symbols used throughout the chapter.

The earlier boosting algorithms were considered very efficient and less prone to overfitting than neural networks in the late eighties and nineties. However, given that weight updating are carried out through gradient descent of a cost function, a number of associated problems emerged: presence of local minima on the error surface, free parameters tuning, control of convergence, etc. A promising boosting algorithm alleviating the latter problems is the *Adaptive Boosting* (AdaBoost) [29].

The AdaBoost algorithm takes as input a labeled training set $(\mathcal{X}, \mathcal{Y}) = \{(\mathbf{x}_1, y_1), \ldots, (\mathbf{x}_n, y_n)\}$, where $\mathbf{x}_i \in \mathbb{R}^N$ and $y_i \in \{-1, +1\}$, $i = 1, \ldots, n$, and calls a *weak* or *base learning algorithm* iteratively, $t = 1, \ldots, T$. At each iteration t, a certain confidence weight $d_i^{(t)}$ is given (and updated) to each training sample $\mathbf{x}_i$. The weights of incorrectly classified samples are increased at each iteration so that the weak learner is forced to focus on the hard samples in the training set. The task of the base learner reduces to find a *hypothesis* $h_t : \mathcal{X} \to \mathcal{Y}$ for the distribution $\mathbf{d}^{(t)} = \{d_i^{(t)}\}_{i=1}^n$. At each iteration, the goodness of a weak hypothesis is measured by its error,

$$\varepsilon_t = P[h_t(\mathbf{x}_i) \neq y_i] = \sum_{i: h_t(\mathbf{x}_i) \neq y_i} d_i^{(t)}. \tag{1}$$

Once the weak hypothesis h_t has been calculated, AdaBoost chooses a parameter

$$\alpha_t = \frac{1}{2} \ln\left(\frac{1 - \varepsilon_t}{\varepsilon_t}\right), \tag{2}$$

which measures the importance assigned to h_t. Note that $\alpha_t \geq 0$ if $\varepsilon_t \leq 1/2$, and that α_t gets larger as ε_t gets smaller. The distribution $\mathbf{d}^{(t)}$ is next updated in order to increase the weight (or importance) of samples misclassified by h_t, and to decrease the weight of correctly classified samples [29]

$$d_i^{(t+1)} = d_i^{(t)} \exp\left(-\alpha_t y_i h_t(\mathbf{x}_i) \right) \frac{1}{Z_t}, \tag{3}$$

where Z_t is a *normalization* constant chosen so that $\mathbf{d}^{(t+1)}$ be a valid distribution, i.e., $\sum_{i=1}^n d_i^{(t)} = 1$ for $d_i^{(t)} \geq 0$, $i = 1, \ldots, n$.

As a consequence, the algorithm tends to concentrate on difficult samples, which reminds somewhat the *support vectors* of SVMs, and that are called here *support patterns*. The final hypothesis (decision) is a weighted majority vote of the T weak hypotheses where α_t are the weights assigned to h_t. Consequently, for a test instance $\mathbf{x}_*$, the weak hypothesis h_t yields a prediction $h_t(\mathbf{x}_*) \in \mathbb{R}$ whose sign is the predicted label given by

$$\hat{y}_* = f_{\boldsymbol{\alpha}}(\mathbf{x}_*) = sgn\left(\sum_{t=1}^T \alpha_t h_t(\mathbf{x}_*) \right) \tag{4}$$

and whose magnitude $|h_t(\mathbf{x}_*)|$ gives a measure of confidence in the prediction, which is specially interesting in some applications.

The main advantage of AdaBoost is that in many cases it increases the overall accuracy of the weak learner. AdaBoost tends to exhibit no overfitting when the data are noiseless, and being a combination procedure, it usually reduces both the variance and the bias of the classification function [30, 31]. However, as illustrated in many applications, the standard algorithm of AdaBoost does not offer good performances in the presence of noise in the data, thus suggesting that *regularization* becomes necessary. In the next section we introduce a regularized version of the AdaBoost algorithm.

3 Regularizing the AdaBoost Algorithm

This section analyzes the soft margin regularized AdaBoost for classification problems. First, we motivate the need for regularizing the standard AdaBoost equations. Then, we present the formulation for the regularized AdaBoost and analyze its main properties. Finally, we show the theoretical relationship between the regularized AdaBoost and the soft margin properties of support vector machines.

3.1 Regularization, Soft Margin and Overfitting

As stated before, and seen in many real applications, AdaBoost rarely overfits in the low noise regime. However, this problem is posed for low signal-to-noise ratios. Intimately related to this fact is the way AdaBoost performs

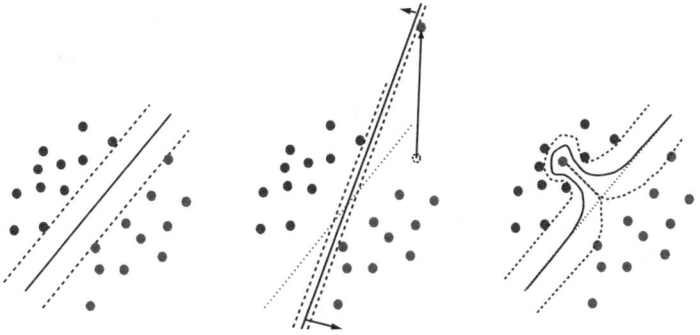

Fig. 2. The problem of finding a maximum margin "hyperplane" on reliable data (*left*), data with outlier (*middle*) and with a mislabeled example (*right*). The *solid line* shows the resulting decision boundary, whereas the *dashed line* marks the margin area. In the *middle* and on the *left* the original decision boundary is plotted with *dots*. The hard margin implies noise sensitivity, because only one example can spoil the whole estimation of the decision boundary. Figure adapted from [23]

learning. Basically, the algorithm concentrates its efforts on the most difficult examples to learn, and thus AdaBoost asymptotically achieves a hard margin separation (see [26, 32] for details). Developing hard margin algorithms is a sub-optimal strategy in noisy environments because the algorithm can focus on noisy irrelevant samples thus skewing the classification boundary (see Fig. 2 for an illustrative example). As a direct consequence, regularization becomes necessary to alleviate this bias, which can be of dramatic consequences in the presence of high noise levels.

From Fig. 2, it becomes clear that AdaBoost (and any other algorithm with large hard margin) is noise sensitive and maximizing the smallest margin in the case of noisy data can (and probably will) lead to sub-optimal results. Therefore, we need to relax the hard margin and allow for a possibility of mistrusting the data. This could be done for instance by imposing a trade-off between the margin maximization and the influence that a given sample has on the final decision boundary. However, introducing an additional (and rather intuitive) parameter is not a good solution. In the following section, we introduce a regularized version of the AdaBoost based on the re-formulation of AdaBoost as a hard margin classifier and then translate it to the *soft margin* concept.

3.2 Soft Margin Regularized AdaBoost

A possibility to reformulate a hard margin algorithm in order to develop a *soft margin algorithm* is to use the soft margin rather than the actual margin. In our specific case, one can develop the previous AdaBoost algorithm using the soft margin regularization as follows. Note from (1) and (3) that the standard

AdaBoost algorithm presented before minimizes the exponential cost function, which can be written as a function of the actual margin as follows [23]:

$$G[f_{\hat{\alpha}}] = \sum_{i=1}^{n} \exp\left\{ -\rho_i(\hat{\alpha}) \sum_{j=1}^{J} \hat{\alpha}_t \right\} , \tag{5}$$

where $\rho_i(\hat{\alpha})$ is the margin defined as a function of the *normalized* function value and the label for sample $\mathbf{x}_i$:

$$\rho_i = y_i \frac{f_{\hat{\alpha}}(\mathbf{x}_i)}{\sum_{t=1}^{T} \alpha_t} , \tag{6}$$

where $\alpha_t \geq 0$, $t = 1, \ldots, T$ are the hypothesis coefficients of the (non-negative) combined hypothesis f. Therefore, it is easy to see that if the margin is positive, the sample is classified correctly.

In order to regularize the AdaBoost algorithm, Rätsch et al. [23, 33] proposed to modify the loss function in (5) by replacing the actual margin ρ by a soft margin $\hat{\rho}$. The new loss function thus becomes

$$\begin{aligned} G_{reg}[f_{\hat{\alpha}}] &= \sum_{i=1}^{n} \exp\left\{ -\hat{\rho}_i(\hat{\alpha}) \sum_{j=1}^{J} \hat{\alpha}_j \right\} \\ &= \sum_{i=1}^{n} \exp\left\{ -[\rho_i(\hat{\alpha}) + C\mu_i] \sum_{j=1}^{J} \hat{\alpha}_j \right\} , \end{aligned} \tag{7}$$

where the *regularization parameter* C tunes the hard margin minimization and the misconfidence given by μ_n, which measures the *influence* of samples on the combined hypotheses. In particular, we used the following confidence measurement:

$$\mu_i^{(t)} = \sum_{r=1}^{t} \frac{\hat{\alpha}_t}{\sum_{r'=1}^{t} \hat{\alpha}_{r'}} \hat{d}_i^{(r)} , \tag{8}$$

where $d_i^{(t)}$ represents the sample weighting. Intuitively, the regularization factor $\mu_i^{(t)}$ for a given hypothesis t is equivalent to the average weight of a sample computed during the learning process.[2]

The idea behind this parametrization is as follows. An example that is very often misclassified (i.e., hard to classify correctly) will have a high average weight, i.e., a high influence. For instance, in the noisy case, there is (usually) a high overlap between samples with high influence and samples that are mislabeled or other samples very near to (or only slightly beyond) the decision boundary. This way, if a training sample has a high influence $\mu_i^{(t)}$, then the

---

[2] Note, however, that one could design more sophisticated regularizers according to Tikhonov's regularization theory, in which, rather than using the l_1-norm $\mu_i^{(t)}$, any meaningful regularization operator $\mathbf{P}(\mu_i^{(t)})$ could be used.

margin is increased, but if one maximizes the smallest *soft margin*, one does not force outliers to be classified according to their possibly wrong labels, but allows for some errors. This idea directly induces a *prior* to weight all samples equally, which contradicts with the general idea of AdaBoost algorithms to overweight certain examples. In fact, in this formulation, we impose a trade-off between maximizing the margin and minimizing the influence of outliers. Once we have defined the cost function, it can be demonstrated that an iterative procedure can be used to minimize it [23, 34].

In this setting, the regularized AdaBoost only differs from the standard AdaBoost algorithm in that (2) is replaced with the following equation:

$$\alpha_t = \underset{\alpha_t \geq 0}{\operatorname{argmin}} \sum_{i=1}^{n} \exp\left(- [\rho_i(\alpha)^{(t)}) + C\mu_i^{(t)}] \sum_{r=1}^{t} \hat{\alpha}_t \right), \tag{9}$$

and the weights $\mathbf{d}^{(t)}$ are updated following the rule:

$$d_i^{(t+1)} = d_i^{(t)} \exp\left(- [\rho_i(\alpha)^{(t)}) + C\mu_i^{(t)}] \sum_{r=1}^{t} \alpha_t \right) \frac{1}{Z_t}, \tag{10}$$

instead of using (3). Here, Z_t is again a normalization constant, such that $\sum_{i=1}^{n} d_i^{(t+1)} = 1$. Finally, the decision function (final hypothesis) with weights $\mathbf{d}^{(t)}$ is computed as in the standard AdaBoost (cf. (4)). Note that for $C = 0$ and $h_t \in \{-1, +1\}$ the algorithm is equivalent to AdaBoost. For illustration purposes, we show in Fig. 3 the performance of the standard and regularized

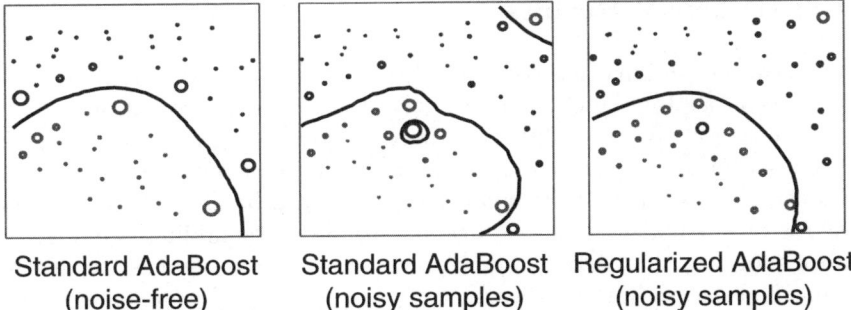

Standard AdaBoost   Standard AdaBoost   Regularized AdaBoost
(noise-free)   (noisy samples)   (noisy samples)

Fig. 3. Illustration of the need of regularized solutions in a toy example. The decision boundary of AdaBoost on a toy dataset (*gray*: positive class, *lightgray*: negative class) without (*left*) and with noise (*middle*). The boundary is considerably changed when adding only three examples (*middle, upper/lower right*). On the *right* the boundary generated by the regularized AdaBoost is plotted. The boundary is almost unchanged compared to the case without noise. The diameter of the points is proportional to the influence of the example. When using regularized AdaBoost, the influences are spread over more examples near the boundary. Figure adapted from [23]

AdaBoost algorithms in a two-class toy example with both "noisy" and "clean" samples.

3.3 SVMs and Regularized AdaBoost Algorithm

SVMs and a particular form of regularized AdaBoost, also known as Arc-GV [23], are explicitly related by observing that any hypothesis set $\mathbf{H} = \{h_j | j = 1, \ldots, J\}$ implies a mapping $\phi(\mathbf{x}) = [h_1(\mathbf{x}), h_2(\mathbf{x}), \ldots, h_J(\mathbf{x})]^T$ and therefore also a kernel $K(\mathbf{x}, \mathbf{y}) = \phi(\mathbf{x})\phi(\mathbf{y})^\top = \sum_{j=1}^{J} h_j(\mathbf{x})h_j(\mathbf{y})$, where $J = dim(\mathcal{H})$. In fact, any hypothesis set $\mathbf{H}$ spans a feature space $\mathcal{H}$, which is obtained by some mapping ϕ and the corresponding hypothesis set can be constructed by $h_j = P_j[\phi(\mathbf{x}_i)]$ [35]. The Reg-AdaBoost [26, 32] can be thus expressed as the maximization of the smallest margin ρ w.r.t. $\mathbf{w}$ and constrained to

$$y_i \left(\sum_{j=1}^{J} w_j h_j[\mathbf{x}_i] \right) \geq \rho, \ \forall i = 1, \ldots, n$$
$$\text{and } \|\mathbf{w}\|_1 = 1. \tag{11}$$

On the one hand, Reg-AdaBoost can be thought as an SVM approach in a high-dimensional feature space spanned by the base hypothesis of some function set (see (11)). It uses effectively an l_1-norm regularizer, which induces sparsity. On the other hand, one can think of SVMs as a "boosting approach" in a high-dimensional space in which, by means of the "kernel trick", we never work explicitly in the kernel feature space.

4 Experimental Results

In this section, we illustrate the performance of the regularized AdaBoost in several scenarios for the classification of satellite images. Specifically, a synthetic multispectral image and a real hyperspectral images acquired with the AVIRIS sensor are considered in the experiments. Different real life scenarios are considered by varying the number of considered bands and training samples. Also, we simulate many additional situations by adding Gaussian and uniform noise to the AVIRIS image in order to test robustness to different amount and nature of the noise source.

4.1 Model Development

The experimental analysis considers an exhaustive evaluation and comparison of SVMs and regularized AdaBoost. In addition, we include the linear discriminant analysis (LDA) as baseline method for comparison purposes. In the case of SVMs, we tested the Gaussian kernel (RBF) and the polynomial kernel, given the good results obtained in previous applications to hyperspectral data [36]. Therefore, only the Gaussian width σ, the polynomial order d, together with the *regularization* parameter C should be tuned. We tried exponentially increase sequences of σ ($\sigma = 1, \ldots, 50$) and C ($C = 10^{-2}, \ldots, 10^{6}$), and tuned d in the range 1–15.

For the case of Reg-AdaBoost algorithm, the regularization term was varied in the range $C = [10^{0}, \ldots, 10^{3}]$, and the number of iterations was tuned to $T = 10$. In this work, we used Radial Basis Functions (RBF) neural networks as base classifiers for the AdaBoost method, and the width and centers of the Gaussians were computed iteratively in the algorithm (see [11] for details on the adaptive algorithm). Therefore, only the number of hidden neurons (N_h) in the hidden layer has to be tuned with regard the base learner.

In all cases, we developed one-against-all schemes to deal with the multiclass problem for all considered classifiers [37]. This approach consists in solving a problem of K classes by constructing K classifiers, each one designed to classify samples of a given class. A winner-takes-all rule across the classifiers is then applied to classify a new sample. Complementary material (MATLAB source code, demos, and datasets) is available at http://www.uv.es/gcamps/reg_ab/ for those interested readers.

4.2 Synthetic Multispectral Image Classification

In this first classification problem, we compare the performance in the three-class multispectral problem proposed in [22]. We generated 3,330, 1,371 and 3,580 training samples for each class, and a four-dimensional spectral image was simulated through the generation of multivariate Gaussian distributions $\mu_1 = [0000]$, $\mu_2 = [1100]/\sqrt{2}$, $\mu_3 = [1.0498, -0.6379, 0, 0]$, and the common variance-covariance matrix $\sigma^2 I$, where I represents the identity matrix. The test set contained 8,281 samples generated in the same way.

Comparison of the proposed method with the standard AdaBoost method, SVMs, and LDA algorithm was carried out for $\sigma^2 = 1$ and $\sigma^2 = 4$. The eight-fold cross-validation procedure was used to tune the parameters using the training set, and the average test results for 50 realizations are shown in Table 2. Several conclusions can be extracted. First, regularized AdaBoost outperforms the LDA (gain: 14.66%), SVMs (gain: 4.39%), and the unregularized algorithm (gain: 3.55%) in low-noise regime, and also in high noise regime (11.90% over LDA, 5.40% over SVM, and 6.22% gain over AdaBoost). Second, in the case of high noise regime ($\sigma^2 = 4$), differences become lower but still in this complex situation, the proposed method obtains a remarkable advantage over the rest of the methods. Finally, it is worth noting that AdaBoost and SVMs yield very similar results in both scenarios.

4.3 Real Hyperspectral Image Classification

This section provides an experimental assessment of the regularized AdaBoost algorithm for the classification of the well-known AVIRIS'92 Indian Pines image. In this set of experiments, we will focus on robustness to noise with different input dimensions, stability and regularization.

Table 2. Results in the toy example validation set[a]

Method	$\sigma^2 = 1$			$\sigma^2 = 4$		
	PARAMS.	OA (%)	κ	PARAMS.	OA (%)	κ
LDA	–	47.22	0.288	–	39.91	0.111
SVM	$C = 12.44, \sigma = 10$	57.49	0.306	$C = 10.20, \sigma = 22$	46.41	0.170
AdaBoost [22]		58.33	0.359		45.59	0.172
Reg-AB	$C = 46.42, N_h = 3$	**61.88**	**0.366**	$C = 1, N_h = 3$	**51.81**	**0.177**

[a] Several accuracy measures are included: users, producers, overall accuracy (OA (%)), and kappa statistic (κ) different classifiers: LDA, SVM with RBF kernel, AdaBoost in [22] and Reg-AdaBoost presented in this work. The column "PARAMS." gives some information about the best free parameters for the models. The best overall scores for each class are highlighted in bold face font

Data Collection and Setup

The AVIRIS Indian Pines scene was gathered in 1992 over a mixed agricultural/forested area in NW Indiana, early in the growing season. The dataset represents a very challenging land-cover classification scenario, in which the primary crops of the area (mainly corn and soybeans) were very early in their growth cycle, with only about 5% canopy cover. Discriminating among the major crops under this circumstances can be very difficult (in particular, given the moderate spatial resolution of 20 m). This fact has made the scene a challenging benchmark to validate classification accuracy of hyperspectral imaging algorithms. The data is available online (along with detailed ground-truth information) from `http://dynamo.ecn.purdue.edu/~biehl/MultiSpec`. The whole scene consists of 145×145 pixels and 16 ground-truth classes, ranging in size from 20 to 2,468 pixels.

We removed 20 noisy bands covering the region of water absorption, and finally worked with 200 spectral bands (see [38] for full details). In order to assess performance in different input dimensions, we will work with (1) this subset of 200 bands, and (2) a more reduced set of 9 bands of a mixed agriculture/forestry landscape selected in [38]. In the following section we extensively compare LDA, SVMs (RBF and polynomial kernels) and the Reg-AdaBoost method proposed here. Since the AdaBoost is a particular solution ($C = 0$) of the reg-AdaBoost, we will not further consider including its results.

Numerical Comparison

Table 3 shows the results obtained when training the models in the different datasets (200 and 9 spectral bands). In the case of working with 200 spectral features, the best overall accuracy (OA (%)) and κ values are provided by the SVM-RBF, closely followed by the Reg-AdaBoost, yielding an overall accuracy equal to 93.85% and 91.67%, respectively. In the case of a higher dimensional

Table 3. *Top*: Results in the 200-bands dataset, and *Bottom*: after the selection proposed in [38] (9 input bands)[a]

Method (γ, C, d, N_h)	Users/Producers																OA	κ
	C_1	C_2	C_3	C_4	C_5	C_6	C_7	C_8	C_9	C_{10}	C_{11}	C_{12}	C_{13}	C_{14}	C_{15}	C_{16}		
200 Spectral bands																		
LDA	80.72	68.35	90.76	76.31	98.66	100	96.71	100	83.4	66.75	83.99	100	90.62	78.31	91.3	86.14	80.57	0.781
	79.08	64.92	56.84	90.05	85.15	81.25	98.74	90.91	67.34	85.55	72.39	98.11	96.24	72.2	91.3	81.57		
SVM-RBF ($10^3, 599.48, -, -$)	89.88	89.93	94.12	96.79	99.46	92.31	100	90	90.04	94.16	96.08	99.04	97.85	83.6	97.83	93.15	**93.85**	0.930
	94.19	91.69	88.19	99.59	99.2	100	99.59	90	91.75	91.2	93.04	100	95.5	89.27	95.74	94.93		
SVM-POLY ($-, 10^5, 9, -$)	85.44	81.06	90.76	96.39	99.19	92.31	100	90	80.71	91.16	93.14	99.04	96.92	85.19	97.83	91.15	90.58	0.892
	89.93	87.56	87.8	97.56	99.19	100	99.59	90	85.49	84.96	91.64	99.04	96.04	86.56	95.74	93.19		
Reg-AB ($-, 46.416, -, 17$)	88.07	87.53	74.79	95.18	98.92	92.31	99.18	40	89	92.62	92.16	99.04	97.08	82.01	93.48	85.82	91.67	0.905
	90.71	92.88	81.65	99.16	97.87	100	97.18	80	88.45	87.44	92.46	100	95.46	86.11	95.56	92.42		
9 Spectral bands																		
LDA	64.63	53.48	66.39	58.23	86.29	84.62	82.3	80	60.79	43.71	66.99	99.04	85.08	42.86	84.78	70.73	63.42	0.592
	63.23	39.96	29.04	70.39	86.29	12.79	97.56	10.81	58.37	72.54	48.24	76.3	88.06	57.45	90.70	58.70		
SVM-RBF ($10^3, 10^5, -, -$)	73.93	66.43	79.83	93.98	96.51	76.92	97.94	50	74.27	86.7	84.64	98.08	96.92	60.85	93.48	80.35	83.79	0.814
	81	80.06	77.87	96.69	93.49	90.91	96.36	83.33	83.45	75.82	80.94	97.14	92.38	74.19	95.56	85.96		
SVM-POLY ($-, 10^5, 9, -$)	64.36	48.2	63.87	90.36	94.62	76.92	99.59	0	57.26	78.91	58.5	98.08	97.69	53.44	93.48	69.14	75.01	0.712
	68.24	74.17	68.47	89.29	89.8	83.33	93.8	NaN	80.47	61.97	61.51	94.44	91.1	71.13	95.56	NaN		
Reg-AB ($-, 464.16, -, 35$)	79.75	76.74	68.07	92.77	94.89	84.62	98.35	20	83.20	85.64	84.64	98.08	94	60.85	93.48	79.82	**85.22**	0.831
	80.87	83.99	74.31	94.67	91.93	100	97.55	66.67	82	81.99	83.01	93.58	93	65.71	97.73	85.80		

[a] Several accuracy measures are included: users, producers, overall accuracy (OA (%)) and kappa statistic (κ) in the test set for different kernel classifiers: Linear Discriminant Analysis (LDA), SVMs with RBF kernel (SVM-RBF) and with polynomial kernel (SVM-POLY), and regularized AdaBoost (Reg-AdaBoost). The column "PARAMS." gives some information about the final models. The best scores for each class are highlighted in bold face font

space (9 bands), the proposed Reg-AdaBoost method clearly outperforms the SVM (with polynomial kernel) and the LDA, and produces better results than the SVM with RBF kernel.

When looking at producers/users accuracies, it is noteworthy that kernel-based methods obtain higher scores on classes $C3$, $C4$, $C5$, and $C9$, and that the most troublesome classes are $C1$, $C6$, $C7$, and $C8$. This has been also observed in [39], and can be explained because grass, pasture, trees, and woods are homogeneous areas which are clearly defined and labeled. Contrarily, corn and soybean classes can be particularly misrecognized because they have specific sub-classes (no till, min till, clean till).

Analysis of Model Stability

Let us analyze the critical issue of tuning the free parameters in the classifiers. For this purpose, we have represented in Fig. 4 the accuracy surface in the validation set as a function of the free parameters of all classifiers.

From Fig. 4 one can see that the behaviour of SVM-RBF is similar for all cases, and that higher accuracy is obtained as the penalization C and γ increase. However, in the case of SVM-POLY, is much more important the value of the regularization parameter compared with the degree of the kernel. Better results are obtained for $d > 6$, which turns to be more clear as the number of available bands is decreased. With regard the Reg-AdaBoost, despite the irregular *plateau* of the error surface, one can see that as the number of spectral channels is decreased, the error surface becomes smoother for a wide range. Also, and very important, is the fact that tuning the value of C is not so critical as in the case of SVMs. In fact, good results are obtained

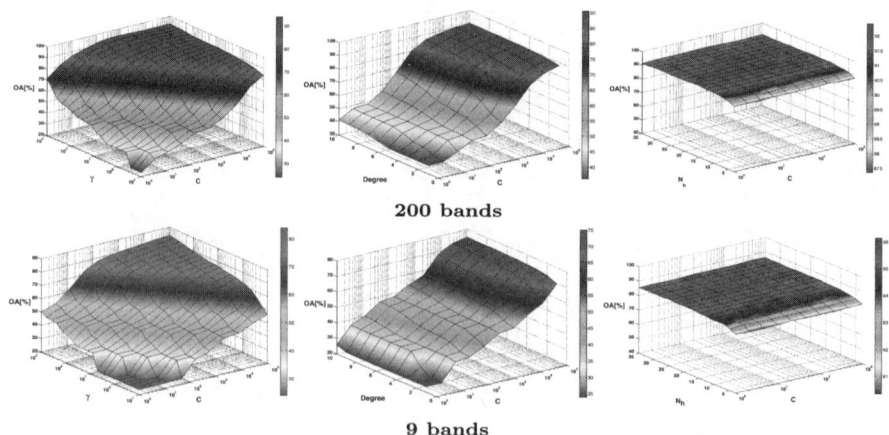

Fig. 4. Overall accuracy in the validation set as a function of the free parameters using (*left*) SVM-RBF (*middle*) SVM-POLY and (*right*) Reg-AdaBoost for different number of spectral bands (200 and 9)

for almost all values of C and a proper value of N_h. This confirms the ease of use of the proposed Reg-AdaBoost.

Robustness to Noise

Let us analyze the issue of robustness to noise in the case of remote sensing images. Figure 5 shows the evolution through different Signal-To-Noise Ratios (SNRs) of the overall accuracy of the models in the test set when either Gaussian or uniform noise is added to the spectral channels.

Once again, we varied the SNR between 4 and 40 dB, and 50 realizations were averaged for each value. Several conclusions can be extracted. First, in the case of Gaussian noise, we can see that Reg-AdaBoost outperforms the SVM approaches in almost all noisy situations and number of available bands. This gain is more noticeable in the high noise level range (SNR < 10 dB). Both RBF and polynomial kernels perform similarly. All methods outperform LDA by large. Note that in the case of Gaussian noise, the loss function is more appropriate and thus the Reg-AdaBoost algorithm shows a better behaviour in this situation. Second, in the case of uniform noise, SVM with RBF kernels shows a better performance in general terms (due to its cost function) but show inferior results to Reg-AdaBoost in the high noise regime

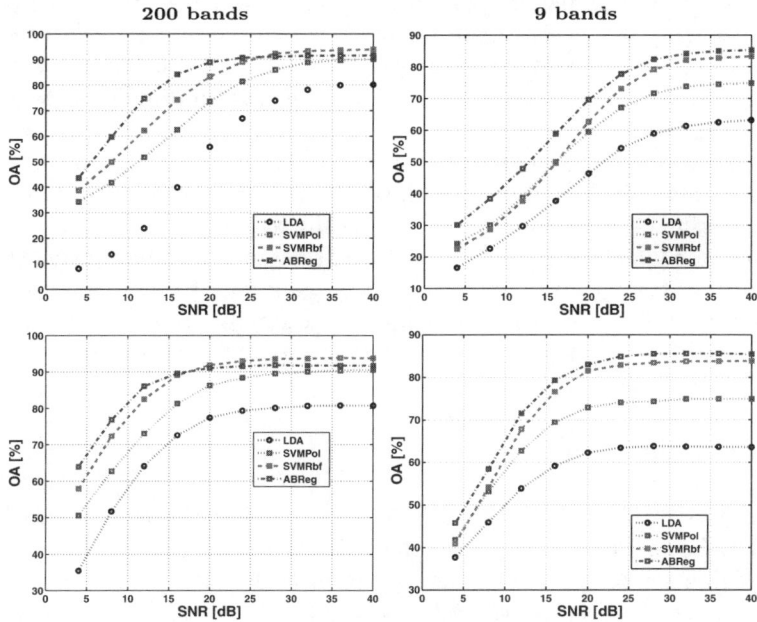

Fig. 5. Overall accuracy (OA (%)) in the validation set vs different amounts of (*top*) Gaussian or (*bottom*) uniform noise added to input features for the different number of spectral bands (200 and 9)

(SNR $<$ 20 dB) and in all noise regimes for low dimensional spaces (9 bands). In these situations, the LDA performs poorly again.

Classification Maps

Figure 6 shows the best classified images with all models in all different situations (number of bands). The numerical results shown in Table 3 are confirmed by inspecting these classification maps, where some classes are better identified by the proposed Reg-AdaBoost algorithm, and smoother classification maps are obtained, more noticeable for the minority classes, class borders, and mainly with low number of bands. These results confirm that the presented approach of regularizing the AdaBoost (smoother classification functions) and soft margin (more versatile classification boundaries) is beneficial in high-dimensional and high noise level scenarios.

5 Discussion and Conclusions

In this chapter, we have presented a soft margin (regularized) AdaBoost method for hyperspectral image classification. The method has been theoretically motivated by looking at the problems observed in the standard AdaBoost algorithm, particularly in the context of high dimensional possibly noisy problems. Extensive experimental evaluation support the theoretical properties presented. Summarizing, this method can be useful in low to moderate signal-to-noise ratio environments and when input dimension is relatively high.

Future work will consider integrating contextual information in the methodology. Certainly, inclusion of spatial, contextual or even textural information in the classifiers is of paramount important in image classification in general and in hyperspectral image classification in particular given the high variability of the spectral signature. An elegant way to include multi-source information in kernel-based classifiers was originally presented in [40], which could be easily extended to the classifiers presented in this chapter.

Acknowledgments

The authors would like to thank Prof. Landgrebe for providing the AVIRIS data, Dr. C. C. Lin for providing the libSVM software for SVMs, and Prof. Günnar Raetsch for providing the Reg-AdaBoost source code.

This work has been partially supported by the Spanish Ministry for Education and Science under project DATASAT ESP2005-07724-C05-03, and by the "Grups Emergents" programme of Generalitat Valenciana under project HYPERCLASS/GV05/011.

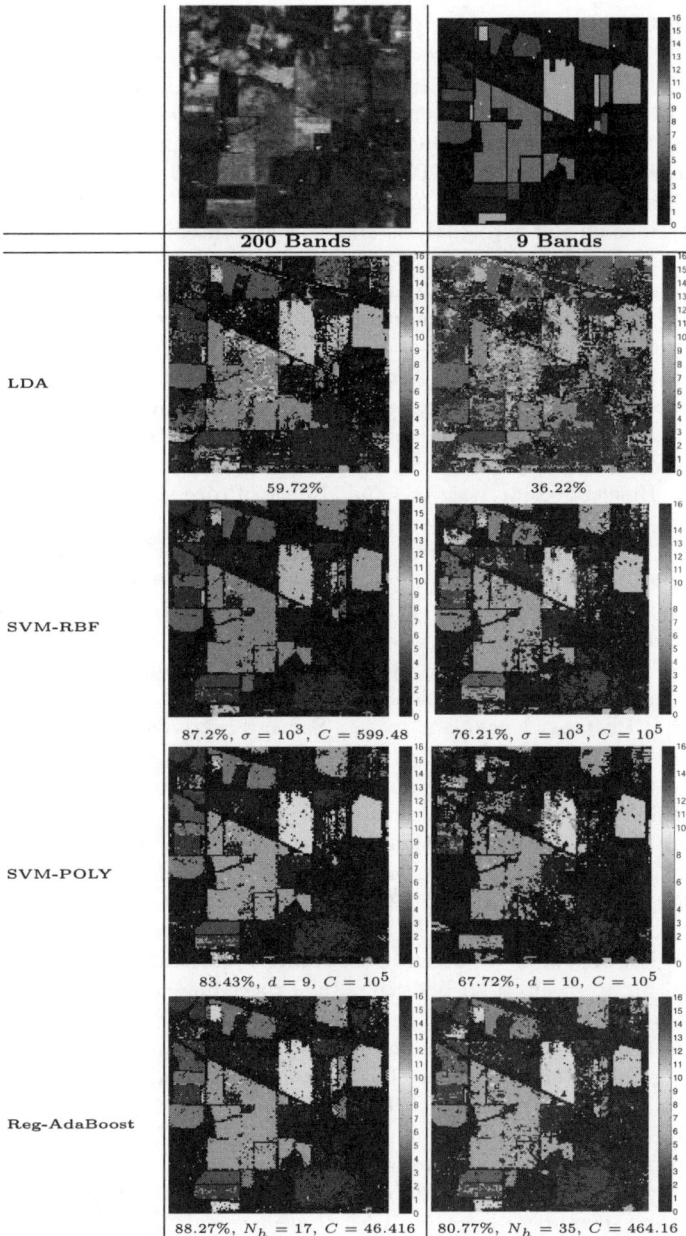

Fig. 6. *Top panel*: Original hyperspectral image: three-channel false color composition and the true classification map. *Bottom panel*: Best thematic maps produced with the LDA, SVM-RBF, SVM-POLY, and Reg-AdaBoost for different number of bands. Model parameters and overall accuracy for the whole image are indicated below each mapping

References

1. Richards, J.A., Jia, X.: Remote Sensing Digital Image Analysis. An Introduction, 3rd edn. Springer, Berlin Heidelberg New York (1999)
2. Hughes, G.F.: On the mean accuracy of statistical pattern recognizers. IEEE Transactions on Information Theory **14** (1968) 55–63
3. Fukunaga, K., Hayes, R.R.: Effects of sample size in classifier design. IEEE Transcations on Pattern Analysis and Machine Intelligence **11** (1989) 873–885
4. Duda, R.O., Hart, P.E., Stork, D.G.: Pattern Classification and Scene Analysis: Part I Pattern Classification, 2nd edn. Wiley, New York (1998)
5. Efron, B., Tibshirani, R.J.: An Introduction to the Bootstrap. Chapman & Hall, New York (1993)
6. Vapnik, V.N.: Statistical Learning Theory. Wiley, New York (1998)
7. Schölkopf, B., Smola, A.: Learning with Kernels – Support Vector Machines, Regularization, Optimization and Beyond. MIT, Cambridge, MA (2002)
8. Camps-Valls, G., Rojo-Álvarez, J.L., Martínez-Ramón, M., eds.: Kernel Methods in Bioengineering, Signal and Image Processing. Idea, Hershey, PA (2006)
9. Camps-Valls, G., Gómez-Chova, L., Calpe, J., Soria, E., Martín, J.D., Alonso, L., Moreno, J.: Robust support vector method for hyperspectral data classification and knowledge discovery. IEEE Transactions on Geoscience and Remote Sensing **42** (2004) 1530–1542
10. Camps-Valls, G., Bruzzone, L., Rojo-Álvarez, J.L., Melgani, F.: Robust support vector regression for biophysical parameter estimation from remotely sensed images. IEEE Geoscience and Remote Sensing Letters **3** (2006) 339–343
11. Camps-Valls, G., Bruzzone, L.: Kernel-based methods for hyperspectral image classification. IEEE Transactions on Geoscience and Remote Sensing **43** (2005) 1351–1362
12. Schapire R.: A brief introduction to boosting. In: Proceedings of the Sixteenth International Joint Conference on Artificial Intelligence (1999)
13. Breiman, L., Friedman, J.H., Olshen, R.A., Stone, C.: Classification and Regression Trees. Wadsworth & Brooks, Pacific Grove, CA (1984)
14. Freund, Y., Schapire, R.: A Decision-theoretic generalization of on-line learning and an application to boosting. In: Proceedings of the Second European Conference on Computational Learning Theory, LNCS (1995)
15. Schwenk, H., Bengio, Y.: Training methods for adaptive boosting of neural networks for character recognition. In: Advances in Neural Information Processing Systems 11, Proceedings of NIPS'98 (1999)
16. Tieu, K., Viola, P.: Boosting image retrieval. In: Proceedings IEEE Conference on Computer Vision and Pattern Recognition (2000)
17. Schwenk, H.: Using boosting to improve a hybrid HMM/neural network speech recognizer. In: Proceedings of IEEE International Conference on Acoustics, Speech and Signal processing (1999) 1009–1012
18. Friedl, M.A., Brodley, C.E., Strahler, A.H.: Maximizing land cover classification accuracies produced by decision trees at continental to global scales. IEEE Transactions on Geoscience and Remote Sensing **34** (1999) 969–977
19. McIver, D.K., Friedl, M.A.: Estimating pixel-scale land cover classification confidence using nonparametric machine learning methods. IEEE Transactions on Geoscience and Remote Sensing **39** (2001) 1959–1968

20. Chan, J.C-W., Huang, C., DeFries, R.: Enhanced algorithm performance for land cover classification from remotely sensed data using bagging and boosting. IEEE Transactions on Geoscience and Remote Sensing **39** (2001) 693–1968
21. Briem, G.J., Benediktsson, J.A., Sveinsson, J.R.: Multiple classifiers applied to multisource remote sensing data. IEEE Transactions on Geoscience and Remote Sensing **40** (2002) 2291–2299
22. Nishii, R., Eguchi, S.: Supervised image classification by contextual AdaBoost based on posteriors in neighborhoods. IEEE Transactions on Geoscience and Remote Sensing **43** (2005) 2547–2554
23. Rätsch, G.: Robust Boosting via Convex Optimization. Ph.D. thesis, University of Potsdam (2001)
24. Quinlan, J.R.: Boosting first-order learning. In: Arikawa, S., Sharma, A., eds.: Proceedings of the 7th International Workshop on Algorithmic Learning Theory. Volume 1160 of LNAI., Springer, Berlin (1996) 143–155
25. Rätsch, G.: Ensemble learning methods for classification (in German). Master's thesis, Department of Computer Science, University of Potsdam (1998)
26. Rätsch, G., Schökopf, B., Smola, A., Mika, S., Onoda, T., Müller, K.R.: Robust ensemble learning. In: Smola, A., Bartlett, P., Schölkopf, B., Schuurmans, D., eds.: Advances in Large Margin Classifiers. MIT, Cambridge, MA (1999) 207–219
27. Freund, Y., Schapire, R.E.: A short introduction to boosting. Journal of Japanese Society for Artificial Intelligence **14** (1999) 771–780
28. Valiant, L.G.: A theory of the learnable. Communications of the ACM **27** (1984) 1134–1142
29. Schapire, R.E.: The strength of weak learnability. Machine Learning **5** (1990) 197–227
30. Breiman, L.: Bagging predictors. Machine Learning **26** (1996) 123–140
31. Tumer, K., Gosh, J.: Linear and order statistics combiners for pattern classification. In: Sharkey, A., ed.: Combining Artificial Neural Nets. Springer, London (1999) 127–162
32. Rätsch, G., Warmuth, M.K.: Maximizing the margin with Boosting. In: Proceedings of the Annual Conference on Computational Learning Theory. Volume 2375 of LNAI., Springer, Sydney (2002) 334–340
33. Rätsch, G., Onoda, T., Müller, K.R.: Soft margins for AdaBoost. Machine Learning **42** (2001) 287–320 also NeuroCOLT Technical Report NC-TR-1998-021, Department of Computer Science, Royal Hollaway, University of London, Egham, UK.
34. Mason, L., Baxter, J., Bartlett, P.L., Frean, M.: Functional gradient techniques for combining hypotheses. In: Smola, A.J., Bartlett, P., Schölkopf, B., Schuurmans, D., eds.: Advances in Large Margin Classifiers. MIT, Cambridge, MA (1999) 221–247
35. Müller, K.R., Mika, S., Rätsch, G., Tsuda, K.: An introduction to kernel-based learning algorithms. IEEE Transactions on Neural Networks **12** (2001) 181–201
36. Gualtieri, J.A., Cromp, R.F.: Support vector machines for hyperspectral remote sensing classification. In: Proceedings of the SPIE, 27th AIPR Workshop (1998) 221–232
37. Rifkin, R., Klautau, A.: In defense of one-versus-all classification. Journal of Machine Learning Research **5** (2004) 101–141
38. Landgrebe, D.: Multispectral data analysis: A signal theory perspective (1998) http://dynamo.ecn.purdue.edu/~biehl/MultiSpec/documentation.html.

39. Melgani, F., Bruzzone, L.: Classification of hyperspectral remote-sensing images with support vector machines. IEEE Transactions on Geoscience and Remote Sensing **42** (2004) 1778–1790
40. Camps-Valls, G., Gómez-Chova, L., Muñoz-Marí, J., Vila-Francés, J., Calpe-Maravilla, J.: Composite kernels for hyperspectral image classification. IEEE Geoscience and Remote Sensing Letters **3** (2006) 93–97

Convolutional Neural Networks for Image Processing with Applications in Mobile Robotics

Matthew Browne, Saeed Shiry Ghidary, and Norbert Michael Mayer

[1] CSIRO Mathematical and Information Sciences, Cleveland Australia,
matthew.browne@csiro.au
[2] Department of Computer Engineering and Information Technology, Amirkabir
University of Technology, Tehran Iran, shiry@ce.aut.ac.ir
[3] Department of Adaptive Machine Systems, Osaka University, Osaka Japan,
norbert@er.ams.eng.osaka-u.ac.jp

Summary. Convolutional neural networks (CNNs) represent an interesting method for adaptive image processing, and form a link between general feed-forward neural networks and adaptive filters. Two-dimensional CNNs are formed by one or more layers of two-dimensional filters, with possible non-linear activation functions and/or down-sampling. Convolutional neural networks (CNNs) impose constraints on the weights and connectivity of the network, providing a framework well suited to the processing of spatially or temporally distributed data. CNNs possess key properties of translation invariance and spatially local connections (receptive fields). The so-called "weight-sharing" property of CNNs limits the number of free parameters. Although CNNs have been applied to face and character recognition, it is fair to say that the full potential of CNNs has not yet been realised. This chapter presents a description of the convolutional neural network architecture, and reports some of our work applying CNNs to theoretical and real-world image processing problems.

1 Introduction

The term convolutional network (CNN) is used to describe an architecture for applying neural networks to two-dimensional arrays (usually images), based on spatially localized neural input. This architecture has also been described as the technique of shared weights or local receptive fields [6–8] and is the main feature of Fukushima's "neocognitron" [1, 2]. Beginning in 1979 with Fukushima's proposal for a translation invariant pattern recognition device [3], the properties and capabilities of CNNs (in their various forms) have been known for some time (see e.g. [4,5]). CNNs are related to time delay neural networks (TDNNs), which have had success in spoken word recognition, signature verification and motion detection. Le Cun and Bengio [9] note three

M. Browne et al.: *Convolutional Neural Networks for Image Processing with Applications in Mobile Robotics*, Studies in Computational Intelligence (SCI) **83**, 327–345 (2008)
www.springerlink.com

architectural ideas common to CNNs: local receptive fields, shared weights, and often, spatial down-sampling. Processing units with identical weight vectors and local receptive fields are arranged in an spatial array, creating an architecture with parallels to models of biological vision systems [9]. A CNN image mapping is characterized by the strong constraint of requiring that each neural connection implements the same local transformation at all spatial translations. This dramatically improves the ratio between the number of effective parameters and training cases, increasing the chances of effective generalization [10]. This advantage is significant in the field of image processing, since without the use of appropriate constraints, the high dimensionality of the input data generally leads to ill-posed problems. CNNs take raw data, without the need for an initial separate pre-processing or feature extraction stage: in a CNN, the feature extraction and classification occur naturally within a single framework.

To some extent, CNNs reflect models of biological vision systems [11]. CNN topology is somewhat similar to the early visual pathways in animals, which is also structured in a feed-forward layer topology. As in the brain, earlier layers ("simple cells") have relatively less shape specificity and rather more spatial specificity [25]. In later layers e.g. "hyper-complex cells") the neurons respond to a broader spatial region, but become increasingly specialised to respond to particular patterns or shapes. While neurons in later layers may become increasingly sensitive to certain patterns, they have the potential to develop insensitivity to other aspects of the stimulus. An important and intrinsic property of both biological and artificial architectures is their ability to become robust to variation in spatial location, micro-feature relation, rotation, and scale: obviously a critical property for recognising objects in the real world.

In the CNN architecture, the 'sharing' of weights over processing units reduces the number of free parameters, increasing the generalization performance of the network. Weights are replicated over the spatial array, leading to intrinsic insensitivity to translations of the input – an attractive feature for image classification applications. CNNs have been shown to be ideally suited for implementation in hardware, enabling very fast real-time implementation [12]. Although CNNs have not been widely applied in image processing, they have been applied to handwritten character recognition [7,12–14,24] and face recognition [10,11,15]. CNNs may be conceptualized as a system of connected feature detectors with non-linear activations. The first layer of a CNN generally implements nonlinear template-matching at a relatively fine spatial resolution, extracting basic features of the data. Subsequent layers learn to recognize particular spatial combinations of previous features, generating 'patterns of patterns' in a hierarchical manner. The features detected by each layer are combined by the subsequent layers in order to detect higher order features.

Once a feature has been detected, its exact location becomes less important. If down-sampling is implemented, then subsequent layers perform pattern recognition at progressively larger spatial scales, but at lower resolution. A CNN with several down-sampling layers enables processing of large spatial arrays, with relatively few free weights. Since all weights are learned, CNNs can be seen as synthesizing their own feature extractor. The weight sharing property has the important further effect of reducing the number of free parameters, thereby reducing the capacity of the machine and improving generalisation.

In the context of image processing, CNNs present a number of clear advantages compared to fully connected and unconstrained neural network architectures. Typical images are large, and without a specialised architecture, the number of free parameters in the network will quickly become unmanageable when presenting input directly to the network . Conventional applications of neural networks may resolve this by relying on extensive pre-processing of the images in order to render them into a suitable form. However, this leads to a hybrid two-stage architecture where much of the "interesting" work is done by the pre-processing stage, which of course will tend to be hard-wired and non-adaptive. Unstructured neural networks have no built-in invariance with respect to translations, or local distortions of the inputs. Indeed, a deficiency of fully connected architectures is that the topology of the input is entirely ignored. Images have a strong 2D local structure that are highly correlated. In general, we argue that a general CNN architecture is more suitable than a generic neural network whenever input information is arranged temporally or spatially.

Figure 1 illustrates the process of "iterated convolution" that is central to CNNs. The output array is generated by convolving a neural filter with adaptive weights with the feature array, which is itself generated by convolution with a second 5×5 filter operating on the input array. The output array is therefore the outcome of two convolution operations (leaving aside the non-linearity in the activation function for the moment). Even without down-sampling, the spatial region to which the pixel in the output array is sensitive grows to a 9×9, illustrating the progressive decrease in spatial sensitivity of later layers. In general, each unit in a layer receives inputs from a set of units located in a small neighborhood in the previous layer. With local connections (receptive fields), neurons can extract elementary visual features such as oriented edges, endpoints, corners. Note that for simplicity of explanation, Fig. 1 omits multiple feature arrays in each layer. Just as a standard neural network requires more than a single neuron in each hidden layer in order to implement arbitrary transformations, CNNs also utilize multiple feature arrays in each layer. Implementations of CNNs often progressively increase the number of feature arrays in later layers, while decreasing the spatial resolution.

Down-sampling of the feature array may be implemented between the convolution layers. For fixed filter sizes, this has the effect of increasing the spatial range of subsequent layers, or conversely reducing the level of spatial resolution. Figure 2 illustrates the effect of convolving two filters with a step size of 2 instead of one on an input array. The number of features encoded in

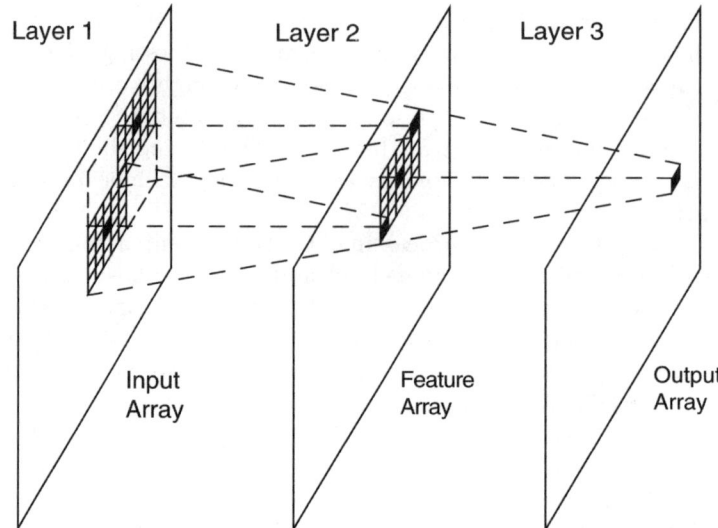

Fig. 1. Architecture of a CNN with a single convolutional neuron in two layers. A 5 × 5 filter at two different translations is shown mapping to shaded pixels in the first feature array. Shaded pixels in turn are part of the local feature information which is mapped to the output (or second feature) array by a second 5 × 5 filter

parallel has increased from 1 to 2, while the spatial resolution has decreased by a factor of 2. Each cell in such a feature map has the same set of afferent weights, but is connected to a square at a unique position in the input array. This point is important enough to reiterate: units in a CNN are organized in planes (feature maps) within which all the units share the same set of weights. This leads to the translation invariance property: if the input image is shifted, the feature map output will be shifted by the same amount, but is otherwise unchanged.

As mentioned previously, CNNs must utilize multiple feature maps in each layer in order to implement complex transformations. Figure 3 shows an example two layer CNN with multiple neurons in each layer, with down-sampling being implemented by the second layer. Also note that because there are three feature arrays in the second hidden layer and two feature arrays in the second layer, a total of six filters are required to map between the two. This should be regarded as a fully interconnected CNN layer. A number of researchers have found that fully interconnected CNN layers can result in too complex a network, and topologies can be pruned by removing filters as desired.

A final architecture diagram should assist in conceptualizing the 2:1 downsampling property of a CNN, which progressively transforms a spatial representation into a totally abstract coding. Figure 4 show an elementary network that maps 8 × 8 input arrays to a single output array via three layers, each consisting of a 2 × 2 weight vector corresponding to a single neuron. The

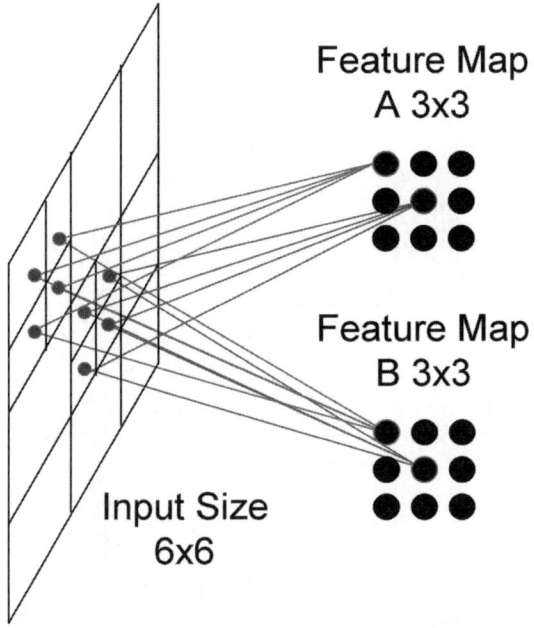

Fig. 2. Illustration of convolution and 2:1 downsampling of two 2×2 filters, generating two 3×3 feature maps from a 6×6 input array. Apart from implementing a local transformation, the spatial resolution has been increased by a factor of 2, while the 'semantic' information has been increased by a factor of 2

feature arrays are formed by convolving the weights vectors with the previous array, using a step size of 2. Because minimal size filters are used, there is no overlap between the input of spatially adjacent neurons at any given layer in the network. In terms of the economical use of free parameters, this represents maximally efficient processing of an image by a neural network. The network has 64 inputs and 84 connections, but due to the constraint of translation invariance, there are only 12 free weights (not including biases). This clearly demonstrates the ability of CNNs to operate on large input arrays with relatively few free parameters. Although the function approximation power of the network shown is trivial since only one array has been included in each layer, processing capacity can be added or removed at will, rather than being linked to the dimension of the input data. As with most neural networks, the exact architecture of a CNN should depend on the problem at hand. This involves determination of: the mixture of convolutional and down-sampling layers (if any), filter-weight sizes, number of neurons, and inter-connectivity between layers.

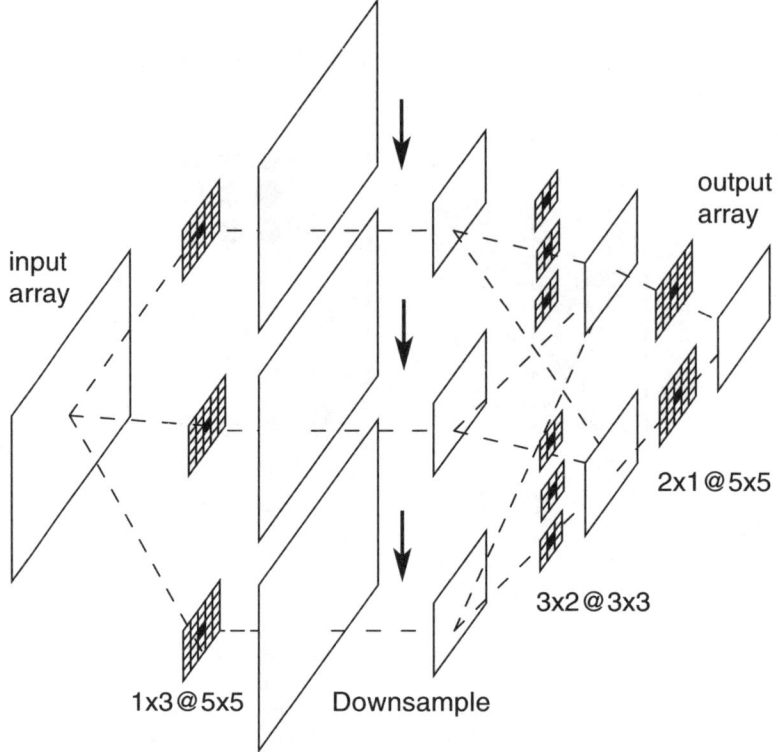

Fig. 3. Architecture of a "fully interconnected" (though not in the conventional neural network sense) CNN with multiple neurons in each convolution layer, and a factor of two translation in the second layer. Note that a total of six 3×3 filters are required to map the three feature maps in the second layer to the two feature maps in the third layer. Thus, the activation of a particular scalar point in the neural array in the third layer is a weighted scalar sum of 27 activations in the previous layer (3 local horizontal points $\times$ 3 local vertical $\times$ 3 arrays). Points within each spatial neural array share the same weights, but have different local inputs

2 Theory

CNNs perform mappings between spatially/temporally distributed arrays in arbitrary dimensions. They appear to be suitable for application to time series, images, or video. CNNs are characterized by:

- Translation invariance (neural weights are fixed with respect to spatial translation).
- Local connectivity (neural connections only exist between spatially local regions).
- An optional progressive decrease in spatial resolution (as the number of features is gradually increased).

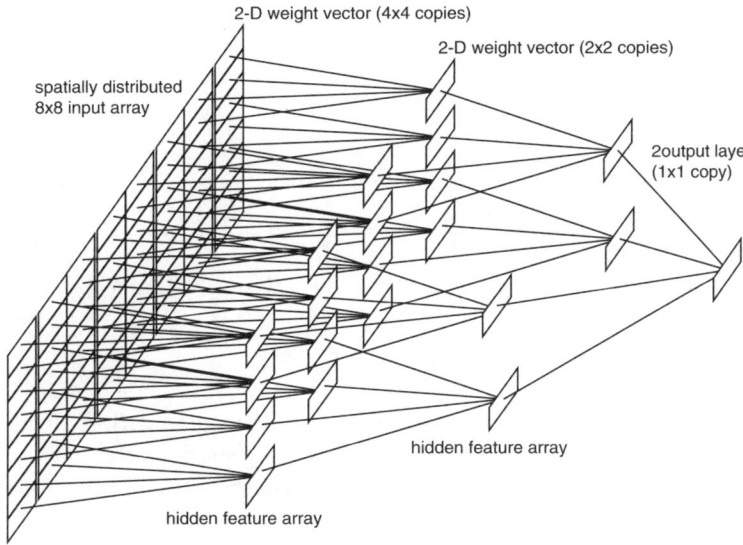

2-D weight vector (4x4 copies)

2-D weight vector (2x2 copies)

spatially distributed
8x8 input array

2output layer
(1x1 copy)

hidden feature array

hidden feature array

Fig. 4. A CNN architecture with non-overlapping receptive fields and minimal (i.e. length 2) filter lengths. Dyadic sampling in each layer progressively transforms an 8×8 spatial array into a single scalar output with no spatial extent

2.1 Formalism

The case of one dimensional input and a single hidden layer CNN is described. Extensions may be then be made to multiple dimensions and multiple layers, with possible operation of downsampling. We wish to obtain the formula for changing the weight and bias parameters of a CNN given a training set of input–output pairs $\xi_{k,r}^{\mu}, \zeta_{i,p}^{\mu}$. The indices i, j, k refer respectively to neuron arrays in the output, hidden, and input layers. In a CNN neurons are 'replicated' with respect to spatial translation, although they share the same weight and bias vectors. Indices p, q, r are used as a spatial index for each layer. A property of CNNs is that translated neurons $h_{j,q}^{\mu}$ receive only local connections from the previous layer – a CNN is not a fully interconnected network (see Fig. 3). When calculating the net input to a particular translated neuron, it is convenient to index the spatially distributed weights separately, using indices s, t, μ. The weights of a CNN are invariant to spatial translation, so it is natural to think of the set of weights $w_{j,k}^{t}, t = \{-T, .., 0, .., T\}$ as a filter that connects input array $\xi_{k,r}^{\mu}$ to feature array $V_{j,q}^{\mu}$. $2T + 1$ is the size of the region surrounding each translation point for which network weights exist, i.e., the filter size. It is constrained to be of an odd-numbered length, so that it is symmetrical about q. A summary of the indices used in the present paper is shown in Table 1.

Table 1. Indexes and array terms, organized with respect to layer and data type

	Output	Hidden	Input
Array label	O	V	ξ
Array index	i	j	k
Spatial index	p	q	r
Weight index	s	t	

Given input pattern μ hidden unit j, q receives net input

$$h_{j,q}^{\mu} = \sum_k \sum_t w_{j,k}^t \xi_{k,q+t}^{\mu} + b_j \tag{1}$$

where the index to ξ_k^{μ} is clamped to spatially local positions, centered at translation q, by setting $r = q + t$. The term b_j refers to the usual constant bias. The neural output forms the hidden feature arrays, produced by the transfer function

$$V_{j,q}^{\mu} = g\left(h_{j,q}^{\mu}\right). \tag{2}$$

The neuron at translation p in the ith array in the output layer receives net input

$$h_{i,p}^{\mu} = \sum_j \sum_s w_{i,j}^s V_{j,p+s}^{\mu} + b_i, \tag{3}$$

where, as before, $s = \{-S, .., 0, .., S\}$, and $2S + 1$ describes the length of the filter in the output layer, and relative indexing has been substituted for absolute indexing: $q = p + s$. Final output of the network is

$$O_{i,p}^{\mu} = g(h_{i,p}^{\mu}) = g\left(\sum_j \sum_s w_{i,j}^s V_{j,p+s}^{\mu} + b_i\right). \tag{4}$$

Although CNN weights have similar optimization rules as standard feedforward neural networks, weight sharing and local input regions add a degree of complexity that demands some care in implementation. For completeness, using the above notation, we briefly restate the standard derivative formulas with respect to the hidden-to-output weights

$$\frac{\partial E}{\partial w_{i,j}^s} = -\sum_{\mu,p} \left(\zeta_{i,p}^{\mu} - g\left(h_{i,p}^{\mu}\right)\right) g'\left(h_{i,p}^{\mu}\right) V_{j,p+s}^{\mu}, \tag{5}$$

and the weight update rules for both layers under gradient descent learning with learning rate η:

$$\Delta w_{i,j}^s = \eta \sum_{\mu,p} \delta_{i,p}^{\mu} V_{j,p+s}^{\mu} \tag{6}$$

$$\Delta w_{j,k}^s = \eta \sum_{\mu,q} \delta_{j,q}^{\mu} \xi_{k,q+t}^{\mu} \tag{7}$$

$$\delta^{\mu}_{i,p} = \left(\zeta^{\mu}_{i,p} - g\left(h^{\mu}_{i,p}\right)\right) g'\left(h^{\mu}_{i,p}\right) \tag{8}$$

$$\delta^{\mu}_{j,q} = \sum_{g}{}'(h^{\mu}_{j,q}) \sum_{i} \delta^{\mu}_{i,q-s} w^{s}_{i,j}. \tag{9}$$

Bias update rules are similar.

2.2 Subsampling

Often when applying CNNs we wish to progressively reduce spatial resolution at each layer in the network. For example, a CNN may be used for classification where an image is mapped to a single classification output. Given fixed filter sizes, reducing spatial resolution has the effect of increasing the effective spatial range of subsequent filters. In a CNN with sub-sampling in each layer, the outcome is a gradual increase in the number of features used to describe the data, combined with a gradual decrease in spatial resolution. Because the change in coordinate system is accomplished in a nonlinear, incremental, hierarchical manner, the transformation can be made insensitive to input translation, while incorporating information regarding the relative spatial location of features. This provides an interesting contrast to methods such as principle components analysis, which make the transition from normal coordinate space to feature space in a single linear transformation.

We can rewrite the previous formulas for calculating the output of the network, given that both layers incorporate spatial sub-sampling. This has been previously accomplished using a separate 'averaging' layer with fixed neural weights [7]. However, it is described below by increasing the shift indexes by a factor of two, thus combining adaptive and downsampling functions. Since the averaging layer in the method of Le Cun [7] may be specified by a 'double shift' layer with a filter size of 2, it may be shown that the present formalism is essentially similar, albeit more general, and allowing for adaption of previously fixed averaging weights.

$$h^{\mu}_{j,q} = \sum_{k}\sum_{t} w^{t}_{j,k}\xi^{\mu}_{k,2q+t} + b_j, \tag{10}$$

$$h^{\mu}_{i,p} = \sum_{j}\sum_{s} w^{s}_{i,j}V^{\mu}_{j,2p+s} + b_i. \tag{11}$$

The output of the network is then

$$O^{\mu}_{i,p} = g\left(\sum_{j}\sum_{s} w^{s}_{i,j}g\left(\sum_{k}\sum_{t} w^{t}_{j,k}\xi^{\mu}_{k,4p+2s+t} + b_j\right) + b_i\right). \tag{12}$$

For a general CNN with N layers, being some combination of non subsampling and subsampling layers, and filter sizes being given by F_n, the local region of input contributing to the output is given by the recursive formulas

$$R_{n+1} = R_n + F_n - 1 \text{ if nonsubsampling,} \qquad (13)$$

$$R_{n+1} = 2(R_n + F_n) - 3 \text{ if subsampling.} \qquad (14)$$

given $R_1 = F_1$. Given fixed filter sizes F_n, it is clear that the input 'window' of CNN may grow rapidly as the number of sub-sampling layers increase. The size of the input array and the total set of sub-sampling layers may be tailored so that the size of the output array is 1×1, or simply scalar. This results in a CNN architecture suitable for image classification tasks. For sub-sampling layers, the weight update rules are identical to those given above, with the shift indices p and q increased by a factor of two.

3 Illustrative Examples

3.1 Synthetic Spatial XOR

Here we shall detail a numerical experiment that is used partly to elucidate the essential properties of CNNs, and partly in order to confirm that CNNs can learn spatially invariant, non-linear filtering systems.

Small 4×4 pixel input arrays were considered. Each input array considered of two 'micro features' with a 2×2 pixel extent, either two pixels in top-left to bottom-right (type A) diagonal arrangement, or two pixel in bottom-left to top-right diagonal arrangement (type B). Each micro-feature was allowed to vary independently over all possible spatial locations in the input space.

To be specific, let $\Phi^{\tau_1}_{\tau_2}[m, n]$ be a 4×4 binary array with a micro-feature located at $[\tau_1, \tau_2], \tau \in \{1, 2, 3\}$ with Φ being equivalent either to class A or B. Then

$$\text{if } \Phi \equiv \text{A}, \ \Phi^{\tau_1}_{\tau_2}[m, n] = \begin{cases} 1 & \text{if } m = \tau_1 \wedge n = \tau_2 | m = \tau_1 + 1 \wedge n = \tau_2 + 1, \\ 0 & \text{otherwise,} \end{cases} \quad (15)$$

$$\text{if } \Phi \equiv \text{B}, \ \Phi^{\tau_1}_{\tau_2}[m, n] = \begin{cases} 1 & \text{if } m = \tau_1 + 1 \wedge n = \tau_2 | m = \tau_1 \wedge n = \tau_2 + 1, \\ 0 & \text{otherwise.} \end{cases} \quad (16)$$

A particular input array ξ was defined as $\xi = \Phi^{\tau_1}_{\tau_2} | \Phi^{\tau_1}_{\tau_2}$ with $|$ denoting an element-by-element OR operation.

Considering input arrays of two micro-features, four combinations are possible: AA, BB, AB, BA. Figure 5 displays class AA and class BB while Fig. 6 displays the combined class AB ∪ BA.

The first task was to test that a CNN could correctly differentiate between classes AA and BB. Casual inspection of Fig. 2 would indicate that the various permutations and interference between the micro features creates a non-trivial problem for a normal neural network. It was expected that a CNN with $2@2 \times 2 : 1@3 \times 3$ filters and 29 free weights was sufficient to differentiate between the

Class AA

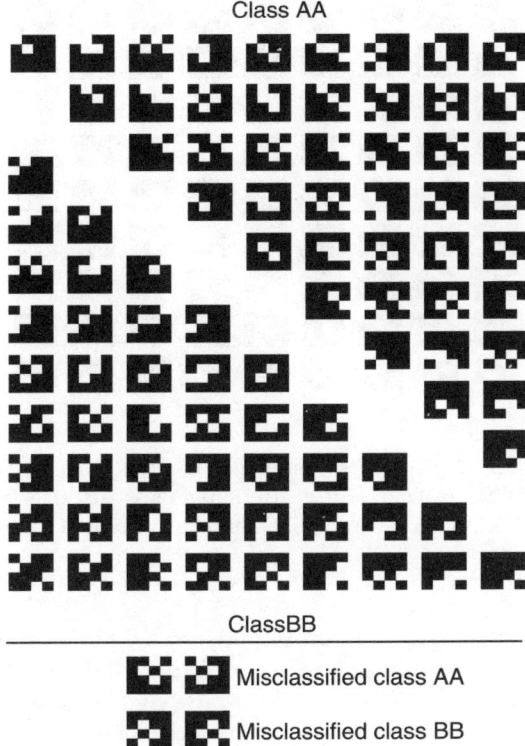

ClassBB

Misclassified class AA

Misclassified class BB

Fig. 5. Enumeration of all inputs comprising homogenous micro-features for the synthetic spatial XOR problem. Differentiation of classes AA and BB requires translation invariance and tolerance of interference between features

two classes. The network successfully classified all inputs except those shown in the lower section of Fig. 5. In these cases, interference between the micro features creates identical patterns. Although the CNN can theoretically learn purely spatial differences, sensitivity to pure translation would have violated the underlying decision rule required to classify the rest of the data set, namely orientation differences.

The second task was more challenging: the CNN was required to differentiate between class AB/BA and class AA/BB. This represents a kind of spatial X-OR problem, with the added difficulty of interference between features. A CNN with 4@2 × 2 : 3@3 × 3 CNN was trained to convergence on this data set. 100% performance was obtained on this problem, demonstrating that a minimal CNN architecture can learn a non-linear translation-invariant mapping in two dimensional feature space. The key property of the first layer of the CNN is that only a single template is learnt for all possible translations.

Class AB / BA

Fig. 6. Enumeration of all inputs comprising heterogenous micro-features for the synthetic spatial XOR problem. Differentiation of class AB∪BA from class AA∪BB represents a spatial XOR problem requires implementation of a translation invariant nonlinear spatial filter system

The following layer integrates information across different templates and spatial translations, enabling approximation of the X-OR function. Thus, CNN units de-couple detection of a features shape and location. In conventional spatial feature extraction (principle component 'eigenfaces' are a prominent example) shape and location are clamped. Thus, a separate representation of a feature must be learnt over the entire of range spatial locations where it may occur.

3.2 Detecting Road Markers

Figure 7 displays the application of a simple CNN to a toy problem of detecting road markers from a camera image mounted above an intersection. It may be seen that the feature arrays in layer 1 capture simple features, while the feature arrays in the second hidden layer capture have more complex responses. The final output of the system does a good job of detecting road markers despite low degrees of freedom, a high degree of noise, and the presence of many distracters such as cars or road signs. Note that certain high-contrast line segments on the bus and sign represent a potential source of confusion for the road-marker detector. This is overcome by recognising the "context" of the bus in terms of its distinctive characteristics at a larger scale. This information utilized in the final output array to effectively rule out "road-marker-like" objects that occurred within the bus and sign objects.

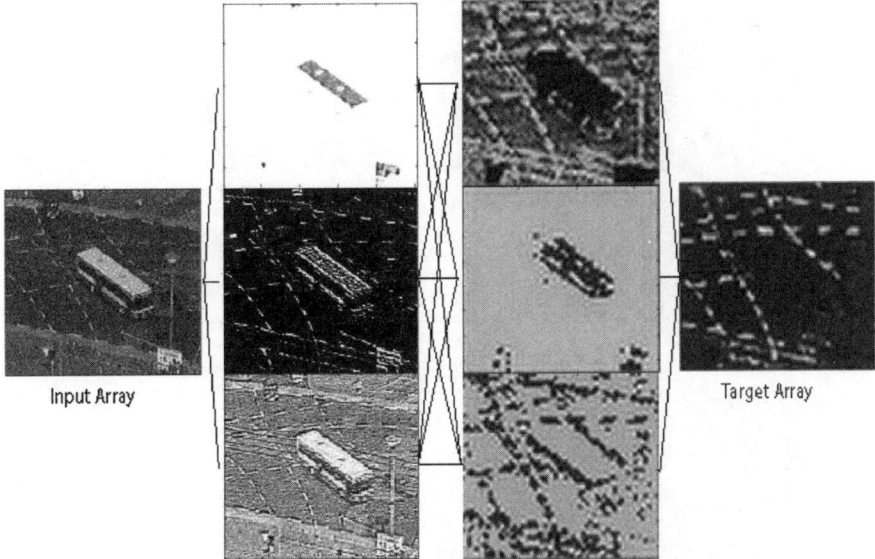

Fig. 7. Input, feature, and output arrays of a convolution network applied to detecting road markers. The highlighted zones in the output array correspond relatively well to the actual locations of road markers in the input array. Intermediate representations provide some insight into the structure of the transformation implemented

4 Applications in Mobile Robotics

4.1 Crack Delineation

Introduction

This application of a CNN involves an applied problem in mobile robotics. In particular, the visual system of a KURT2 autonomous mobile robot [16] designed for sewer inspection and equipped with an infra-red video camera. The task of the robot is the on-line detection of cracks and other faults in sewer pipes. The cracks are defined by a relatively distinctive space-frequency structure. However, they are often embedded in a variety of complex textures and other spurious features. Lighting and reflection effects present an additional source of difficulty. The role of the CNN is to implement a pre-processing stage, eliminating spurious features, noise and lighting artifacts from the image before passing to a separate crack-detection stage. A CNN with relatively little sub-sampling, and a large output array was implemented, yielding output similar to a filtering of the original image.

Detailed context of the image processing system on an autonomous mobile sewer inspection robot are provided in [17–19]. The task of the system is autonomous detection and characterization of cracks and damage in sewer

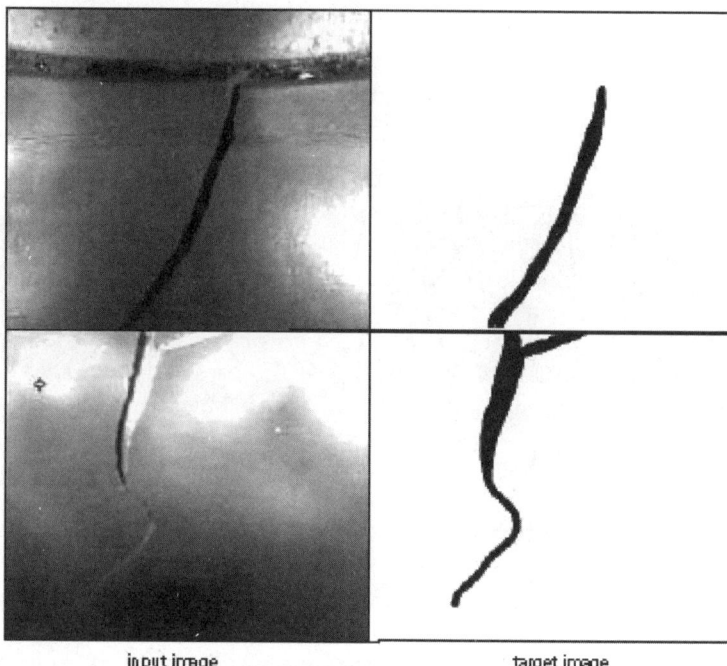

input image   target image

Fig. 8. Example input and target images for large cracks on a concrete pipe. Note the horizontal feature in the upper left image is a pipe joint, not a crack. Differentiating between pipes and joints, accounting for shadows and lighting effects are significant challenges for a detection/filtering system

pipe walls. The robot scans the pipe wall using a monochrome CCD camera, which is digitally converted at a resolution of 320 × 240 pixels per frame. The task of the CNN is to perform filtering of the raw pixel data, identifying the spatial location of cracks, enabling subsequent characterization of the length, width, etc of the damage. Figure 8 displays a sample input frame, along with the ideal output of the CNN. Although the cracks are easily identifiable by eye, the image processing task is quite complex, as variability in lighting and orientation, width and type of crack, along with the presence of other crack-like structures (such as joins between pipe sections), combine to make a challenging computer vision task.

Method

A representative data set of 37 frames from an on-line recording session were manually classified for training and validation of the network. A training data set was generated using 20 of the images, sampling 100 crack and 200 non-crack pixel locations, yielding a total of 6,000 input-target pairs. Although all pixel locations had associated targets, not every pixel was used for training because: (a) computational expense (b) the low proportion of 'crack' to 'clean' training samples tended to bias the network towards classifying all samples as 'clean'. Alternatively, it would be possible to use a biased learning procedure, where the error generated by the rarer class would be weighted in inverse proportion to the ratio of occurrence.

The CNN architecture used involved a total of five layers: a single input and output map, and three hidden layers. The filter sizes used in all cases were 5×5, and the common activation function used was a log-sigmoid. The number of feature maps used in the three hidden layers was, from input to output: $4, 3, 2$. Thus, the number of neural weights to be optimized was $(5^2 + 1)(1\cdot4 + 4\cdot3 + 3\cdot(2 + 2)\cdot1) = 624$ while the input to the network was a square sub-region of the input image, with side lengths of $(((5\cdot2 + 4) + 4) + 4) = 68$ pixels, yielding a total of 4,624 pixel inputs to the network. These figures are indicative of the degrees-of-freedom saving achieved by the weight sharing method. Training was conducted using standard weight updated rules, as described about, for 10,000 epochs, using a learning rate $\eta = 0.05$, requiring approximately two hours of compute time. The network was implemented in C++.

Discussion

Approximately 93% of pixels in the validation set were correctly classified by setting a simple hard-max selection function of the output layer. However, as the current application relates to an image processing/filtering task, with non-independence between nearby input-target pairs, numerical results (i.e. percentage of pixels correctly/incorrectly classified) are less informative than graphical results. Figure 9 displays three example frames, that were representative of the data set, including crack present, no crack and pipe joint, and crack and joint together. The network appears to have successfully ignored the presence of joints, and attenuated lighting effects while enhancing the cracks. In the context of the application to sewer pipe defect detection and characterization, the output was profitably used by a subsequent crack detection algorithm.

The present application of a general CNN architecture to a 'real-world' problem results in a practical video pre-processing system of some interest to the civil robotics community. CNNs may be expected to achieve significantly better results than standard feed-forward networks for this kind of task because they impose appropriate constraints on the way the function mapping is learnt. The key characteristics of local connectivity, and translation invariant weight sharing, are appropriate when the input data is spatially or temporally distributed. In addition, implementing down-sampling allows the network to progressively trade-off resolution for a greater input range.

There are some practical recommendations for this form of image filtering application. Training using images results in an over abundance of training sample points. It appears that, in order to maximize the available computational resources, that not all pixel-centre points of the training set should

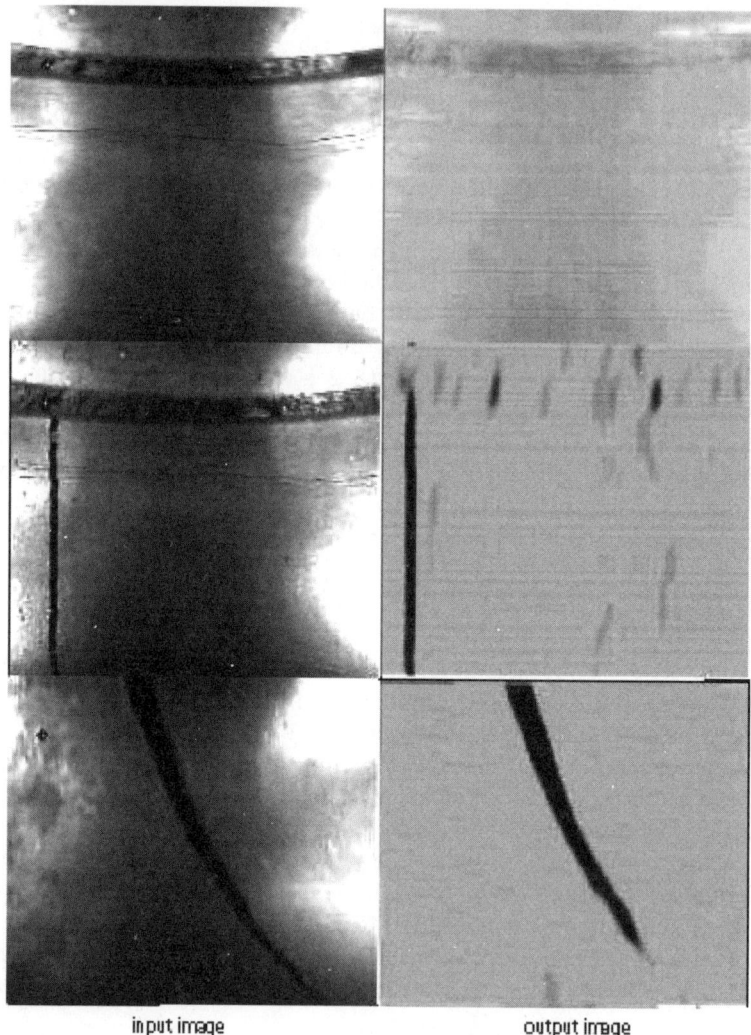

input image   output image

Fig. 9. Example input and CNN output frames Note that because of the 68 pixel window required by the network, the output image represents is subregion of the input

be used. Rather, a representative sub-sample would be more appropriate. In the present application, this might mean over-sampling joints as opposed to flat-wall regions, or thin cracks as opposed to thick cracks. This may be done manually by hand-coding multiple targets, and sampling each one equally. Alternatively, some statistical criteria might be developed for selection of a representative data subset.

4.2 Landmark Detection

Introduction

Autonomous sewer robots must navigate independently the sewer pipe system using only available information from sensor systems. The task of accurately detecting and classifying relevant landmarks and features in the environment is an essential part of the navigation routines. Because video is often used for the inspection work, performing detection of cracks and other faults, it is useful if the same data can be utilized for landmark detection.

Paletta, Rome and Platz [23] used a multi-stage system for landmark detection based on video data. This involved an attention controller, pre-processing, feature extraction, probabilistic interpretation, and final classification. Early components of the system operate according to fixed a priori rules while latter components are data-driven and adaptive. A related work for sewer pipe navigation include procedures for navigating under uncertainty [22]. Other pattern recognition approaches designed for use in autonomous sewer inspection include application of neural architecture to segmentation of pipe joints [21] and reconstruction of a 3D model of the interior of the pipe based on video data [20].

In the context of the navigation of a robot in an uncertain environment, the use of CNNs for video processing is consistent with the autonomous agent perspective [23], holding that interpretation of sensor images should be learnt from experience, and modeling of features of interest takes place in terms of their appearance to the agent. This is therefore a strong data-based perspective: analytical approaches towards constructing an objective model of the objects in question are rejected in favor of methods that directly learn from experienced sensor data. As an entirely trainable system, pre-processing and feature detection, transformation and classification modules are integrated into a single adaptive CNN architecture. Convolutional neural networks (CNNs) form the theoretical basis that makes such an approach possible.

The current project involves video data gathered from a robot fitted with an omnidirectional camera with the task of navigating and inspecting civil sewer pipe systems. Effective navigation requires that the robot be capable of detecting pipe inlets and pipe joints using the same sensor data used for fault detection, namely the omnidirectional video image data. Sewer pipes represent a relatively impoverished environment in terms of the types of stimuli generally encountered, making an video-based landmark detection method feasible. After detection of landmarks, a topological map of the sewer system was constructed, and used to navigate autonomously within the system. The aim of the present study was to train an image processing system that would be capable of recognising the three most important state-locations for navigation: pipe-inlets, pipe-joints, and clear pipe sections.

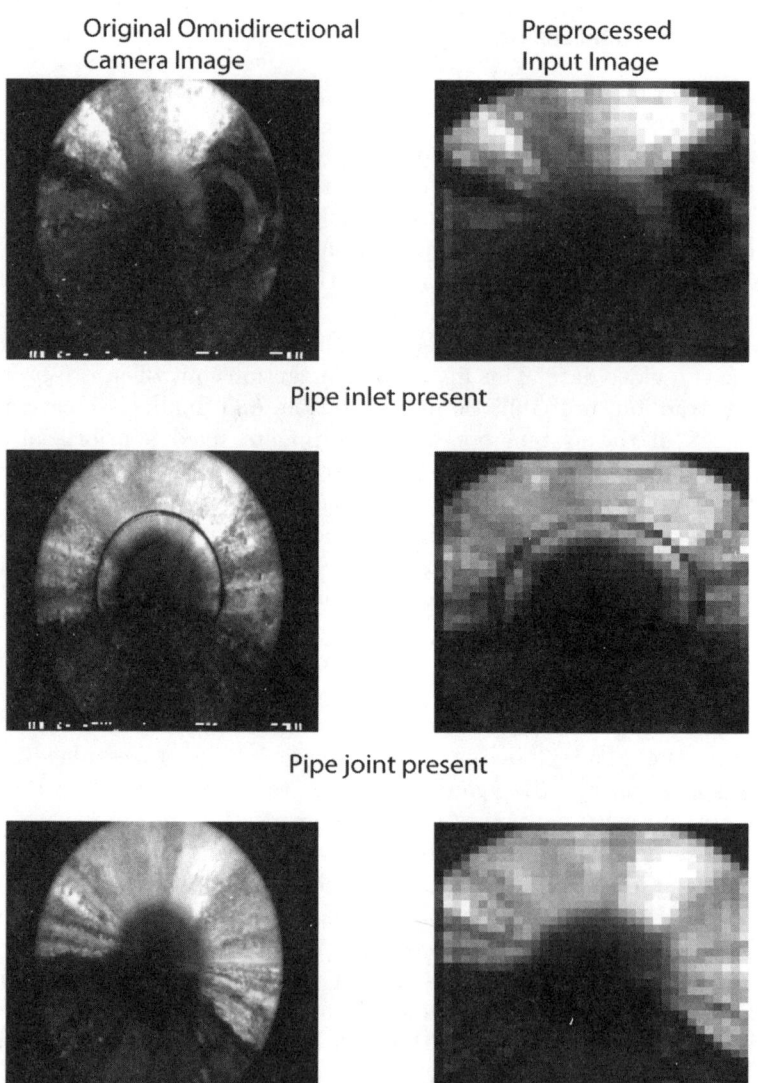

Original Omnidirectional
Camera Image

Preprocessed
Input Image

Pipe inlet present

Pipe joint present

Clear pipe

Fig. 10. Examples of three classes of omnidirectional camera images for classification by the CNN, corresponding to the three important navigational regions
encountered by the sewer robot

Method and Results

Figure 10 displays examples of each sensor data class. The CCD original
camera image frames were truncated and down sampled to arrays of 36 ×
36 pixels. It was found that this was approximately the maximum amount

Table 2. Specification of the CNN architecture used for omni-directional CCD processing

Layer	1	2	3	4	5	6
Filter size	5	2	2	2	2	2
Map size	32	16	8	4	2	1
Downsampling	No	Yes	Yes	Yes	Yes	No
Transfer fun	Tan	Tan	Tan	Tan	Tan	Log

of downsampling that could be achieved while preserving enough resolution for detection of the relatively finely structured pipe joints. Standard 1:255 valued black and white pixel intensities were normalized across frames to lie within a range of 0:1. Dirt, reflections, and changes in reflectance represented a challenge in this classification task. However, the objects to be detected had a relatively simple structure. The main challenge for classification of this kind of spatial input by machine learning methods is the size of the input ($36^2 = 1{,}296$ inputs) and variability in the scale and location of features as they pass through the robot's visual field. Table 2 shows the relatively simple architecture used for classification of the camera frames. Only one feature map per layer was used to detect pipe joints and pipe inlets, respectively. The 'inlet detector' and the 'joint detector' subnetworks each consisted of a total 51 free weights, including biases to each 2D weight vector. We note that the input size of the 36×36 was somewhat tailored to the architecture, since application of a 5×5 convolution filter without downsampling results in a 32×32 array, which is convenient for subsampling to a single 1×1 array, corresponding to the class prediction. Tan-sigmoid ($g(x) = 2/(1 + \exp(-2x)) - 1)$) and log-sigmoid ($g(x) = 1/(1 + \exp(-x))$) transfer functions were used.

The networks were trained using 160 manually classified video frames drawn from sample runs through the test pipe, consisting of an approximately equal number of 'joint present', 'inlet present', and 'nothing present' class samples. Training was performed for 1,000 epochs, using backpropagation, with momentum and an adaptive learning rate. After training, mean square error rates of 0.0022 and 0.0016 were obtained, which though relatively uninformative in itself, represents relatively good performance on data in the form of 0/1 targets. Validation was performed on a second set of 840 frames drawn from a video sample through the same pipe. For classification of continuous video data, subsequent frames are not statistically independent, and there also exists some 'grey area' where one class stops and the next starts. Thus, calculation of an overall correct classification rate is rather misleading. Of more practical interest was how the vision system responded as the robot traversed a number of known structures. This is illustrated in Fig. 11, which displays the output of the CNN, along with the actual state of the sewer pipe within the visual field of the robot. We note the uncertainty of the 'joint detector'

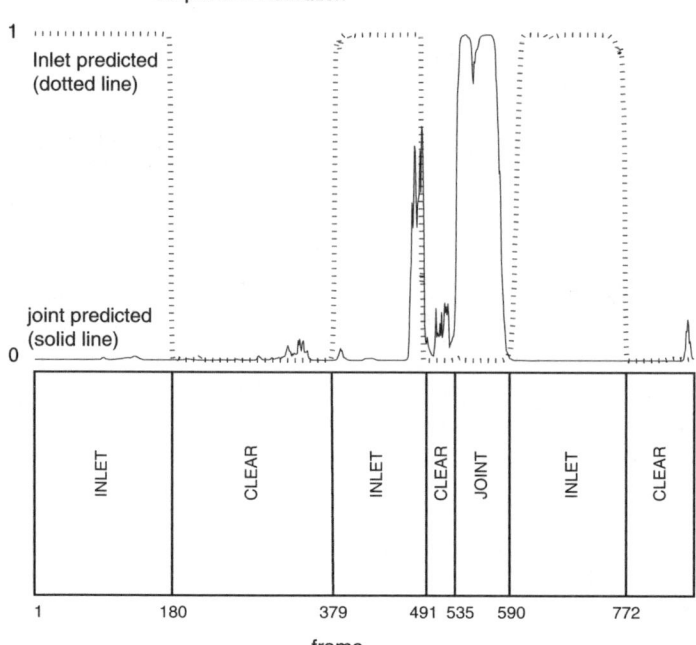

Fig. 11. Performance on continuous video input during a validation run of the robot with trained CNN vision system. Network output of the two landmark detectors is shown along with actual objects in the visual field of the omnidirectional camera

around frame 491 is due to a slightly unusual inlet construction that bears some similarity to a normal pipe joint.

Discussion

In the context of robot navigation, we have found the CNN properties of spatial invariance and weight constraints are necessary for application of machine learning methods to high dimensional image input, where features of interest may occur at a variety of spatial locations. The results in Fig. 11 were a positive indication that CNNs may be effectively applied to detecting navigational landmarks in the sewer-pipe environment. Subsequent field trials showed that the landmark detection system function effectively when sewer pipe width and lighting were similar to that of the test pipe structure. With appropriate thresholding, the activations of the 'inlet detector' and 'joint detector', would result in perfect landmark detection for the validation data. The system is computationally efficient, capable of processing image frames using an on-board processor in real time.

   With respect to the development of an industrial standard landmark-detection system, much work is required to train and test any adaptive landmark detection system with a wide variety of environments and lighting conditions. However, the results of this experiment were a promising indication of the effectiveness of CNNs for robot vision in sewer robots.

5 Conclusions

This chapter has presented a demonstration of the properties of CNNs using artificial data, a toy problem, and the application of the architecture to applied problems in robot vision. CNNs are shown to implement non-linear mappings of features with invariance to spatial translation. More precisely, CNNs decouple the learning of feature structure and feature location, and are therefore well suited to problems where the relative or absolute location of features has a degree of uncertainty.

In order to encourage more research in this field, the machinery of the CNN architecture requires further development. Although basic gradient descent with an adaptive learning rate is adequate (as used in our work), implementation of more advanced optimization techniques (such as Levenburg–Marquardt or conjugate-gradient optimization) is a priority. The basic CNN framework allows a wide variety of possible network architectures: pruning and growing algorithms for specification of the various network parameters are another topic of interest. Finally, although CNNs are an efficient method of applying neural networks to image processing, real-time processing of high definition images with a sophisticated architecture would benefit from development and application of dedicated hardware such as FPGAs.

As yet, the utility of CNNs does not appear to have been fully realized, and applications to a wide variety of data-types and function mapping problems (i.e. physiological recordings, financial time-series analysis, remote sensing) remain to be explored. In particular, through the implementation of 3-D filters, CNNs may represent a computationally feasible method of adaptive video processing. Refinements in the CNN architecture remain to be explored. For example, sequential CNN layers comprising $1 \times K$ and $K \times 1$ filters may be used to learn separable $K \times K$ filter functions. There are clear links between CNNs and finite impulse response filters, adaptive filters, and wavelet transforms, and theoretical work bridging these disciplines would be of significant interest.

References

1. Fukushima K, Miyake S, Ito T (1983) "Neocognitron: a neural model for a mechanism of visual pattern recognition," IEEE Transactions on Systems, Man, and Cybernetics, 13:826–834.
2. Fukushima K (1988) "Neocognitron: A hierachical neural network capable of visual pattern recognition," Neural Networks, 1(2):119–130.
3. Fukushima K (1979) "Neural-network model for a mechanism of pattern recognition unaffected by shift in position," Trans. IECE Japan, 62-A(10):658–665.

4. Lovell DR, Simon D, Tsoi AC (1993) "Improving the performance of the neocognitron," In Leong P, Jabri M (Eds.) Proceedings of the Fourth Australian Conference on Neural Networks, pp. 202–205.
5. Lovell DR, Downs T, Tsoi AC (1997) "An evaluation of the neocognitron," IEEE Trans. on Neural Networks, 8(5):1090–1105
6. Rumelhart DE, Hinton GE, Williams RJ (1986) "Learning internal representation by error propagation," In Rumelhart DE, McClelland JL (Eds.) Parallel Distributed Processing: Explorations in the Microstructure of Cognition, 1:318–362. MIT, Cambridge, MA.
7. Le Cun YB, Boser JS, Denker D, Henderson RE, Howard W, Hubbard W, Jackel LD (1988) "Backpropagation applied to handwritten zip code recognition," Neural Computation, 4(1):541–551.
8. Lang KJ, Hinton GE (1990) "Dimensionality reduction and prior knowledge in e-set recognition," In Touretzky, DS (Ed.) Advances in Neural Information Processing Systems, 178–185. Morgan Kauffman, San Marteo, CA.
9. Le Cun Y, Bengio Y (1995) "Convolutional networks for images, speech, and time series," In Arbib, MA (Ed.) The Handbook of Brain Theory and Neural Networks, 255–258. MIT, Cambridge, MA.
10. Lawrence S, Giles CL, Tsoi AC, Back AD (1997) "Face recognition: A convolutional neural network approach," IEEE Transactions on Neural Networks 8(1):98–113.
11. Fasel B (2002) "Robust face analysis using convolutional neural networks," In Proceedings of the International Conference on Pattern Recognition (ICPR 2002), Quebec, Canada.
12. Sackinger E, Boser B, Bromley J, LeCun Y (1992) "Application of the ANNA neural network chip to high-speed character recognition," IEEE Transactions on Neural Networks, 3:498–505.
13. Le Cun Y (1989) "Generalization and network design strategies," Tech. Rep. CRG-TR-89-4, Department of Computer Science, University of Toronto.
14. Bengio Y, Le Cun Y, Henderson D (1994) "Globally trained handwritten word recognizer using spatial representation, convolutional neural networks, and Hidden Markov Models," In Cowan JD, Tesauro G, Alspector J (Eds.) Advances in Neural Information Processing Systems, 6:937–944. Morgan Kaufmann, San Marteo, CA.
15. Fasel B (2002) "Facial expression analysis using shape and motion information extracted by convolutional neural networks," In Proc. of the International IEEE Workshop on Neural Networks for Signal Processing (NNSP 2002), Martigny, Switzerland.
16. Kirchner F and Hertzberg J (1997) "A prototype study of an autonomous robot platform for sewerage system maintenance," Autonomous Robots, 4(4):319–331.
17. Browne M, Shiry S, Dorn M, Ouellette R (2002) "Visual feature extraction via PCA-based parameterization of wavelet density functions," In International Symposium on Robots and Automation, pp. 398–402, Toluca, Mexico.
18. Browne M, Shiry Ghidary S (2003) "Convolutional neural networks for image processing: an application in robot vision," Lecture Notes in Computer Science, Springer, Berlin Heidelberg New York 2903:641–652.
19. Shiry Ghidary S, Browne M (2003) "Convolutional neural networks for robot vision: numerical studies and implementation on a sewer robot," In Proceedings of the 8th Australian and New Zealand Intelligent Information Systems Conference, 653–665, Sydney, Australia.

20. Cooper TPD, Taylor N (1998) "Towards the recovery of extrinsic camera parameters from video records of sewer surveys," Machine Vision and Applications, 11:53–63.
21. del Solar JR, K-Pen, R (1996) "Sewer pipe image segmentation using a neural based architecture," Pattern Recognition Letters 17:363–368.
22. Hertzberg J, Kirchner F (1996) "Landmark-based autonomous navigation in sewerage pipes," In Proceedings of First Euromicro Workshop on Advanced Mobile Robots (EUROBOT '96):68–73. Kaiserslautern, IEEE Press.
23. Paletta ERL, Pinz A (1999) "Visual object detection for autonomous sewer robots," In Proceedings of 1999 IEEE/RSJ International Conference on Intelligent Robots and Systems (IROS '99), 2:1087–1093. Piscataway NJ, IEEE Press.
24. Simard PY, Steinkraus D, Platt JC (2003) "Best Practices for Convolutional Neural Networks Applied to Visual Document Analysis," In Proceedigs of the Seventh International Conference on Document Analysis and Recognition (ICDAR '03):958–963. Washington, DC, USA.
25. Hubel DH, Wiesel TN (1959) "Receptive fields of single neurons in the cat's striate cortex," Journal of Physiology, 148:574–591.

SVM Based Adaptive Biometric Image Enhancement Using Quality Assessment

Mayank Vatsa, Richa Singh, and Afzel Noore

West Virginia University, Morgantown, WA, USA, mayankv@csee.wvu.edu, richas@csee.wvu.edu, noore@csee.wvu.edu

Summary. The quality of input data has an important role in the performance of a biometric system. Images such as fingerprint and face captured under non-ideal conditions may require additional preprocessing. This chapter presents intelligent SVM techniques for quality assessment and enhancement. The proposed quality assessment algorithm associates the quantitative quality score of the image that has a specific type of irregularity such as noise, blur, and illumination. This enables the application of the most appropriate quality enhancement algorithm on the non-ideal image. We further propose a SVM quality enhancement algorithm which simultaneously applies selected enhancement algorithms to the original image and selects the best quality regions from the global enhanced image. These selected regions are used to generate single high quality image. The performance of the proposed algorithms is validated by considering face biometrics as the case study. Results show that the proposed algorithms improve the verification accuracy of face recognition by around 10–17%.

1 Introduction

Biometrics is the science in which physiological and behavioral characteristics are used to establish an individual's identity [1, 2]. A typical biometric system consists of three stages: data acquisition, feature extraction and matching. In data acquisition, raw data such as face image, fingerprint image or voice data is captured using appropriate hardware. Feature extraction extracts useful and distinct information from the data and finally matching is performed to establish identity by comparing the features extracted from the query image and the reference features stored in the database.

In biometrics, *quality* refers to the intrinsic physical data content. National Institute of Standards and Technology (NIST) defines biometric quality scores as the accuracy with which physical characteristics are represented in a given biometric data [3, 4]. Here the term *quality* is not limited to image resolution, dimension, gray scale/color depth and other acquisition parameters. In ideal

M. Vatsa et al.: *SVM Based Adaptive Biometric Image Enhancement Using Quality Assessment*, Studies in Computational Intelligence (SCI) **83**, 351–367 (2008)
www.springerlink.com © Springer-Verlag Berlin Heidelberg 2008

case, data acquisition should produce good quality images for feature extraction and matching. However, in many real world applications live samples are collected under non-ideal conditions, necessitating preprocessing the non-ideal image to enhance quality and remove different types irregularities present. For instance, face images can have blurriness, noise and illumination variation. Thus quality of a biometric data is an important factor in ensuring robustness of the security system. The quality of images can be improved by capturing the images under ideal conditions by using advanced sensor technologies, and by applying noise removal and image enhancement techniques.

Non-ideal biometric data requires quality assessment and enhancement to improve the quality and remove the irregularities present in the image. In this chapter, we propose a biometric image quality assessment and enhancement algorithm using Support Vector Machine (SVM) [5] learning to address this challenge. In the proposed quality assessment algorithm, we apply Redundant Discrete Wavelet Transform (RDWT) to compute the quality score of an image and then project the score into the hyperspace of multiclass 2ν-SVM to classify and determine the irregularities present in the image. The knowledge of discerning one or more types of irregularities can then be used to selectively apply the 2ν-SVM based enhancement algorithm. The proposed assessment and enhancement algorithms can be applied to any image-based biometric modality such as fingerprint, face, and iris. However, in this chapter we validate the results with face biometrics as the case study. Section 2 presents the literature survey of existing biometric quality assessment and enhancement algorithms. Section 3 presents an overview of 2ν-SVM, Sect. 4 describes the proposed image quality assessment algorithm, and Sect. 5 describes the proposed image quality enhancement algorithm. The case study on face biometrics is presented in Sect. 6 with experimental results discussed in Sect. 7.

2 Literature Review of Biometric Assessment and Enhancement Algorithms

In literature, there are several biometric quality assessment and enhancement algorithms. Most of these algorithms are associated with a particular biometric modality. For example, [6–16] focus on fingerprint quality assessment and enhancement, [17–21] describe iris quality assessment and enhancement, and [22–33] address face quality assessment and enhancement. Assessment algorithms generally give the quality scores but do not specify the type of irregularity present in the image. Enhancing the images without knowing the irregularity present is both difficult and time consuming. Further, applying different enhancement algorithms on the image might also degrade the good quality regions present in the image.

Bolle et al. [6] propose fingerprint image quality assessment using ratios of directional area and non-directional area in fingerprint. Shen et al. [7] use gabor wavelets to identify the blocks with good fingerprint features.

Agrawal [8] describe the local information based fingerprint image quality measure. Ratha and Bolle [9] propose WSQ compressed fingerprint image quality assessment algorithm using wavelet transform. Lim et al. [10] propose spatial domain quality estimation algorithm using eigen values of the covariance matrix obtained from grayscale gradient of fingerprint image. Hong et al. [11] use spatial and frequency modules to compute the quality of fingerprint images. Tabassi et al. [3] propose classification based approach for computing the fingerprint quality score. Chen et al. [12, 13] propose the global quality index captured in spatial frequency domain. Further, there exist different quality enhancement algorithms to remove irregularities from fingerprint images. O'Gorman et al. [14] propose the enhancement algorithm using anisotropic smoothening kernel based contextual filters. Greenberg et al. [15] use structured adaptive anisotropic filter to enhance the quality of fingerprint images. Hong et al. [11] use Gabor kernel for fingerprint enhancement. Chikkerur et al. [16] propose short time fourier transform based fingerprint enhancement algorithm for enhancement at local levels.

In iris biometrics, Daugman [17] measures the high frequency component in fourier spectrum to determine the quality of iris images. Ma et al. [18] used Support Vector Machine based iris quality assessment which classifies fourier spectra into four classes: clear, defocused, deblurred, and occluded. Chen et al. [19] propose the use of continuous wavelet transform to compute the quality index of iris images. Kalka et al. [21] propose Dempster Shafer theory based iris image quality assessment algorithm in which they estimate seven different quality parameters and fuse them to generate a composite quality score. These quality measures improve the recognition performance by incorporating assessment in quality enhancement and recognition. Ma et al. [19] use estimated background illumination to enhance the quality of iris images. Further, Zhang et al. [20] use super resolution based enhancement technique to improve the quality of iris images.

In face recognition, several quality models have been prepared but most of these models handle only the illumination variation. Further, most of these models neither generate quality scores nor enhance the bad quality face images. Models such as illumination cone model [22], spherical harmonic based model [23–25], quotient image based models [26,27], and total variation models [28] normalize face images with varying illumination for improving the feature extraction and recognition performance. Fronthaler et al. [29] propose an automatic image quality assessment algorithm which can be used for face image quality assessment. This approach uses orientation tensor with a set of symmetry descriptions to assess the quality of face images. In face image enhancement, Levine and Bhattacharya [30] propose an algorithm for detecting and removing specularities present in a face image. In this algorithm, a retinex based algorithm first processes face image to remove and detect specularities using Support Vector Machine. Gross and Brajovic [31] propose an image enhancement algorithm for face images affected by illumination variation. This algorithm first eliminates the illumination field in a face image and then enhances by compensating the illumination. In [32], cross channel histogram equalization is used to enhance the quality of face images. Kela et al. [33] use retinex based algorithm to compensate the illumination variation.

Most of these techniques are modeled for a specific biometric modality or a particular type of irregularity. However, in a real world scenario, quality assessment and enhancement algorithms should be generalized to handle different irregularities.

3 Overview of 2ν-Support Vector Machine

There are several algorithms such as Bayesian learning and neural networks which can be trained to learn and perform image quality assessment and enhancement. One of the major drawbacks with neural network based approach is choosing a mapping function between the input data space and the output feature space. The mapping function may be inefficient in determining the local minima consistently. When a large set of training data is used, it could lead to overfitting and hence poor generalization affecting the matching performance. Such learning algorithms have several parameters which are controlled heuristically, making the system difficult and unreliable to use in a real world scenario. Also, traditional multilayer perceptron neural networks suffer from the existence of multiple local minima solutions. To alleviate these inherent limitations, we use Support Vector machine (SVM) [5] based learning algorithms for biometric image quality assessment and enhancement. SVM starts with the goal of separating the data with a hyperplane and extends this to non-linear decision boundaries. Further, SVM training always finds a global minimum by choosing a convex learning bias. SVM is thus a classifier that performs classification by constructing hyperplanes in a multidimensional space and separating the data points into different classes. To construct an optimal hyperplane, SVM uses an iterative training algorithm which maximizes the margin between two classes. However, some researchers have shown that margin maximization does not always lead to minimum classification error [34]. Sometimes the training data points are not clearly separable and are characterized as fuzzy separable data. From biometrics perspective, fuzzy data are more common and SVM which deals with such data can provide probabilistic recommendation to the user. To achieve this, we use dual ν-SVM (2ν-SVM) originally proposed by Chew et al. [35]. 2ν-SVM is an attractive alternative to SVM and offers much more natural setting for parameter selection which is a critical issue in practical applications. Dual ν-SVM (2ν-SVM) is briefly described as follows:

Let $\{x_i, y_i\}$ be a set of N data vectors with $x_i \in \Re_d$, $y_i \in (+1, -1)$, and $i = 1, ..., N$. x_i is the ith data vector that belongs to a binary class y_i. The objective of training 2ν-SVM is to find the hyperplane that separates two classes with the widest margins, i.e.,

$$w(x) + b = 0 \tag{1}$$

subject to

$$y_i \left(w\, \varphi(x) + b \right) \geq (\rho - \psi_i), \ \psi_i \geq 0 \tag{2}$$

to minimize

$$\frac{1}{2}\|w\|^2 - \sum_i C_i(\nu\rho - \psi_i), \tag{3}$$

where ρ is the position of the margin and ν is the error parameter. $\varphi(x)$ is the mapping function used to map the data space to the feature space, and provide generalization for the decision function that may not be a linear function of the training data. $C_i(\nu\rho - \psi_i)$ is the cost of errors, w is the normal vector, b is the bias, and ψ_i is the slack variable for classification errors. Slack variables are introduced to handle classes which cannot be separated by a hyperplane. ν is the error parameter that can be calculated using ν_+ and ν_- which are the error parameters for training the positive and negative classes respectively.

$$\nu = \frac{2\nu_+\nu_-}{\nu_+ + \nu_-}, \ \ 0 < \nu_+ < 1 \ \ and \ \ 0 < \nu_- < 1 \tag{4}$$

Error penalty C_i is calculated as,

$$C = \begin{cases} C_+, & if \quad y_i = +1, \\ C_-, & if \quad y_i = -1, \end{cases} \tag{5}$$

where

$$C_+ = \left[n_+ \left(1 + \frac{\nu_+}{\nu_-} \right) \right]^{-1} \tag{6}$$

$$C_- = \left[n_- \left(1 + \frac{\nu_-}{\nu_+} \right) \right]^{-1} \tag{7}$$

and n_+ and n_- are the number of training points for the positive and negative classes, respectively. 2ν-SVM training can thus be formulated as,

$$max_{(\alpha_i)} \left\{ -\frac{1}{2} \sum_{i,j} \alpha_i\, \alpha_j\, y_i\, y_j\, K(x_i, x_j) \right\}, \tag{8}$$

where

$$0 \leq \alpha_i \leq C_i,$$

$$\sum_i \alpha_i y_i = 0, \tag{9}$$

$$\sum_i \alpha_i \geq \nu,$$

$i, j \in 1, ..., N$, α_i, α_j are the Lagrange multipliers and kernel function is

$$K(x_i, x_j) = \varphi(x_i)\varphi(x_j) \tag{10}$$

2ν-SVM uses iterative decomposition training based on optimization algorithm originally proposed by Chew et al. [35]. This optimization algorithm can be seen as pairwise decomposition method which breaks the problem to a two variable decision problem and solves the subproblem analytically. Applying the optimization algorithm thus leads to reduction in the computational complexity.

4 Biometric Image Quality Assessment

The proposed biometric image quality assessment algorithm uses redundant discrete wavelet transform. Usually, Discrete Wavelet Transform (DWT) [36, 37] is used for image based operations such as image fusion, denoising and quality measure because DWT preserves different frequency information and allows good localization both in time and spatial domain. However, one of the major limitations of DWT is that the transformation does not provide shift invariance. This causes a major change in the wavelet coefficients of the image/signal even for minor shifts in the input image/signal which leads to inaccurate data processing. Researchers have proposed several approximation techniques to overcome the shift variance of DWT. One approach is known as Redundant DWT (RDWT) [36–40]. RDWT can be considered as an approximation to DWT by removing the down-sampling operation from traditional critically sampled DWT to produce an over-complete representation. The shift variance characteristic of DWT arises from the use of down-sampling, while RDWT is shift invariant since the spatial sampling rate is fixed across scale [38–40]. Along with shift invariance, the transform not only captures some notion of the frequency content of the input by examining it at different scales, but also captures the temporal content. Another important aspect of using RDWT is per-subband noise relationship [38]. Fowler showed that the distortion in the original image from noise in a single RDWT subband depends only on the decomposition scale at which the subband resides and is independent of the other subbands. As shown in Fig. 1, high frequency content exists along edges, and low frequency content exists where little or no edges occur. Therefore, each RDWT subband provides information about where different types of edges exist. Coefficients in the subbands are large for edges, and zero or close to zero for non-edge regions. To determine the image quality, we need to find the edge information in the image along with blurriness, smoothness, and noise present in the image.

Let I be an image of size $n \times n$. I is decomposed to 3-level RDWT using Daubechies-9/7 (Db9/7) mother wavelet [41]. For wavelet based image processing operations, Db9/7 is considered to be the best mother wavelet [41]. As mentioned earlier, quality assessment requires different details of images such as edges, frequency and temporal content, and per-subband noise relationship. Hence, the image is decomposed to 3-levels of RDWT to compute

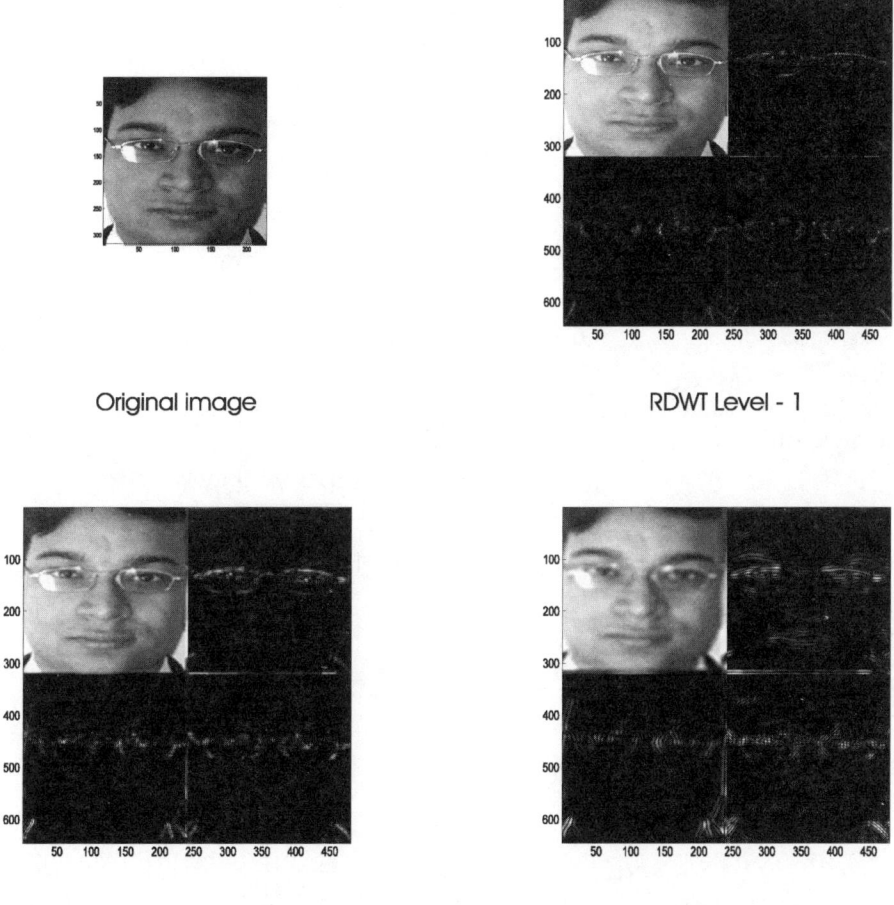

Original image

RDWT Level - 1

RDWT Level - 2

RDWT Level - 3

Fig. 1. 3-level RDWT decomposition of a face image

the image quality at different resolution levels. Equation (11) represents the decomposition of image I,

$$[I_{Ai}, I_{Hi}, I_{Vi}, I_{Di}] = RDWT(I),\tag{11}$$

where $i = 1, 2, 3$ represents the level of decomposition and A, H, V and D represent the approximation, horizontal, vertical and diagonal bands respectively. Approximation and detailed bands at each decomposition level are used to compute the quality factor of the bands. Let Q_A, Q_H, Q_V, and Q_D be the quality factor for the approximation, horizontal, vertical and diagonal bands respectively. The quality factor for each band is computed using (12)–(15).

$$Q_A = \sum_{i=1}^{3} \sum_{k,l=1}^{n} I_{Ai}(k,l), \tag{12}$$

$$Q_H = \sum_{i=1}^{3} \sum_{k,l=1}^{n} I_{Hi}(k,l), \tag{13}$$

$$Q_V = \sum_{i=1}^{3} \sum_{k,l=1}^{n} I_{Vi}(k,l), \tag{14}$$

$$Q_D = \sum_{i=1}^{3} \sum_{k,l=1}^{n} I_{Di}(k,l), \tag{15}$$

Further, these quality factors are combined using (16) to compute the quality score Q of image I,

$$Q = \frac{m_A Q_A + m_H Q_H + m_V Q_V + m_D Q_D}{m_A + m_H + m_V + m_D} \tag{16}$$

where m_A, m_H, m_V, and m_D are the weight factors computed using the following equations:

$$m_A = \sum_{i=1}^{3} \frac{1}{1 + \sum_{k,l=1}^{n} \nabla(I_{Ai}(k,l))}, \tag{17}$$

$$m_H = \sum_{i=1}^{3} \frac{1}{1 + \sum_{k,l=1}^{n} \nabla(I_{Hi}(k,l))}, \tag{18}$$

$$m_V = \sum_{i=1}^{3} \frac{1}{1 + \sum_{k,l=1}^{n} \nabla(I_{Vi}(k,l))}, \tag{19}$$

$$m_D = \sum_{i=1}^{3} \frac{1}{1 + \sum_{k,l=1}^{n} \nabla(I_{Di}(k,l))}, \tag{20}$$

where i represents the level of decomposition and ∇ represents the gradient operation. The weight factors ensure proper weight assignment to all bands depending on the information contained in each of the bands.

The next step after computing the quality score is to determine the class of the score, i.e. whether the score belongs to a good image or is affected with a specific type of irregularity. To perform this classification, we use 2ν-SVM based multiclass classification. Let the number of irregularities under consideration be n. This gives the total number of possible combinations of irregularities which can be present in an image as $2^n - 1$. The total number of classes including good quality class thus becomes 2^n. These 2^n classes are classified using multiclass implementation of 2ν-SVM [42]. The process is divided into two steps, multiclass 2ν-SVM training and multiclass 2ν-SVM classification.

4.1 Multiclass 2ν-SVM Training for Quality Assessment

The training process involves constructing a decision function for 2^n classes using the independent and identical distributed samples. We use multiclass implementation of 2ν-SVM to construct the decision function for classifying 2^n classes by learning the training quality scores. To solve 2^n class learning, we construct a piecewise linear separation of 2^n classes with 2ν-SVM. The optimization function in (8) is optimized for all 2^n classes. For $c = 2^n$, (8) is written as,

$$max_{\alpha_i}^c \left\{ -\frac{1}{2} \sum_{i,j} \alpha_i^c \, \alpha_j^c \, y_i^c \, y_j^c K(x_i^c, x_j^c) \right\}. \tag{21}$$

Labeled training quality scores are used to learn 2ν-SVM for multiclass hyperplane generation using the optimization function expressed in (21).

4.2 Multiclass 2ν-SVM Classification for Quality Assessment

Multiclass 2ν-SVM is trained to classify the score using labeled training quality scores. Trained multiclass 2ν-SVM is then used to classify the quality score of input biometric image. We project the quality score Q of input biometric image into the multiclass high dimensional space and obtain the output class label depending on the distance of Q from different class boundaries. The output class label represents irregularities present in the image.

5 Biometric Image Quality Enhancement

Quality assessment algorithm described in Sect. 4 computes the quality class of the input image. This quality class can be used to enhance the image accordingly. If the image contains only one irregularity such as noise, or blur, then we can apply the standard enhancement algorithm for that particular irregularity. However, the input image may contain multiple irregularities such as noise with blur, or illumination variation with noise. The challenge in enhancing such images is to locally segment the affected regions from the image and apply appropriate enhancement algorithm. Finding regions of a degraded image can be challenging, time consuming, and not pragmatic.

To address these challenges, we propose an image enhancement algorithm that globally applies different enhancement techniques to an image. The enhancement algorithms are selected depending on the quality class of the input image obtained from the quality assessment algorithm. Each resulting image contains regions that are enhanced by specific algorithms. A support vector machine (SVM) based learning algorithm identifies the good quality regions from each of the globally enhanced images and synergistically combines these good quality enhanced regions to form a single image. The combined image contains high quality feature-rich enhanced regions that can be used by biometric recognition algorithms for improving the recognition performance. The proposed biometric image quality enhancement algorithm is divided into two steps, 2ν-SVM training, and 2ν-SVM classification and integration.

5.1 2ν-SVM Training

2ν-SVM is trained to classify input pixels of the image as good or bad. Different regions from selected reference input and enhanced images are labeled as good or bad and these labeled regions are used for training the 2ν-SVM. Training images are decomposed to l levels by Discrete Wavelet Transform (DWT) which gives $3l$ detailed subbands and one approximation band. The activity level of wavelet coefficients is computed over a small window by treating each coefficient separately and then averaging the wavelet coefficients for the window under consideration. The activity level is computed for different regions in the detailed subbands and approximation band of the images. It is then provided as input to the 2ν-SVM for training and determining the quality. The output of training is label G or 1 if the coefficient is good, and B or 0 if the coefficient is bad.

5.2 2ν-SVM Classification and Integration

Trained 2ν-SVM is used to classify the pixels from the original input image and the globally enhanced images, and combine them to generate a new feature-rich image. The enhanced images along with the original input image are decomposed to l levels using the DWT. 2ν-SVM classifier is then used to classify the coefficients of different bands as good or bad. A decision matrix $Decision$ is generated to store the quality of each coefficient in terms of G and B. At any position (x, y), if the 2ν-SVM output $O(x, y)$ is positive then that coefficient is labeled as G; otherwise it is labeled as B.

$$Decision(x, y) = \begin{cases} G, \; if \; O(x, y) > 0, \\ B, \; if \; O(x, y) < 0, \end{cases} \tag{22}$$

This operation is performed on all the images and the decision matrix is generated corresponding to every image. At every location (x, y), average of coefficients having label G is obtained and the coefficients with label B are discarded. In this manner, the enhanced approximation band and detailed subbands are generated. A single feature-rich high quality enhanced image is obtained by applying inverse DWT on the enhanced bands.

6 Generation of High Quality Face Image

The quality assessment and enhancement algorithms described earlier are applied on a face image to generate good quality feature-rich face image. In face biometrics, images can have two types of irregularities: image based

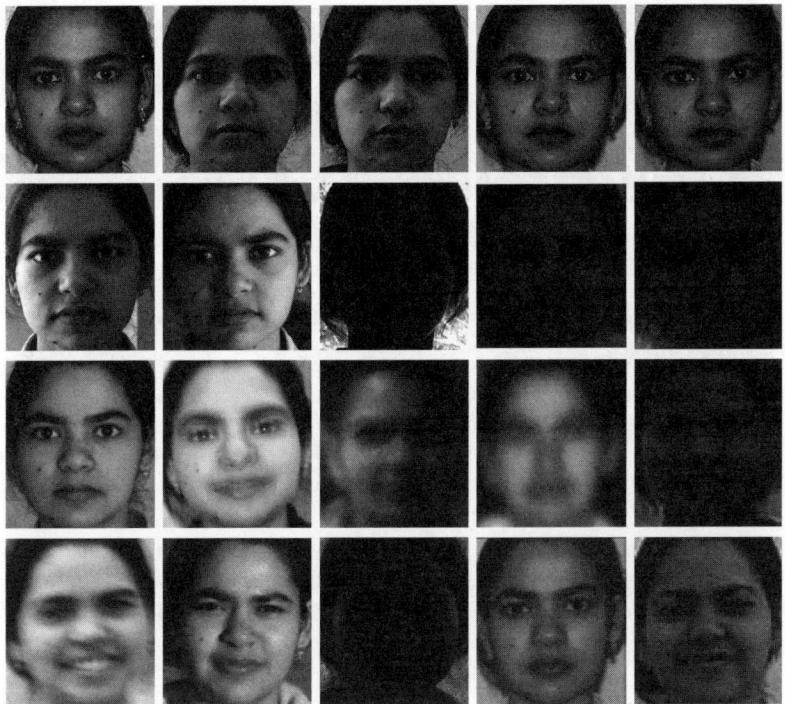

Fig. 2. A subset of face images of an individual in the database

irregularities and intrinsic irregularities. Typical image based irregularities that occur during capture of a face image in an unconstrained environment are blur, noise, and improper illumination as shown in Fig. 2. Intrinsic irregularities are pose and expression variation which cannot be removed by enhancement algorithms. In this case study, we consider only image based irregularities namely blur, noise and improper illumination along with their possible combinations which leads to seven irregularity classes present in the image. Thus including the good class, there are total eight quality classes. Input image is classified into one of the eight classes using the proposed quality assessment algorithm. If the input image is affected by a single irregularity then the corresponding standard enhancement algorithm is applied. However, if the image is affected by multiple irregularities then the set of corresponding enhancement algorithms is applied and the proposed 2ν-SVM enhancement algorithm is used to generate the enhanced image.

For example, if the input face image is affected by all three types of irregularities then *deblur* function [43] is used to remove blur from the image, *denoise* function [44] removes noise present in the image, and finally *retinex* function [45] is used for illumination correction in the face images.

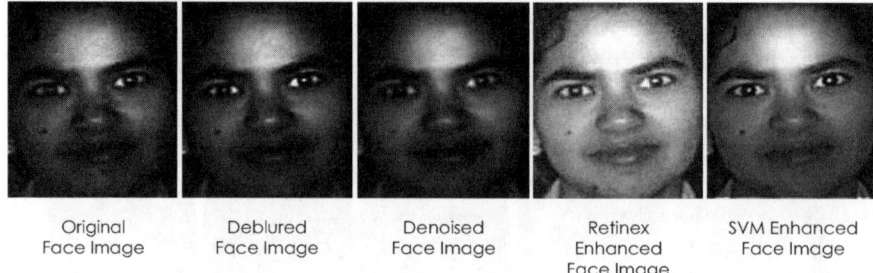

Original	Deblured	Denoised	Retinex	SVM Enhanced
Face Image	Face Image	Face Image	Enhanced	Face Image
			Face Image	

Fig. 3. Generation of single SVM enhanced face image from different globally enhanced face images

$$F_1 = deblur(F),$$
$$F_2 = denoise(F), \qquad (23)$$
$$F_3 = retinex(F).$$

Original face image F, and enhanced images F_1, F_2, and F_3 are given as input to the trained 2ν-SVM to generate the feature-rich and quality enhanced face image. Figure 3 shows an example of the original face image, different globally enhanced images, and the SVM enhanced face image.

7 Experimental Results

This section is divided into four subsections. First subsection describes the database and the face recognition algorithms used for validation of the proposed quality assessment and enhancement algorithms. Further, the experimental results are discussed in Sects. 7.2–7.4. Section 7.5 presents the computational complexity of the proposed algorithms.

7.1 Face Database and Recognition Algorithms used for Evaluation

To validate the performance of the proposed quality assessment and enhancement algorithm, we prepared a face database that consists of 4,000 frontal face images from 100 individuals with five images for each irregularity class. From these 4,000 face images, we used 1,600 face images for training both quality assessment and enhancement algorithms and the rest of the 2,400 face images are used for testing. Training dataset thus comprises of two images from each irregularity class ($2 \times 8 \times 100$) and test dataset consists of three images from each irregularity class ($3 \times 8 \times 100$) for every individual. Figure 2 shows a subset of the images of an individual from the database.

To evaluate the performance of the proposed quality assessment and enhancement algorithms, we selected four face recognition algorithms.

- *Principal component analysis* (PCA) [46] is an appearance based face recognition algorithm which computes the principal eigen-vectors from the training face images. The query image is projected in the eigen space formed by the training sets to extract the appearance based facial features. Matching is done using the Euclidean distance measure.
- *Fisher Linear Discriminant Analysis* (FLDA) [47] is an appearance based face recognition algorithm. It computes a set of vectors that maximizes the interclass variation and minimizes the intraclass variation between images. The feature vectors are matched using Mahalanobis distance.
- *Local feature analysis* (LFA) [48] is one of the most widely used face recognition approach which can accommodate some changes in facial expression. LFA extracts a set of geometrical metrics and distances from facial images and uses these features as the basis for representation and comparison.
- *Texture feature* [49] based face recognition algorithm extracts phase features from face image using 2D log Gabor wavelet and matches them using hamming distance based matching algorithm.

7.2 Performance of Proposed Quality Assessment Algorithm

This subsection presents the experimental results of the proposed quality assessment algorithm described in Sect. 4. First, we evaluated the performance of three kernels: linear, polynomial and RBF. These kernels are used in SVM learning and classification in (21). These three kernels can be expressed as,

Linear kernel:
$$K(x_i, x_j) = x_i^T x_j. \tag{24}$$

Polynomial kernel:
$$K(x_i, x_j) = (\gamma x_i^T x_j + r)^d, \qquad \gamma, r > 0. \tag{25}$$

RBF kernel:
$$K(x_i, x_j) = exp(-\gamma ||x_i - x_j||^2), \qquad \gamma > 0. \tag{26}$$

In our experiments, for polynomial kernel, parameters $r = 1$, $\gamma = 1$ and $d = 2$ provide best results and for RBF kernel, $\gamma = 4$ provides best results. These parameters are obtained empirically by computing the classification results for different combination of parameters. For quality assessment, RBF kernel performs best followed by polynomial and linear kernels. Table 1 shows the experimental results for the proposed quality assessment algorithm. For each irregularity class, there are 300 test cases. Results are shown in terms of the classification accuracy, i.e., the percentage of correctly classified cases. Table 1 shows that the proposed assessment algorithm correctly identifies all the good quality images and the images containing single irregularity. It identifies some of the images with multiple irregularities incorrectly thus giving the overall classification accuracy of 99.38%.

Table 1. Performance of the proposed image quality assessment algorithm

Irregularity class	Number of correctly classified images	Classification accuracy (%)
Good quality	300	100.00
Noisy	300	100.00
Blurriness	300	100.00
Illumination variation	300	100.00
Noise + Blurriness	297	99.00
Noise + Illumination variation	296	98.67
Blurriness + Illumination variation	297	99.00
Noise + Blurriness + Illumination variation	295	98.33
Total	2,385	99.38

Table 2. Performance evaluation of the proposed quality assessment and enhancement algorithm using face recognition algorithms (verification accuracy at 0.001% False Accept Rate)

Recognition Algorithm	Accuracy without quality assessment and enhancement (%)	Accuracy with proposed quality assessment and enhancement (%)	Improvement in accuracy (%)
PCA [46]	53.18	70.34	17.16
FLDA [47]	56.45	72.93	16.48
LFA [48]	69.52	81.57	12.05
Texture features [49]	88.38	98.21	9.83

We further compare the performance of the proposed SVM based quality assessment algorithm with Multilayer Perceptrons (MLP). Similar to 2ν-SVM, the quality scores are first generated using RDWT and then trained using MLP to compute the classification accuracy. We found that MLP based approach yields the classification accuracy of 86.51% which is 12.87% less than 2ν-SVM based assessment approach. This experiment shows that SVM outperforms MLP in the classification of quality scores.

7.3 Performance of Proposed Quality Enhancement Algorithm

The performance of the proposed quality enhancement algorithm is evaluated using four existing face recognition algorithms, Principal component analysis [46], Fisher linear discriminant analysis [47], Local feature analysis [48] and texture features [49]. Similar to performance evaluation of the linear, polynomial and RBF kernel for quality assessment, we evaluate the performance for enhancement using all three kernels. In this experiment, the irregularity class of face images determined by 2ν-SVM and RBF based assessment algorithm is used to enhance the images. The enhancement is performed with all three kernels and verification accuracy for the enhanced images is computed using PCA based face recognition algorithm. Results show that the images enhanced using RBF kernel with $\gamma = 4$ provide better accuracy compared to the linear and polynomial kernels. We therefore choose the RBF kernel for the remaining experiments.

To show the efficacy of the proposed quality enhancement algorithm, we computed the face verification accuracy without using the proposed quality assessment and enhancement algorithms and then with the proposed algorithms. Table 2 shows that the maximum improvement in accuracy is 17.16%, which is obtained for PCA based face recognition algorithm. The significant improvement in verification accuracies shown in Table 2 shows the

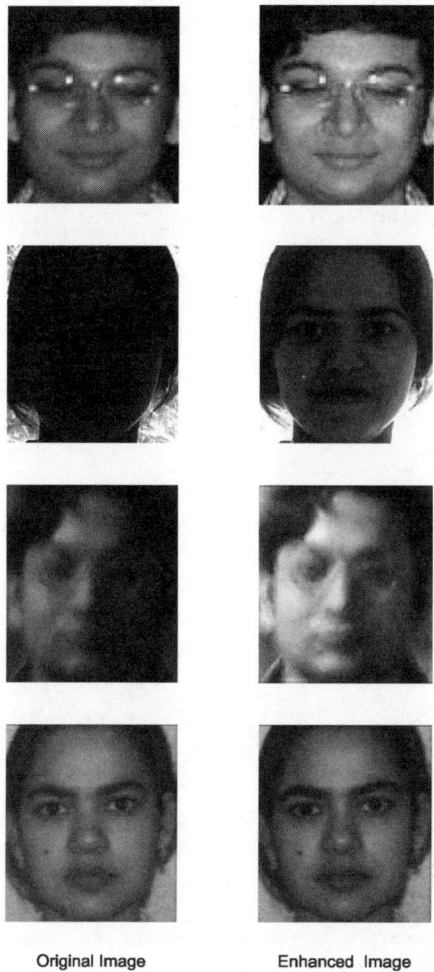

Original Image   Enhanced Image

Fig. 4. Enhanced face images generated by the proposed SVM image enhancement algorithm. *Left* column shows original images with multiple irregularities and *right* column shows corresponding enhanced images

effectiveness of the proposed image quality assessment and enhancement algorithms. Figure 4 shows sample results of the proposed quality enhancement algorithm when the input face image is affected by multiple irregularities.

7.4 Comparison with Existing Quality Enhancement Algorithms

The proposed quality assessment and enhancement algorithms are compared with three existing quality enhancement algorithms used in face biometrics. The three existing algorithms we used are Histogram equalization [32], Multiscale retinex [33] and illumination correction algorithm [31]. These three approaches are termed as Histogram, Retinex and SI respectively. Figures 5–8, show the Receiver Operating Characteristics (ROC) plots for comparison using PCA, FLDA, LFA and texture feature based face verification algorithms. ROC plots of these face verification algorithms demonstrate that the proposed assessment and enhancement algorithms improve the verification performance by at least 3.5% compared to existing enhancement algorithms. The existing algorithms provide comparatively low performance because they correct only the illumination variations present in images and irregularities due to noise and blur are not corrected. On the contrary, the proposed algorithm compensates for noise, blur, and illumination irregularities, thus yielding higher accuracy.

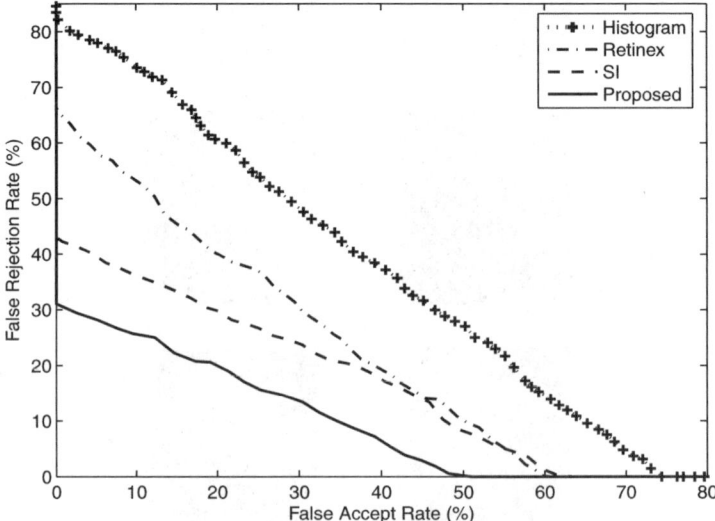

Fig. 5. ROC plots comparing the performance of the proposed quality assessment and enhancement algorithm with existing image quality enhancement algorithms using PCA based face verification algorithm [46]

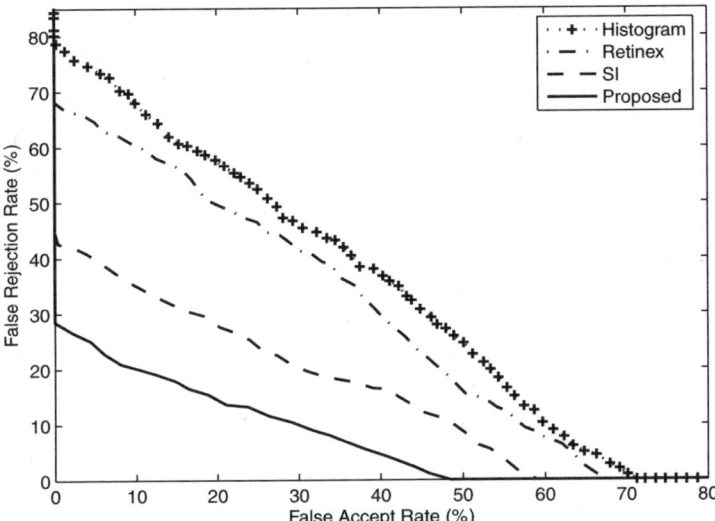

Fig. 6. ROC plots comparing the performance of the proposed quality assessment and enhancement algorithm with existing image quality enhancement algorithms using FLDA based face verification algorithm [47]

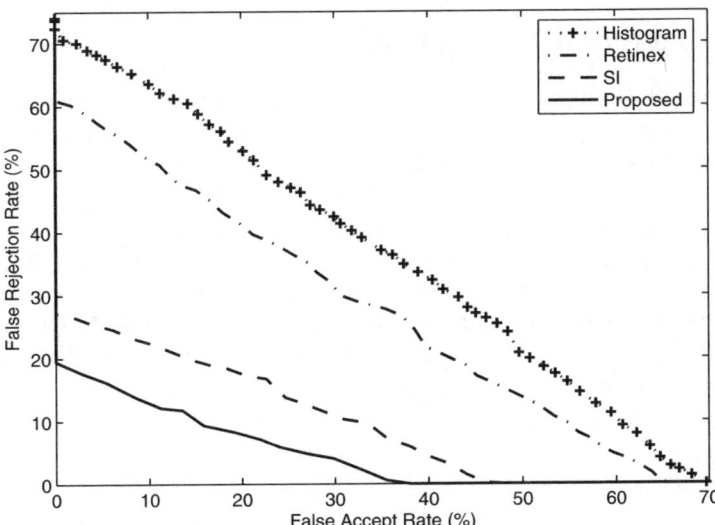

Fig. 7. ROC plots comparing the performance of the proposed quality assessment and enhancement algorithm with existing image quality enhancement algorithms using LFA based face verification algorithm [48]

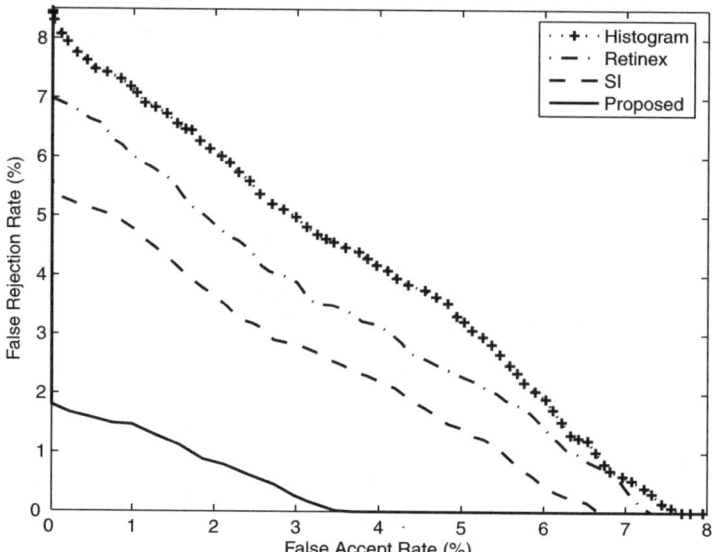

Fig. 8. ROC plots comparing the performance of the proposed quality assessment and enhancement algorithm with existing image quality enhancement algorithms using texture based face verification algorithm [49]

7.5 Computational Time Complexity

The proposed quality assessment and enhancement algorithms can be used as preprocessing algorithms in a biometrics system. It is desirable in a real time system that the preprocessing algorithm should not increase the time complexity. Using a computer with 3.0 GHz, 1 GB RAM, the lower and upper bounds of time complexity of the proposed algorithm ranges from 2 to 9 s, respectively depending upon the irregularity present. If the quality of image is good and requires no enhancement then the computational time is 2 s which is required to assess the quality. If the input image is affected by all three irregularities, i.e. noise, blur and illumination variation, then the total time taken by quality assessment and enhancement algorithm is 9 s.

8 Summary

Many biometric technologies using different modalities will be deployed in the future for authenticating individuals. Proliferation of these technologies leads to new challenges that should collectively address scalability, accuracy, and speed issues for meeting performance requirements in real-time applications. Furthermore, images such as fingerprints, face, or iris maybe captured under non-ideal conditions, with or without the cooperation of an individual. This chapter proposes intelligent learning techniques to associate a quantitative quality score of an image with a specific type of irregularity. The irregularity in an image can be due to a single dominant factor or a combination of factors such as noise, blur, and illumination. Identifying the specific nature of irregularity using SVM enables the most appropriate enhancement algorithms to be applied. For enhancing images that have multiple irregularities, the problem is exacerbated because localization of affected regions is non trivial under normal operational time constraints. This chapter describes an approach to simultaneously apply selected enhancement algorithms to the original image. SVM is trained to recognize and select the best quality pixels from the globally enhanced images to generate a single high-quality feature-rich image. The proposed approach is validated using face images as a case study. The performance is validated using well known face recognition algorithms, with and without quality assessment and image enhancement. Experimental results show that using the proposed intelligent quality assessment and enhancement techniques, the face recognition accuracy is improved by at least 10%.

References

1. Wayman J., Jain A. K., Maltoni D., Maio D. (2005) Biometric systems: technology, design and performance evaluation, Springer, Berlin Heidelberg New York.
2. Jain A. K., Bolle R., Pankanti S. (Eds.) (1999) Biometrics: personal identification in networked society, Kluwer Academic, Dordretch.
3. Tabassi E., Wilson C., Watson C. (2004) Fingerprint image quality. In: NIST Research Report NISTIR7151.
4. Hicklin A., Khanna R. (2006) The role of data quality in biometric systems. Technical Report, Mitretek Systems.
5. Vapnik V. N. (1995) The nature of statistical learning theory. Springer, Berlin Heidelberg New York.
6. Bolle R. M., Pankanti S. U., Yao Y. S. (1999) System and methods for determining the quality of fingerprint images. In: United States Patent Number US5963656.
7. Shen L., Kot A., Koo W. (2001) Quality measures of fingerprint images. In: Proceedings of Audio- and Video-based Biometric Person Authentication 266–271.
8. Agrawal M. (2006) Multi-impression enhancement of fingerprint images. Master Thesis. West Virginia University.
9. Ratha N., Bolle R. (1999) Fingerprint image quality estimation. In: IBM computer science research report RC21622.
10. Lim E., Jiang X., Yau W. (2002) Fingerprint quality and validity analysis. In: Proceedings of International Conference on Image Processing 1:469–472.
11. Hong L., Wan Y., Jain A. (1998) Fingerprint image enhancement: algorithms and performance evaluation. In: IEEE Transactions on Pattern Analysis and Machine Intelligence 20(8):777–789.
12. Chen Y., Dass S., Jain A. (2005) Fingerprint quality indices for predicting authentication performance. In: Proceedings of Audio- and Video-based Biometric Person Authentication 160–170.

13. Aguilar J. F., Chen Y., Garcia J. O., Jain A. (2006) Incorporating image quality in multi-algorithm fingerprint verification. In: Proceedings of International Conference on Biometrics 213–220.

14. O'Gormann L., Nickerson J. V. (1989) An approach to fingerprint filter design. In: Pattern Recognition, 22(1):29–38.

15. Greenberg S., Aladjem M., Kogan D., Dimitrov I. (2000) Fingerprint image enhancement using filtering techniques. In: International Conference on Pattern Recognition 3:326–329.

16. Chikkerur S., Cartwright A., Govindaraju V. (2005) Fingerprint Image Enhancement Using STFT Analysis, International Workshop on Pattern Recognition for Crime Prevention, Security and Surveillance, Lecture Notes in Computer Science, Springer Berlin/Heidelberg, Volume 3687/2005, Pages 20–29.

17. Daugman J. (2001) Statistical richness of visual phase information: update on recognizing persons by iris patterns. In: International Journal on Computer Vision 45:25–38.

18. Ma L., Tan T., Wang Y., Zhang D. (2003) Personal identification based on iris texture analysis. In: IEEE Transactions on Pattern Analysis and Machine Intelligence 25(12):1519–1533.

19. Chen Y., Dass S., Jain A. (2006) Localized iris image quality using 2-D wavelets. In: Proceedings of International Conference on Biometrics 373–381.

20. Zhang G., Salganicoff M. (1999) Method of measuring the focus of close-up image of eyes. In: United States Patent Number 5953440.

21. Kalka N. D., Zuo J., Dorairaj V., Schmid N. A., Cukic B. (2006). Image Quality Assessment for Iris Biometric. In: Proceedings of SPIE Conference on Biometric Technology for Human Identification III, 6202:61020D-1-62020D-11.

22. Belhumeur P. N., Kriegman D. J. (1998) What is the space of images of an object under all possible lighting conditions? In: International Journal on Computer Vision 28(3):245–260.

23. Basri R., Jacobs D. (2000) Lambertian reflectance and linear subspaces. NEC Research Institute Technical Report 2000-172R.

24. Ramamoorthi R., Hanrahan P. (2001) On the relationship between radiance and irradiance: determining the illumination from images of a convex Lambertian object. In: Journal of Optical Society of America 18(10):2448–2459.

25. Zhang L., Samaras D. (2003) Face recognition under variable lighting using harmonic image exemplars. In: Proceedings of International Conference on Computer Vision on Pattern Recognition 1:19–25.

26. Shan S., Gao W., Cao B., Zhao D. (2003) Illumination normalization for robust face recognition against varying lighting conditions. In: Proceedings of International Workshop on Analysis and Modeling of Faces and Gestures 157–164.

27. Shashua A., Riklin-Raviv T. (2001) The quotient image: class-based re-rendering and recognition with varying illuminations. In: IEEE Transactions on Pattern Analysis and Machine Intelligence, 23(2):129–139.

28. Chen T., Yin W., Sean X. Z., Comaniciu D., Huang T. S. (2006) Total variation models for variable lighting face recognition. In: IEEE Transactions on Pattern Analysis and Machine Intelligence 28(9):1519–1524.

29. Fronthaler H., Kollreider K., Bigun J. (2006) Automatic image quality assessment with application in biometrics. In: Computer Vision and Pattern Recognition Workshop 30–30.

30. Levine M. D., Bhattacharya J. (2005) Detecting and removing specularities in facial images. In: Computer Vision and Image Understanding 100:330–356.
31. Gross R., Brajovic V. (2003) An image pre-processing algorithm for illumination invariant face recognition. In: Proceedings of Audio- and Video Based Biometric Person Authentication 10–18.
32. King S., Tian G. Y., Taylor D., Ward S. (2003) Cross-channel histogram equalisation for colour face recognition. In: Proceedings of International Conference on Audio- and Video-Based Biometric Person Authentication 454–461.
33. Kela N., Rattani A., Gupta P. (2006) Illumination invariant elastic bunch graph matching for efficient face recognition. In: Proceedings of International Conference on Computer Vision and Pattern Recognition Workshop 42–42.
34. Chen P.-H., Lin C.-J., Schölkopf B. (2005) A tutorial on nu-support vector machines. In: Applied Stochastic Models in Business and Industry 21:111–136.
35. Chew H. G., Lim C. C., R. E. Bogner. (2004) An implementation of training dual-nu support vector machines. In: Qi, Teo, and Yang, (Eds.) Optimization and Control with Applications, Kluwer, Dordretch.
36. Daubechies I. (1992) Ten lectures on wavelets. In: Society for Industrial and Applied Mathematics.
37. Rioul O., Vetterli M. (1991) Wavelets and signal processing. In: IEEE Signal Processing Magazine 8(4):14–38.
38. Hua L., Fowler J. E. (2001) RDWT and image watermarking. In: Technical Report MSSU-COE-ERC-01-18, Engineering Research Center, Mississippi State University.
39. Fowler J. E. (2005) The redundant discrete wavelet transform and additive noise. In: IEEE Signal Processing Letters 12(9):629–632.
40. Cao J.-G., Fowler J. E., Younan N. H. (2001) An image-adaptive watermark based on a redundant wavelet transform. In: Proceedings of the International Conference on Image Processing 2:277–280.
41. Antonini M., Barlaud M., Mathieu P., Daubechies I. (1992) Image coding using wavelet transform. In: IEEE Transactions on Image Processing 1(2):205–220.
42. Weston J., Watkins C. (1999) Support vector machines for multi-class pattern recognition. In: Proceedings of the 7th European Symposium on Artificial Neural Networks 219–224.
43. Kang S. K., Min J. H., Paik J. K. (2001) Segmentation-based spatially adaptive motion blur removal and its application to surveillance systems. In: Proceedings of International Conference on Image Processing 1:245–248.
44. Malladi R., Sethian J. A. (1995) Image processing via level set curvature flow. In: Proceedings of National Academy of Sciences 92:7046–7050.
45. Land E. H., McCann J. J. (1971) Lightness and retinex theory. In: Journal of the Optical Society of America 61:1–11.
46. Turk M., Pentland A. (1991) Eigenfaces for recognition. In: Journal of Cognitive Neuroscience 3:72–86.
47. Belhumeur P. N., Hespanha J. P., Kriegman D. J. (1997) Eigenfaces vs. fisherfaces: recognition using class specific linear projection. In: IEEE Transactions on Pattern Analysis and Machine Intelligence 19(7):711–720.
48. Penev P., Atick J. (1996) Local feature analysis: A general statistical theory for object representation. In: Network: Computation in Neural Systems 7:477–500.
49. Singh R., Vatsa M., Noore A. (2005) Textural feature based face recognition for single training images. In: IEE Electronics Letters 41:23–24.

Segmentation and Classification of Leukocytes Using Neural Networks: A Generalization Direction

Pedro Rodrigues[1], Manuel Ferreira[2], and João Monteiro[3]

[1] Instituto Politécnico de Bragança, Campus de Santa Apolónia, Apartado 134, 5301-857 Bragança, Portugal, pjsr@ipb.pt
[2] Escola de Engenharia, Universidade do Minho, Campus de Azurém 4800-058 Guimarães, Portugal, mjf@dei.uminho.pt
[3] Escola de Engenharia, Universidade do Minho, Campus de Azurém; 4800-058 Guimarães, Portugal, Joao.Monteiro@dei.uminho.pt

Summary. In image digital processing, as in other fields, it is commonly difficult to simultaneously achieve a generalizing system and a specialized system. The segmentation and classification of leukocytes is an application where this fact is evident. First an exclusively supervised approach to segmentation and classification of blood white cells images is shown. As this method produces some drawbacks related to the specialized/generalized problems, another process formed by two neural networks is proposed. One is an unsupervised network and the other one is a supervised neural network. The goal is to achieve a better generalizing system while still doing well the role of a specialized system. We will compare the performance of the two approaches.

1 Introduction

The automatic identification and the automatic counting of human leukocytes, from blood, represent a factor of combativity and a factor of quality for the modern laboratories of clinical analyses. However, the leukocyte types that the current equipment is able to manage are restricted to few classes. Moreover, that equipment needs a set of expensive chemicals to reach its goals. Thus, nowadays, there are situations where the identification and differential counting operations have to be made by the human eye using a microscope. Those situations can be correlated with disease cases. Some abnormal cells may be observed only in specific diseases, while other may be seen occasionally in normal peripheral blood. The number of different white cells that can come out is relatively huge, more than 20, and this increases the difficulty of obtaining a large functional process. Even restricting the set of white cells to a normal group, Fig. 1, the task, when seen by artificial vision, is not straightforward. The cell images might be in a very random aspect. The intensity,

P. Rodrigues et al.: *Segmentation and Classification of Leukocytes Using Neural Networks: A Generalization Direction*, Studies in Computational Intelligence (SCI) **83**, 373–392 (2008)
www.springerlink.com

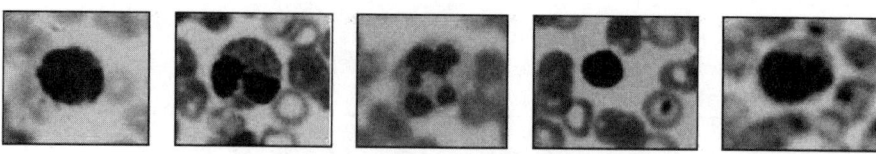

Fig. 1. Different types of leukocytes

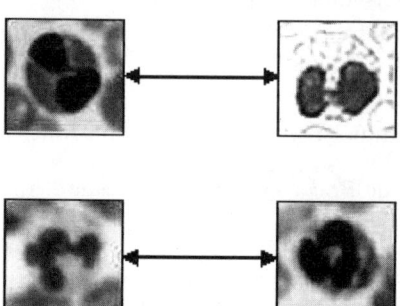

Fig. 2. Each row shows cells of the same type. The geometrical differences among cells of the same class can be higher

the color, the natural artifacts, the relative statistical position among cells, the adjacency among cells, all these features can be very different from one sample to another sample. The cytoplasm geometry and the nucleus geometry are very diverse, even if the cells belong to the same class, Fig. 2. The texture of the nucleus and the texture of the cytoplasm are present in some images but in other images are blurred. A few issues of this problem are overcome by manipulating the physical and visual acquisition process and having care on the staining slides process. The majority of these issues are inherent in the blood itself. In fact, all these issues involve enormous tasks in order to be run in machine vision. They take, mainly, two directions. First, it is difficult to figure out all the mechanism that does the visual job as correctly as the humans do. And second, to obtain a little improvement it is necessary to widely increase the computational effort. We will show, mainly, a solution to try to minimize the first problem. The base for that is prosecuted through neural networks. First we will analyze an approach that uses, in neural terms, exclusively, feedforward neural networks trained by a backpropagation algorithm [1] (BPNN). We will see the positive and the negative aspects that this process discloses when applied on the white cells images. Then, we will show a hybrid model that tries to reduce some problems that were found in the previous solution. Support vector machine (SVM) [2–4] and pulse-coupled neural networks (PCNN) [5,6] are integrated in such way that some negative aspects that arise with the first model are minimized.

This visual process is split into two different mechanisms. One mechanism must detect the pixels that are forming the region cell-segmentation. The second part is served by the first part. It has the task of extracting useful information from the segmented region. Finally a classifier is fed with the extracted information and it will decide in which cell class the region must be considered -classification. Both approaches have these two parts. The main difference lies in the form the data are delivered to the supervised neural networks. In the first case, the segmentation and the classification are exclusively supported by a supervised mechanism. The other method has an unsupervised part that cooperates with the supervised part. The junction of an unsupervised mechanism with a supervised mechanism is important to overcome the problems that are related with the strong variability of these images. The unsupervised mechanism will aggregate pixels that are forming features in multiple directions. The same high level feature can appear on the cells even if the cells have different colors. An exclusive supervisor method tends to learn this color instead of learning the real feature behind the color. Thus, if there appears a cell with a different color, when compared with the training cells colors, the exclusive supervised methods can fail. The addition of an unsupervised method improves the generality capacity of the system. This chapter begins with a presentation of a couple of techniques that exist to solve the problem. The presentation of our solutions is done in two parts. The first one is about a method that uses a supervised neural network to produce the segmentation and the classification of leukocytes. In the second part there is shown a method that uses an unsupervised neural network and a supervised neural network to reach the same goal. In the following section the results of each approach are presented. At the end, a discussion and conclusions are expressed.

2 Current Techniques

The companies that currently make equipment in this area generally solve the problem without image processing. Those methods are very accurate, but, with the exception of an extra of one or two classes, only the five common cells classes can be classified. They can be based in methods such as radiofrequency, direct current resistance, forward and side scatter of laser light, and fluorescence [7]. Normally these techniques require the addition of reagents. The usual form that makes possible the classification of any type of cells is still human vision and a microscope. It allows a diagnostic approach to blood disorders using blood film examination. For instance, an immature cell in the granulocytic series, occurring normally in bone marrow but not in the blood. During blood film examination, the individual types of white blood cells are enumerated. This is referred to as the differential count. Trying to put this ability into a machine has been very hard. Some researches [8] have shown very good results using an unsupervised methods in the segmentation process. They assume some color saturation relations with the leukocytes and the rest of the elements. They assume also that the nuclei of the leukocytes have a similar hue on their cytoplasms. If for some fact these assumptions drop, the mechanisms can fail, as they have exemplified. A similar problem can be deduced from our first approach: the supervised neural networks assume, during the training process, arbitrary relations to store differential information among different regions. Thus, if the images to be processed are agreeing with training images, the process works at a high level of performance, but if the images to be processed and the training images mismatch, the process fails too. Our second approach tries to avoid creating these primary assumptions, but as we will see the goal is not reached in some directions. There are other researchers that are using image processing on fluorescence images by exploring other visual features [9]. The recent works that were studied still suffer in multiple ways from the issues of the variability present in natural cell images. To improve the generalization capacity, in [10] is presented a segmentation model that does not use the color information.

3 Approach − I (BPNN)

3.1 Backpropagation Neural Network

Feedforward Backpropagation is, probably, the artificial neural system most widely used in pattern recognition applications. In the majority of the supervision applications this neural network performs well and shows good computational performance. These were the initial presuppositions that led us to choose this network. The backpropagation algorithm used for training feedforward networks is based in the gradient descent. This method allows us to set the weights of the neurons automatically in a way to minimize the error between the target pattern set and the output pattern set. The neural network after the end of the training must map the input vector in an output vector without error. This neural network can consist in a several layers, generally, at least one hidden layer and one output layer. Each neuron has an activation function, in most cases a sigmoid function. The output value of each neuron can be represented by

$$Y_{(l,n)} = f\left(\sum_{m=1}^{M_{l-1}} W_{(l,n,m)}\, X_{(l-1,m)} + b_{(l,n)}\right), \tag{1}$$

where n identifies the neuron within the l layer. W is the weight placed between the neuron n of the layer l and the output neuron m of the previous layer, or in the case where this layer is the input layer, m will indicate an input node of this neural net. X are the values of the neurons in the previous layer. b represents the biases of the neuron n at the layer l. M is the number of neurons for each layer. $f(.)$ is the activation function that can be the sigmoid function,

$$f(v) = \frac{1}{1 + e^{-v}}. \tag{2}$$

Learning in a backpropagation network is in two steps. First each input pattern is presented to the network and propagated forward to the output. Second, the gradient descent is used to minimize the total error on the patterns in the training set. In gradient descent, weights are changed in proportion to the negative of an error derivative with respect to each weight,

$$\Delta W_{(l,\,n,\,m)} \propto -\frac{\partial E}{\partial W_{(l,\,n,\,m)}}, \tag{3}$$

where E is,

$$E = \frac{1}{2} \sum_{i} \sum_{n} (T_{(i,n)} - Y_{(i,n)})^2, \tag{4}$$

where Y is the activation of output unit n in response to pattern i and T is the target output value for unit n and pattern i.

3.2 The Segmentation Stage (BPNN)

A square scanning window is placed, which is centered over a pixel in the image, Fig. 3. In this way the window acquires the central pixel value and all the values of the neighbouring pixels. At the training step, this supplies the input pattern to the learning algorithm. To extract the rest of the patterns, the sampling window scans the entire image. To reduce the dimension of the neural net, only a part of the pixels that are inside the sampling window is extracted. The feedforward network used was trained by the backpropagation

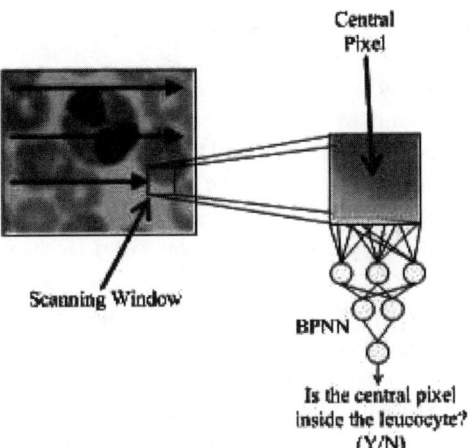

Central
Pixel

Scanning Window

BPNN

Is the central pixel
inside the leucocyte?
(Y/N)

Fig. 3. In the segmentation part the goal is to recognize patterns where the central pixel is inside of one white cell region

algorithm. The goal is to recognize the pattern where the central pixel is inside of one white cell region. The output patterns are indirectly formed manually. This enforces that a cell expert paints a set of images with a singular color that indicates the pixel is in the region cell. After the learning process ends, the system must be able to indicate if any pixel in an image belongs to a white cell region. Thus, all the pixels that make that statement true will be marked in the result image.

Architecture Used at Segmentation Stage (BPNN)

The modes and parameters that best performed for the segmentation were:

- Dynamic range to initialize the weights = $[-0.02, 0.02]$;
- Initializing mode: Nguyen–Widrow;
- Activation function: Bipolar Sigmoid (for all neurons);
- Weights update mode: Delta-Bar-Delta (Batch);
- $\alpha = 0.002$ (Delta-Bar-Delta parameter);
- $\beta = 0.7$ (Delta-Bar-Delta parameter);
- $\kappa = 0.0001$ (Delta-Bar-Delta parameter);
- $\gamma = 0.4$ (Delta-Bar-Delta parameter);
- Hidden Layers: 1;
- Neurons in the input layer: 243;
- Neurons in the hidden layer: 4;
- Training patterns: 95,354 (each pixel produces a pattern);
- Sampling window size: 25×25;
- Sampling pattern inside the sampling windows: on-off-off-on-off-off...
- Pixel format: HSL.

3.3 The Classification Step (BPNN)

An isolation process, based in the chain code, extracts each set of those pixels, previously segmented, that are forming one region cell. They are now numerically described by the Hu descriptors [11], geometrical primary descriptors and texture descriptors. These numbers are inputs to a new set of feedforward neural network. Each of them is training to recognize a white cell class, Fig. 4. Improvement of the generalization was tried by using a cross validation schema.

As the majority of the leukocytes may show two significant separable parts, the nucleus and the cytoplasm, it is important to split them to improve and simplify the classification process. A set of heuristics is used to analyze the histogram outline. Typically the cell histogram consists of two large modes, one will be the pixel occurrence of the cytoplasm and the other one will be the pixels occurrence of the nucleus. The threshold value must be between the peaks of these modes, Fig. 5. The heuristics conveys to a threshold value that

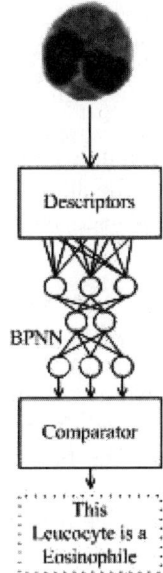

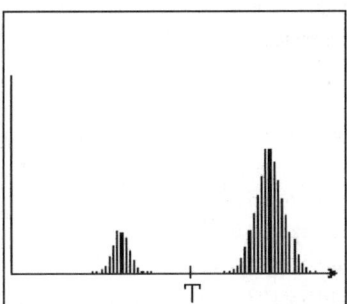

Fig. 4. The classification part. The classifier, after being trained, should be able to "announce" the class of white cell that is present at its inputs

Fig. 5. Typical histogram of a leukocyte region

separates the pixels of the nucleus from the pixels of the cytoplasm. The histogram outline is first smoothed and then it is scanned, with the gradient information being observed by the heuristics to try not to select a minimum that is not between the two modes. In most cases this method is enough to produce satisfactory results. But if the cytoplasm or the nucleus have a notorious granularity, some pixels in both regions can have the same values. Moreover, some times the cytoplasm and the nucleus have a low contrast between them. If these issues happen, the simple use of global threshold produces some mistakes. Thus, in future works this part needs to be improved.

The main heuristics:

- Is the negative gradient low enough to be the new threshold value considered? (if the negative gradient is high, the new threshold value can be in the wrong valley.)
- Has the relative long slope of one large mode already been overcome?
- After first mode descended, is the slope positive again and for many iterations? If yes, we are now at the second mode and the process must finish with the minimum found.

After separating the nucleus from the cytoplasm, it is necessary to extract features from each one. The features are numerically coded by the Hu descriptors, geometrical primary descriptors and texture descriptors. They are invariant to translation, rotation and some of them are invariant to scale change. These features are extracted from the gray nucleus regions, from the gray cytoplasm regions, from the binarization of the nucleus and from the binarization of the cytoplasm. The set of features that were used and given to the neural network inputs are present in Table 1.

Shape Descriptors – Spatial Moments With Invariance to Translation and Rotation

- Row moment of inertia;
- Column moment of inertia;
- Aspect;
- Spread;
- Hu descriptor 1 normalized (nh1);
- Hu descriptor 2 normalized (nh2);
- Hu descriptor 3 normalized (nh3);
- Hu descriptor 4 normalized (nh4);
- Hu descriptor 5 normalized (nh5);
- Hu descriptor 6 normalized (nh6).

The Hu descriptors tend to be very small and to have an enormous dynamic range. This could be negative to the neural process. Thus, the Hu descriptors are normalized as shown in [12].

Texture Descriptors

To represent numerally the texture were used:

- Mean deviation;
- Standard deviation;
- Skewness;
- Kurtosis;
- FNC – First neighbour Contrast [12];
- SNC – Second neighbour Contrast [12].

Table 1. Relation of the used descriptors

	Nucleos (binary)	Cytoplasm (binary)	Nucleos (gray)	Cytoplasm (gray)	Whole cell (gray)
Mq(2,0)	X	X			X
Mq(0,2)	X	X			X
Aspect	X	X			X
Spread	X	X			X
nh1	X	X			
nh2	X	X			
nh3	X	X			
nh4	X	X			
nh5	X	X			
nh6	X	X			
Mean deviation			X	X	
Standard deviation			X	X	
Skewness			X	X	
Kurtosis			X	X	
FNC			X	X	
SNC			X	X	
$N_u I$			X		
$C_y I$				X	
$N_u A$	X				
$C_y A$		X			
$N_u P$	X				
$C_y P$		X			
IR	X	X			
AR	X	X			
PR	X	X			
Circularity	X	X			

Primary Descriptors

Six geometrical primary descriptors were used:

- The average intensity of the nucleus ($N_u I$);
- The average intensity of the cytoplasm ($C_y I$);
- The area of the nucleus ($N_u A$);
- The area of the cytoplasm ($C_y A$);
- The perimeter of the nucleus ($N_u P$);
- The perimeter of the cytoplasm ($C_y P$);
- Circularity.

These descriptors were related through the followings ratios:

$$IR = \frac{N_u I}{C_y I},$$

(5)

$$AR = \frac{C_y A}{N_u A},$$

(6)

$$PR = \frac{N_u P}{C_y P}.$$

(7)

Architecture Used at Classification Stage (BPNN)

The modes and parameters that performed best for the classification were:

- Dynamic range to initialize the weights = $[-0.05 , 0.05]$;
- Initialization mode: Nguyen–Widrow;
- Activation function: bipolar sigmoid (for all neurons);
- Weights update mode: Delta-Bar-Delta (Batch);
- $\alpha = 0.0035$ (Delta-Bar-Delta parameter);
- $\beta = 0.7$ (Delta-Bar-Delta parameter);
- $\kappa = 0.0009$ (Delta-Bar-Delta parameter);
- $\gamma = 0.5$ (Delta-Bar-Delta parameter);
- Hidden Layers: 1;
- Neurons in the input layer: 46;
- Neurons in the hidden layer: 12;
- Training patterns: 276;
- Patterns to cross validation: 48;
- Rejection thresholds: $K = 0.15$ and $Q = -0.3$.

Comparator

To minimize mistakes in the classification, a rejection factor was added. The neural net output domain is $[-1; 1]$. There are two values to control the rejection. If one of the set of classification neural nets has its output in 1, this neural net is indicating the class for which it was trained. On other hand, if the neural net gives the value -1 it means this cell does not belong to that class. As the number of classification neural nets is equal to the number of cell classes to classify, it is important to observe the output value of each neural net. This is done in the comparator block. Thus, if none of the output values is higher than the Q threshold, the solution will be rejected with an "Unknown". If the two higher outputs have values where their difference is lower than the K threshold, then the rejection message will be "Uncertain". These values are figured out analyzing the receiver operating characteristic (ROC) curves to the neural net outputs.

   The main conclusion after using this approach was that the performance on this method is very dependent, especially, on the training set. Obviously, if the segmentation fails the classification fails too. So, considering the segmentation step, we can find a set of problems. If the images, which we have submitted to the evaluation process after the training, are similar, in brightness, in color, in cell position statistics, then the process reaches a relatively good succeeded level. Roughly, 81% of the cells are classified correctly. But those are natural images. Thus, it is very difficult to guarantee the stability of the future images in relation to the training pattern set. If we try to increase the size of the training set, the problems with the training itself start increasing. The common issues related to "the curse of the dimensionality" of the neural network arise intensely. Staying stuck at local minimums and the exponential growth of the operations number make this solution impractical. Another problem about the use of the neural networks in this scenario and in this way was the great difficulty in enforcing the learning of the neural network over a particular feature. In other words, it is extremely difficult to select a training set to give essential features to the neural network. Normally to presume that the neural network will extract the essential features is an improbable presupposition.

4 Approach − II (PCNN-SVM)

The proposal of the second approach, the hybrid approach, is to try to overcome several problems of the first approach.

4.1 PCNN

One characteristic of the pulsed-coupled neural network is to be able to extract essential visual information from a given image. Regions composed by different patterns at different scales, as textures and sets of subregions made of many intensities, are identified at different interactions. The PCNN process is unsupervised. Therefore, we don't need to look out for the essential features of a region, neither does a neural network supervisor have to do that. This was the main reason to choose the PCNN as the unsupervision method. The pulsed coupled neural network is based on the Eckhorn model [5]. A PCNN neuron has the Feeding and the linking compartment. Its synaptic weights, M and W, respectively control the communication with the neighbouring neurons. Only the Feeding compartment receives the input stimulus (pixel value) S. The neuron output is represented by Y. V_F and V_L adjust the global influences of the external neurons in each compartment respectively. The memory effect magnitude is controlled by the α_F, α_L and δ_n. The value of the Feeding compartment at the iteration n is determined by

$$F_{ij}[n] = e^{\alpha_F \delta_n} F_{ij}[n-1] + S_{ij} + V_F \sum_{kl} M_{ijkl} Y_{kl}[n-1]. \qquad (8)$$

The value of the Linking compartment at the iteration n is determined by

$$L_{ij}[n] = e^{\alpha_L \delta_n} L_{ij}[n-1] + V_L \sum_{kl} W_{ijkl} Y_{kl}[n-1]. \qquad (9)$$

The state of these compartments is combined by

$$U_{ij}[n] = F_{ij}[n]\{1 + \beta L_{ij}[n]\}. \tag{10}$$

to obtain the internal state of the neuron and where β adjusts the interference between both compartments.

The internal state is compared with a dynamic value to produce the output Y,

where Θ is the dynamic threshold described by

$$\Theta_{ij}[n] = e^{\alpha_\Theta \delta_n}\Theta_{ij}[n-1] + V_\Theta Y_{ij}[n]. \tag{11}$$

The memory effect of Θ is controlled by α_Θ and δ_n. The participation of the fired value Y in Θ is adjusted by V_Θ.

The state of each PCNN neuron in a neural network is dependent on the external stimuli that are represented by the input pixel values. It is also dependent on the output state of the neighbouring neurons and on its own internal state. i.e., a neuron that has fired recently reduces its sensibility to fire immediately again. These features enable the neuron fire patterns to show, at different times, important regions of the image. Those regions are features of the image in diverse domains, such as different textures, gray levels or aggregate subregions.

4.2 SVM

Given a training set of instance-label pairs $(x_i, y_i), i = 1, \ldots, l$ where $x_i \in \Re^n$ and $y \in \{1, -1\}^l$, the support vector machine requires the solution of the following optimization problem:

$$\min_{w,b,\xi} \quad \tfrac{1}{2}w^T w + C \sum_{i=1}^{l} \xi_i$$
$$\text{subject to} \quad y_i\left(w^T \phi(x_i) + b\right) \geq 1 - \xi_i, \; \xi_i \geq 0 \tag{12}$$

A SVM network maps the training vectors x_i in a higher dimensional space by the function ϕ. Thus, the SVM just needs to find the linear separated hyperplane with the maximal margin in this higher dimensional space. C is the penalty parameter of the error term ξ, $C > 0$.

$K(x_i, x_j) \equiv \phi(x_i)^T \phi(x_j)$ is the kernel function. There are innumerous kernel function types; some can fit better a specific application than others. We used the Adatron kernel [13, 14].

4.3 Zernike Descriptors

The Zernike moments perform better than the Cartesian moments in terms of noise resilience and information redundancy and furthermore they allow reconstruction capability (Note that this subsection is derived largely from [15].). The Zernike orthogonal moments [16] are constructed using a set of complex polynomials which form a complete orthogonal basis set defined on the unit disc $x^2 + y^2 \leq 1$. They are represented as:

$$A_{mn} = \frac{m+1}{\pi} \iint_{x^2+y^2 \leq 1} f(x,y) \left[V_{mn}(x,y) \right]^* dx \, dy, \tag{13}$$

where $m = 0, 1, 2, \ldots, \infty$, $f(x,y)$ is the function being described, $*$ denotes the complex conjugate and n is an integer, depicting the angular dependence and subject to the conditions:

$$m - |n| = even, \quad |n| \leq m. \tag{14}$$

The Zernike polynomials $V_{mn}(x,y)$ expressed in polar coordinates are:

$$V_{mn}(r, \theta) = R_{mn}(r) \exp(jn\theta), \tag{15}$$

where (r, θ) are defined over the unit disc and $R_{mn}(r)$ is the orthogonal radial polynomial, defined as:

$$R_{mn}(r) = \sum_{s=0}^{\frac{m-|n|}{2}} (-1)^s F(m,n,s,r), \tag{16}$$

where:

$$F(m,n,s,r) = \frac{(m-s)!}{s! \left(\frac{m+|n|}{2} - s \right)! \left(\frac{m-|n|}{2} - s \right)!} r^{m-2s}. \tag{17}$$

Translation and scale invariance can be achieved by normalising the image using the Cartesian moments prior to calculation of the Zernike moments [17] and the absolute value of a Zernike moment is rotation invariant.

4.4 Segmentation Stage (PCNN-SVM)

The whole process is quite similar to the previous approach. A sampling window scans the image pixel by pixel, but now the pattern that is extracted from it is supplied to the inputs of the PCNN. After some iterations, which could reach the 20th or more, the PCNN has segmented different subregions of the image. Each iteration represents a new segmentation, Fig. 6. They are images with the size of the sampling window. All the iterations form a segmentation set and this could be seen as a signature [18]. The whole set is a huge dimension. To reduce it we have tried to use Zernike descriptors [17,19] since normally they have a better behavior over the Hu descriptors. As the SVMs do not have the same problems as the feedforward/backpropagation neural networks (may become stuck in a local minimum) and show a high

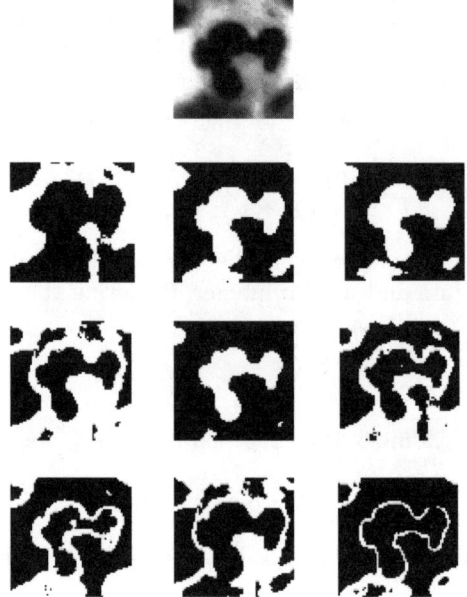

Fig. 6. The first image is an example of one image sampled by the sampling windows over a blood cells image. The rest of the images are the PCNN outputs along several iterations

generalization ability in many real situations, the descriptors results are now the inputs for a SVM.

The SVM is trained as in the segmentation first approach. The difference is that now the PCNN and the Zernike are preceding transforms. We can see this relation in the Fig. 7.

The steps of the algorithm for the segmentation training part are explained next

1. Scan the training image to be segmented, pixel by pixel, in a defined sequence. It can be from the top to the bottom and from the right to the left.
2. For each pixel in the image, consider a set of neighbouring pixels. This set can be extracted from a sampling window centrally placed over the pixel in decision. The size of this window, for this problem, was 51×51. In other problems the size must be adapted.
3. Each captured subregion in the previous steps is now processed by the pulsed coupled neural network. The PCNN will give a signature about each subregion. In this case, the PCNN will produce signatures of these subregions in 20 iterations.
4. The PCNN outputs construct, in each iteration, a new binary subregion. These binary subregions are numerically described by Zernike descriptors.

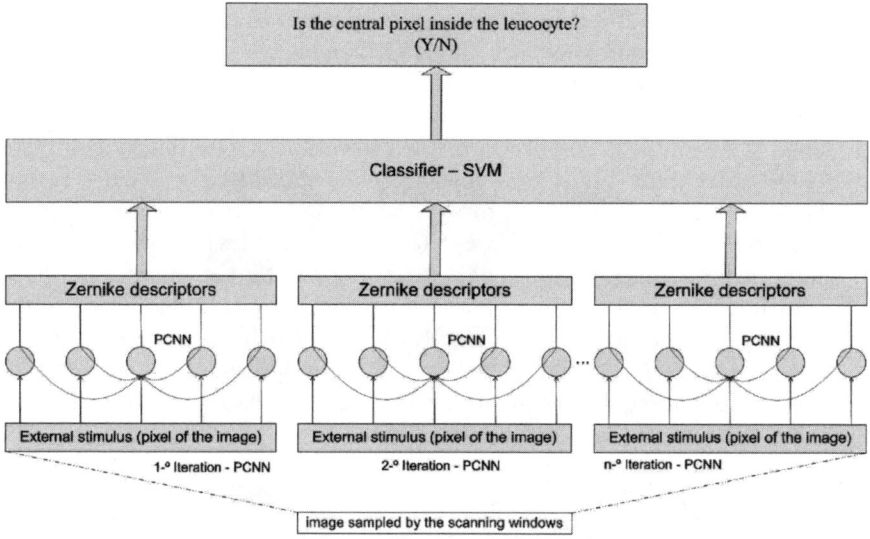

Fig. 7. The main components of the segmentation mechanism (PCNN-SVM)

5. Until this step each image pixel to be segmented will have a numerical sequence with a length that is given by the number of PCNN iterations times the number of descriptors used. In the leukocytes problems six Zernike descriptors were used. Each one of those numerical sequences will be a training pattern.
6. The numerical sequences are normalized, one being the standard deviation and the mean value being zero.
7. The training patterns are presented at the inputs of the support vector machine. The output patterns are formed by a unique variable. The value for this variable could be one of the two possible values. One value hints that the central pixel belongs to the white cell. On the other hand the complementary value refers that the central pixel is outside of a white cell region.
8. The last step will be repeated until all the training patterns are run through and until the SVM reaches convergence.

The steps of the algorithm for the segmentation pos-training part are explained next

1. Scan the image to be segmented, pixel by pixel, in a defined sequence. It can be from the top to the bottom and from the right to the left.
2. For each pixel in the image, consider a set of neighbouring pixels. This set can be extracted from a sampling window centrally placed over the pixel in decision. The morphology and the size of the sampling windows have to be the same that were chosen for the training process.

3. Each captured subregion in the previous steps is now processed by the pulsed coupled neural network. The PCNN will give a signature about each subregion. The PCNN configuration has to be the same as the training part.
4. The PCNN outputs construct, in each iteration, a new binary subregion. These binary subregions are numerically described by Zernike descriptors. They are the Zernine descriptors that were used in the training stage.
5. Until this step each pixel of image to be segmented will have associated a numerical sequence with a length that is given by the number of PCNN iterations times the number of descriptors used. In the leukocytes problem six Zernike descriptors were used. Each one of those numerical sequences will be a pattern to be analyzed by the supervised method.
6. The numerical sequences are normalized, the standard deviation being the value one and the mean value being zero.
7. The patterns are presented at the inputs of the support vector machine. The output patterns are formed by a unique variable. The value for this variable can be one of the two possible values. One value hints that the central pixel belongs to the white cell. On the other hand the complementary value refers that central pixel is outside of a white cell region.
8. The whole process will be finished when all the pixels of the image to be segmented are classified.

Table 2 contains the parameterizations that were used in the PCNN (segmentation) and Table 3 contains the general parameterizations that were used in the segmentation stage.

4.5 The Classification Stage (PCNN-SVM)

At the classification problem the changes are identical to the changes made in the segmentation (PCNN-SVM) problem relatively to approach I. Before applying the descriptors over the segmented regions we apply the PCNN. The signature result is numerically coded by the Zernike descriptors. Several SVMs are now used as classifiers of the type of cell, one for each cell type,

Table 2. PCNN configuration for the segmentation stage

PCNN components	Settings
Kernel M	5×5; type $1/r$
Kernel W	5×5; type $1/r$
V_f	0.1
V_l	0.1
V_t	33
ß	0.1
A_f	10
A_l	6
A_t	18

Table 3. General configuration for the segmentation stage (PCNN-SVM)

Components	Settings
Pixel format	Gray − 8 bits
Sampling window	Square − 51×51
PCNN iterations	20
Descriptors	The first six Zernike descriptors: (0,0); (1,1); (2,0); (2,2); (3,1); (3,3)
Supervised net	SVM
The number of SVM inputs	240 − 20 iterations × six complex values descriptors
SVM Kernel	Adatron
Number of SVMs	1
Training patterns	95,354

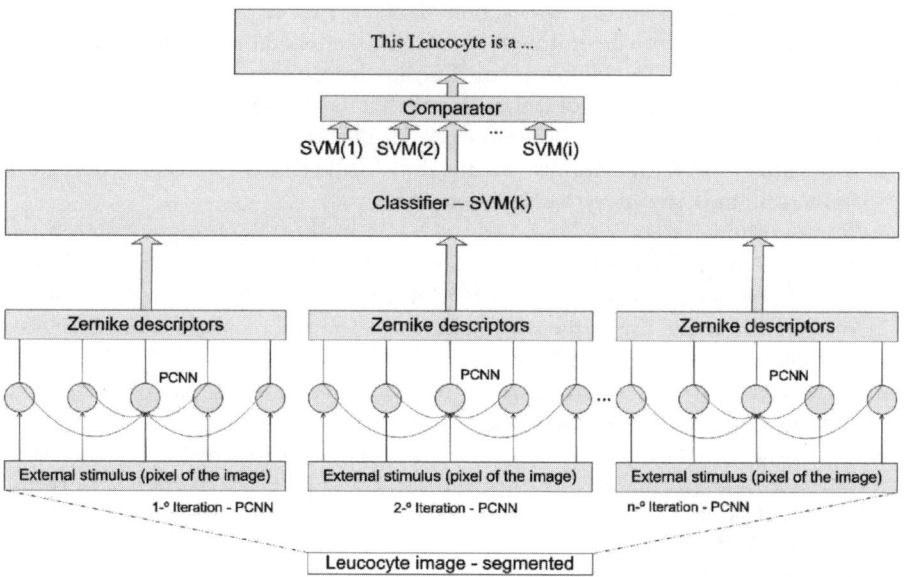

Fig. 8. The main components of the classification mechanism

Fig. 8. The PCNN has the task of splitting the cell in subregions in agreement with the essential features. Thus, the specific descriptors that were used in approach I are no longer needed, nor is the process to split the nucleus from the cytoplasm.

The algorithm steps for the classification in training part are explained next

1. Scan the pixels of the total image, that were segmented in the segmentation task, in a defined sequence. It can be from the top to the bottom and from the right to the left.

2. When the first pixel of a leukocyte that was previously marked by the segmentation process is found, a chain code method is used to mark the cell contour pixels.
3. Mark again all the pixels that are inside of the cell contour. This allows us to mark pixels that are inside the cell but were not detected by the segmentation process.
4. All the cell pixels are extracted from the marked region. Each region is processed by each PCNN in several iterations. At this stage, the number of iteration was set to 40. This number allows us to obtain a good information set of cell partition in order for it to be possible to produce the classification of the cell.
5. The PCNN outputs set is, in each iteration, a new binary subregion. These binary subregions are now numerically described by the Zernike descriptors.
6. For each cell the length of the numerical sequence that will describe the region is given by the number of PCNN iterations times the number of descriptors used. In this problem Zernike descriptors were used until their sixth order. Each one of those numerical sequences will be a training pattern to be learned by the supervised method.
7. The numerical sequences are normalized, the standard deviation being the value one and the mean value being zero.
8. Each training pattern is given to the SVM inputs. Each input pattern has a paired output pattern. The number of SVM is equal to the number of leukocytes to classify. Each SVM is trained in a way to signalize in its output when the input pattern belongs to the cell class that was assigned to it.
9. The last step will be repeated until all the training patterns are run through and until the SVM reaches the convergence.
10. To optimize the decision mechanism of the comparator block, the ROC curves are analyzed to figure out the threshold values to the SVM outputs, as was done in approach I.

The algorithm steps for the classification in the post-training part are explained next

1. Scan the pixels of the total image, which were segmented in the segmentation task, in a defined sequence.
2. When the first pixel of a leukocyte that was previously marked by the segmentation process is found, a chain code method is used to mark the cell contour pixels.
3. Mark again all the pixels that are inside of the cell contour. This allows us to mark pixels that are inside the cell but were not detected by the segmentation process.
4. All the cell pixels are extracted from the marked region. Each region is processed by a PCNN in several iterations.

Table 4. PCNN configuration for the classification stage

PCNN components	Settings
Kernel M	7×7; type $1/r$
Kernel W	7×7; type $1/r$
V_f	0.5
V_l	0.8
V_t	20
ß	0.2
A_f	10
A_l	1
A_t	5

5. The PCNN outputs set are, in each iteration, a new binary subregion. These binary subregions are now numerically described by the Zernike descriptors.
6. For each cell the length of the numerical sequence that will describe the region is given by the number of PCNN iterations times the number of descriptors used. Each one of those numerical sequences will be a pattern to be analyzed by the supervised method.
7. The numerical sequences are normalized, the standard deviation being the value one and the mean value being zero.
8. Each pattern is given to the SVM inputs. This SVM has to be already trained. Each SVM will signalize in its output if the input pattern belongs to the cell class that was assigned to it.
9. Now, in the comparator block, if no SVM output is higher than the Q threshold, the solution will be rejected with an "Unknown". If the two higher outputs have values where their difference is lower than the K threshold, then the rejection message will be "Uncertain".

Table 4 contains the parameterizations that were used in the PCNNs (classification) and Table 5 contains the general parameterizations that were used in the classification stage.

The comparator block works like the first approach comparator.

5 Results

The main purpose of the second approach when compared to the first approach is to increase the generalization capacity, which is the weak point in the first approach. This is evident in the segmentation process. We tried to use gray pixels with the segmentation, in the first approach, but the results were very weak, even during the training step. Then, we decided to make the comparison using colored pixels in the first solution. The second solution demands a very high computational effort. In fact, one image with 129×110 pixels was processed in 1 s at the first approach and spent 217 s in the second approach.

Table 5. General configuration for the classification process (PCNN-SVM)

Components	Settings
Pixel format	Gray – 8 bits
Input pixels	All the pixels belonging to a segmented cell
PCNN iterations	40
Descriptors	The first six Zernike descriptors: (0,0); (1,1); (2,0); (2,2); (3,1); (3,3).
Supervised net	SVM
The number of SVM inputs	480 – 40 iterations × six complex values descriptors
SVM Kernel	Adatron
Number of SVMs	5 (the number of classes of cells that were considered)
Training patterns	276
Rejection thresholds	$K = 0.10$ and $Q = -0.4$.

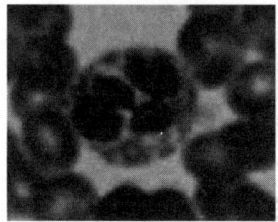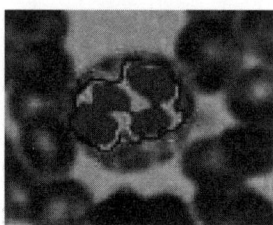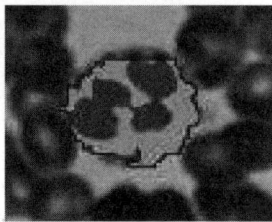

Fig. 9. Comparison between the segmentation results of approach I and approach II for a Neutrophile

This was obtained in a 3 Ghz Pentium 4 PC. This is due mainly to the PCNN process, as a PCNN task is executed by each pixel that appears in the image to be segmented. Then to see the performance of this second method and to minimize the computation time, which is the major weak point of this method, we have performed the tests with gray pixels.

5.1 Segmentation

In Figs. 9–11 the first column has the original images. The second column has images that are the result of the first approach. The last column has the images from the second approach. All of them were out of the training set and their aspects were very random. In the two first cases the cytoplasm is completely segmented by the PCNN-SVM methods but the BPNN method fails, segmenting just partially the cytoplasm. The number of test images was 227. 77% of the test images were correctly segmented using the first approach. 86% of the test images were correctly segmented using the second approach. It is difficult to compare these results with other studies because the images databases are not the same. In any case, the segmentation results that are announced in [8] have a 95% success rate.

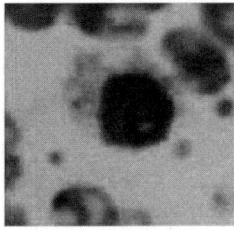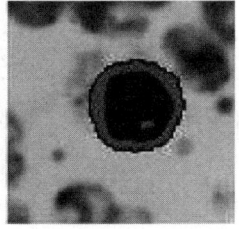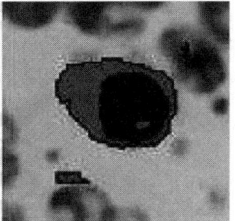

Fig. 10. Comparison between the segmentation results of approach I and approach II for a Lymphocyte

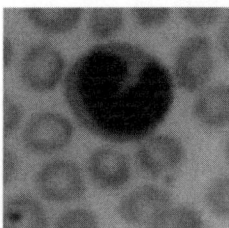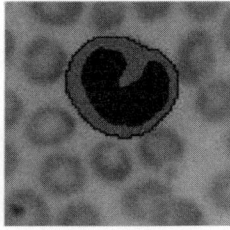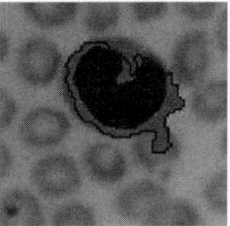

Fig. 11. Comparison between the segmentation results of approach I and approach II for a Monocyte

Table 6. Confusion matrix for the classification process with the BPNN

	Basophile	Eosinophile	Lymphocyte	Monocyte	Neutrophile	Unc.	Unk.
Basophile	4			1			
Eosinophile	1	38		4	4	2	2
Lymphocyte			55	6	1	3	
Monocyte	1	1	3	19	2	2	3
Neutrophile	2	10		4	61	8	4
Uncertain	1			3		12	2
Unknown		4	1	3	2	4	22

5.2 Classification

To produce the classification results we used the white cell regions classified by a haematologist. Those results are presented in a confusion matrix. Table 6 is concerned with the first approach and Table 7 is concerned with the second approach. In the first column we have the cell classes announced by the haematologist and in the first row there are the classes announced by the system. Only the five normal white cells classes were considered. The Unc. (Uncertain) label means that the two higher matching classes were considered similar to the analyzed region. The Unk. (Unknown) means that the analyzed region does not seem to correspond to any class. Table 8 shows the fraction of correct answers for the BPNN method and for the PCNN-SVM method,

Table 7. Confusion matrix for the classification process with the PCNN-SVM

	Basophile	Eosinophile	Lymphocyte	Monocyte	Neutrophile	Unc.	Unk.
Basophile	4			1			
Eosinophile	1	41		2	4	1	2
Lymphocyte			57	5		3	
Monocyte	1		1	22	2	2	3
Neutrophile	1	7		4	70	6	1
Uncertain	1			2		11	2
Unknown		2	1	3	2	3	20

Table 8. The fraction of correct answers

	BPNN (%)	PCNN-SVM (%)	Number of images
Basophile	80.0	80.0	5
Eosinophile	80.8	85.4	51
Lymphocyte	88.7	91.9	65
Monocyte	73.0	84.6	31
Neutrophile	79.2	85.3	89

as well as the number of cells classified. To obtain the classification results we selected satisfactory segmented cells for both approaches. The rates were figured out from test data set. The approach I and the approach II rates were calculated using the same images. In [8] is announced 75–99% correct classification using 13 cell types.

6 Conclusions and Discussion

Approach II showed its greater ability to segment cell images with gray pixels over the first approach. Since color is a significant help to make the correct detection (but it does not help the generalization), the first method has problems even during the training step if it is given gray pixel levels. This fact exposes the difficulty of the supervised neural net extracting the general features that are necessary to have success in the segmentation of images that are not in the training set. The color information only demonstrates itself to be a good feature if the future images do not change in relation to the training images and when the segmentation method is not supervised. The PCNN is able to correctly segment many subregions even in gray level pixels. The set of these subregion forms a better way to allow the generalistic capacity. Then, the SVM only needs to learn the generalistic features that are supplied to it by the PCNN. For these reasons, the second method has shown a better performance when images that have different aspects from the images of the training set were processed. We think that the results could be improved by giving to the PCNN color pixels too. On the other hand, it is difficult to control the previous segmentation made by the PCNN. The question is: is the subregion set figured out by the PCNN always the best set that defines the generalistic features? The answer is no. Although the subregion set can be extended to increase the PCNN iterations, hoping to achieve new generalistic features, there is not any certainty about it. We believe that a feedback mechanism applied to the PCNN bringing information from the supervised method could give clues to the PCNN to take the good direction. Another problem with the PCNN is the sequence order, where the same segmented subregion could appear in a different position in the sequence when the images to be segmented are different. This disturbs the signature that was learned by the SVM. A system to recognize each subregion extracted by the PCNN, in a isolated way, would be necessary. Figure 11 shows a failed case using method II. Replacing the PCNN by other unsupervised mechanisms could be a good solution. Self-organizing neural networks and multiresolution segmentation are just a few of a large set of possible choices. Certainly, it was possible to select a type of image where both methods always fail or to select types of images where both methods will perform always correctly. In a real situation we never know: images will appear that will fit the two cases (success or failure), and in this situation the second approach is in the better position to minimize the mistakes.

The average rate of correctly classified cells increases until 86.9%. The average rate of the first method was 81.31%. For these calculations the "Uncertain" and the "Unknown" cases were not considered. As this count is differential, it is possible to reject the doubtful cases and we shall do it. The second process is not a perfect method but it is better than the first one when related to the generalization problem. Thus we can say that the second method has a generalist potential over the first method.

When these results are compared with other studies they appear to be weak, but as we have said it is difficult to make the comparison because the images are not from the same set.

The first approach needs a very completed training set to avoid failures. And that is not easy to reach. The second approach is not so dependent on the training set but it obliges a very large computational effort. The first method has a much better time performance than the PCNN based method. In fact, the main responsibility for this lies with the PCNN and consequently the number of PCNN iterations. This kind of neural network is easily paralleled on distributed computers, on FPGAs and on other similar techniques [20, 21]. This makes practicable the implementation of this method for a useful system. We believe that the fusion of unsupervised with supervised methods must be pursued to try to overcome the problems of the variability in natural cell images.

References

1. Rumelhart, D.E., Hinton, G.E., Williams, R.J.: Learning representations by back-propagating errors. Nature **323** (1986) 533–536
2. Boser, B., Guyon, I., Vapnik, V.N.: A training algorithm for optimal margin classifiers (1992)

3. Cortes, C., Vapnik, V.: Support-vector networks. Machine Learning **20** (1995) 273–297
4. Schölkopf, B., Burges, C.J.C., Smola, A.: Introduction to support vector learning. In: Advances in kernel methods. MIT Press, Cambridge, MA (1999) 1–15
5. Eckhorn, R., Reitboeck, H.J., Arndt, M., Dicke, P.: Feature linking via synchronization among distributed assemblies: Simulations of results from cat visual cortex. Neural Comp. (1990) 293–307
6. Johnson, J.L.: Pulse-coupled neural nets – translation, rotation, scale, distortion, and intensity signal invariance for images. Applied Optics **33** (1994) 6239–6253
7. de Jonge, R., Brouwer, R., van Rijn, M., van Acker, B.A.C., Otten, H.J.A.M., Lindemans, J.: Automated analysis of pleural fluid total and differential leukocyte counts with the sysmex xe-2100. Clin. Chem. Lab. Med. **44** (2006) 1367–1371
8. Ramoser, H., Laurain, V., Bischof, H., Ecker, R.: Leukocyte segmentation and classification in blood-smear images. In: Engineering in Medicine and Biology Society, 2005. IEEE-EMBS 2005. 27th Annual International Conference of the 2005. 3371–3374
9. Hirono, T., Yabusaki, K., Yamada, Y.: The development of leukocyte counter using fluorescence imaging analysis. In: Biophotonics, 2004. APBP 2004. The Second Asian and Pacific Rim Symposium on. (2004) 213–214
10. Nilsson, B., Heyden, A.: Model-based segmentation of leukocytes clusters. In: Pattern Recognition, 2002. Proceedings. 16th International Conference on 2002, Volume 1. (2002) 727–730
11. Hu, M.K.: Visual pattern recognition by moment invariants. Information Theory, IEEE Transactions on **8** (1962) 179–187
12. Masters, T.: Signal and Image Processing with Neural Networks. Wiley, New York (1994)
13. Anlauf, J.K., Biehl, M.: The adatron – an adaptive perceptron algorithm. Europhys. Lett. (1989)
14. Cristianini, N., Campbell, C.: The Kernel–Adatron algorithm: a fast and simple Learning Procedure for Support Vector Machines (1998)
15. Shutler, J.D., Nixon, M.S.: Zernike velocity moments for description and recognition of moving shapes. (2002)
16. Teague, M.R.: Image analysis va the general theory of moments. J. Optical Soc. Am. (1917–1983) **70** (1980) 920–930
17. Khotanzad, A., Hong, Y.: Invariant image recognition by zernike moments. Pattern Anal. Machine Intell., IEEE Trans. **12** (1990) 489–497
18. Johnson, J.L.: The signature of images (1994)
19. Zernike, F.: Beugungstheorie de schneidenverfahrens und seiner verbesserten form, der phasenkontrastmethode. Physica (1934)
20. Kinser, J., Lindblad, T.: Implementation of pulse-coupled neural networks in a cnaps environment. IEEE Trans. Neural Netw. **10** (1999) 584–590
21. Lindblad, T., Kinser, J.M.: Image processing using pulse-coupled neural networks. Springer, Berlin Heidelberg New York (2005)

A Closed Loop Neural Scheme to Control Knee Flex-Extension Induced by Functional Electrical Stimulation: Simulation Study and Experimental Test on a Paraplegic Subject

Simona Ferrante, Alessandra Pedrocchi, and Giancarlo Ferrigno

Department of NITLAB Bioengineering, Politecnico di Milano, via Garofalo 39, 20133 Milano, Italy, simona.ferrante@polimi.it

Summary. Functional electrical stimulation (FES) is a well established technique for the rehabilitation of neurological patients. In this study we proposed a controller, called Error Mapping Controller (EMC), for the neuromuscular stimulation of the quadriceps during the knee flex–extension movement. The EMC is composed by a feedforward inverse model and a feedback controller, both implemented using neural networks. The training of the networks is conceived to avoid to a therapist and a patient any extra experiment, being the collection of the training set included in the normal conditioning exercises. The EMC philosophy differs from classical feedback controllers because it does not merely react to the error in tracking the desired trajectory, but it estimates also the actual level of fatigue of the muscles. The controller was first developed and tested in simulation using a neuro–musculo–skeletal model and then in some experimental sessions on an untrained paraplegic patient.

1 Introduction

Functional electrical stimulation (FES) is a well established technique for the rehabilitation of neurological patients. The concept is to apply electrical current to excitable tissue in order to activate the intact lower motor neurons using implanted or superficial electrodes. Appropriate electrical stimuli can in fact elicit action potential in the innervating axons and the strength of the resultant muscle contraction can be regulated by modulating the stimulus parameters [1]. Therefore, a crucial point for the wide spreading of these applications is the development of an efficient and robust control system to regulate the stimulator current in order to allow stimulated limbs to achieve the desired movements.

Among all the FES applications, the control of knee position of a free-swinging leg by means of quadriceps stimulation has become a good benchmark for studying and testing the efficiency of FES control strategies [2–7].

S. Ferrante et al.: *A Closed Loop Neural Scheme to Control Knee Flex-Extension Induced by Functional Electrical Stimulation: Simulation Study and Experimental Test on a Paraplegic Subject*, Studies in Computational Intelligence (SCI) **83**, 397–415 (2008)
www.springerlink.com   © Springer-Verlag Berlin Heidelberg 2008

The importance of this movement is crucial in the training of the quadriceps muscle which plays a fundamental role in the main motor activities, such as standing up, sitting down, walking, cycling, standing posture, and climbing stairs. In addition, this simple movement can be performed in a very safe and comfortable condition for the patients and can be used to strengthen the quadriceps muscle daily before carrying on complex functional tasks.

The challenge of controlling the quadriceps stimulation, even in a single joint movement, is made more difficult by the non-linear and time-varying nature of the muscle contraction dynamics, and the presence of external and internal disturbances. In general, when FES is applied to the neuromuscular system, the strength of the resulting muscle contraction can be regulated by modulating frequency, amplitude and duration of the current stimulus pulse. To obtain the temporal summation it is possible to change the stimulation frequency. This is a good procedure that is able to modulate the force produced. Instead, the spatial summation is gained by increasing the stimulation intensity (i.e., the amplitude or the pulse width (PW)). In this way, a higher electrical charge is delivered to the neuromuscular system and consequently, the motor units with higher activation thresholds are recruited. The fibers recruitment characteristic has frequently been represented by a static saturation-type non-linearity because initially, as stimulation intensity starts increasing, the activation threshold of very few motor units will be reached. This phase is followed by a range of stimulation intensities where increasing numbers of motor units are recruited. Finally, at high levels of stimulation, it is possible that all motor units associated with the target nerve are activated so that a further increase in stimulation level does not elicit increased muscle response [6]. The non-linearity of the system depends also on other factors, including electrode placement and location, physiological properties of the muscle, its innervation and electrode properties [6].

Another crucial problem to face in FES applications is the muscular fatigue that is a time-variable phenomenon very complex to be adequately compensated especially using surface stimulation. In fact, the artificial contraction induced by using superficial electrodes amplifies the problem of fatigue for three reasons:

- Muscular fibres are activated synchronously and thus it is impossible to reproduce the fibers turn-over;
- An higher frequency is required to reach the artificial tetanic contraction (20–30 Hz instead of the 10 Hz for the natural tetanic contraction);
- The fibers are recruited in the opposite order with respect to the natural contraction; the fast fatiguable fibers being the first to activate, because of their lower activation threshold.

All these findings are very important and must be taken into consideration in the design of control systems for stimulated muscle responses. Namely, the controller should be able to cope properly with fatigue, trying at the beginning of the session to track a desired trajectory of the knee and, after the occurrence of fatigue, to prolong the movement as much as possible, even if the maximum extension decreased. This is really important because knee flex-extension is intended as a training movement to be executed in order to reach a good muscle conditioning level. In addition, the controller has to be calibrated on a single subject and even on a single session of each subject and it has to be able to react promptly against disturbaces. Finally, the controller has to be very easy to use because the probability of a real widespread use of the controller in clinical practice as well as the probability of being accepted by many patients strongly depend on the short preparation and on the easiness of the exercise procedures.

The first controllers developed for the knee flex-extension FES application were open-loop systems in which the pre-specified stimulation parameters were determined by trial-and-error [2]. The open-loop systems have been successful in demonstrating the feasibility of using electrical stimulation and they have helped to understand that the trial-and-error method is time-consuming and unreliable. Moreover, the open-loop systems, by definition, can not react to disturbances and can not cope with the occurrence of muscular fatigue properly. In order to overcome these problems, a feedback signal must be used to modify stimulus parameters in real time. Among the traditional closed-loop controllers, proportional integrative derivative (PID) are widely used in engineering, as well as in biomedical applications, including control of FES [3,4,8]. In a recent study performed on knee flex-extension induced by FES [5], it was shown that the good tracking performance of PID controllers was offset by a considerable time-lag between reference and actual joint angle, which became more marked when exercises were protracted in time. In addition to the these open- and closed-loop control systems, model-based control systems have been used to control FES knee flex-extension [3, 9]. This kind of control schemes included a neuro-musculo-skeletal model of the system to be controlled. Unfortunately, the large quantity of parameters required for the identification of the system to control is very difficult to determine experimentally [10] and moreover, a re-calibration of the identified parameters is required for each subject to maintain optimal performance.

The physiological model could be replaced by a non-linear black-box model, such as an artificial neural network (ANN). Like their biological counterparts, ANNs can learn from examples and generalise and thus they can successfully cope with the complexity and non-linearity of systems [11]. Chang et al. suggested combining an ANN (acting as an inverse model of the system to be controlled) with a closed-loop, fixed-parameter PID feedback controller, thereby making adjustments for residual errors, due to disturbances and fatigue [7]. This approach was very promising and represents one of the main sources on which the controller developed in this chapter is based. Abbas et al. [12–14] proposed a control system which utilised a combination of adaptive feedforward and feedback control techniques. The feedforward adaptive controller was a pattern generator/pattern shaper (PG/PS) in which PG generated a stable oscillatory rhythm and a single layer neural network

(PS), placed in cascade to PG, provided stimulation to the muscles. A fixed-parameter proportional derivative (PD) feedback controller was included to correct the action of the feedforward controller in case of disturbances. This neural controller showed good performance in both simulation and experimental sessions. However, the authors did not report in details the efficacy of the controller in tracking fatigue. In addition, even if PG/PS could be used with many patterns, this could strongly decrease the efficiency and velocity of the adaptive controller, being the architecture of PS multiplied by the number of patterns.

In a previous study of our research group, an adaptive control system (NEURADAPT) based on ANNs was developed [15]. This control scheme included an inverse neural model of the stimulated limb in the feedforward line and a neural network trained on-line to learn PID behaviour in the feedback loop. Despite the encouraging results, the drawback was that the desired output on the adaptive feedback neural network was based on the PID action and this made the identification phase slower and produced a considerable time-lag between the reference and actual joint angle.

The aim of the study presented in this chapter was to develop a neural controller that is able to face properly the fatigue effect and that is simple to use in a clinical environment avoiding extra complex procedures to the therapist and the patient. The neural controller here proposed is called error mapping controller, EMC in the following. The novelty of EMC is that it is able to provide a fatigue compensation not only based on tracking performance, but also able to understand if the tracking is feasible. When the level of fatigue is so high that it is not possible to reach the desired extension EMC does not over-stimulate the system provoking an extra increase of fatigue but tries to obtain a movement prolonged in time as much as possible in order to enhance the rehabilitative effect of training. This chapter includes a simulation study in which the EMC was developed and validated and an experimental study in which the controller was tested on one paraplegic patient.

2 The EMC: Physiological Rationale

In this study the ANNs were used to calculate the relationship between a biomedical motor signal, that is the angular trajectory of the knee, and the artificial motor command able to generate the desired movement, that is, the stimulation delivered to the quadriceps muscle. In particular, the ANNs had to estimate the shape of the stimulation signal to deliver to the subject in order to produce a desired knee angular trajectory of the knee. To reach this aim we developed a physiologically inspired controller, EMC, which comprised two different levels of ANNs. The first ANN worked as an inverse model, ANNIM, that represented the stable anthropometrical and strength characteristics of the limbs and the second one, NeuroFeedback (NF), was a feedback controller able to predict the correction of the motor command, on the basis of the

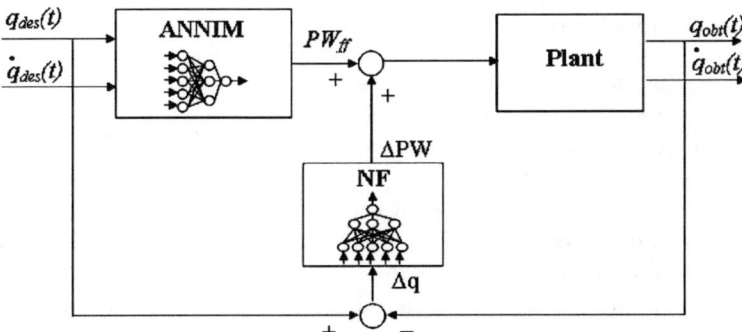

Fig. 1. EMC structure

current error in the angular trajectory and the current fatigue level (Fig. 1). The EMC scheme was thought to replicate the way in which the central nervous system produces a motor task in order to perform any movement, as proposed and discussed by Wolpert et al. [16]. In fact, we started building a motor command sequence on the basis of the known characteristics of our motor apparatus. This can be called a "baseline motor programme" and it is stored, learning from examples like the ontogenic development of a human being. In the same way, ANNIM was trained off-line, learning from examples collected from the system to be controlled (plant in simulation or subject in the experimental study) in the nominal conditions. In the motor control, in addition to the baseline motor programme, exists a short term memory, which is able to face occurring variations of our actuators because of a wide range of events, such as a trauma temporarily limiting the range of motion of one joint or fatigue due to prolonged exercises. This second control is strongly dependent on the current sensory feedback signals coming to the central nervous system once the movement is started. In a similar way, a feedback controller was designed to compensate errors occurring within the exercise and to modify the baseline motor programme (inverse model) depending on its error.

3 Design of EMC in Simulation

In order to design the controller three crucial steps were required. First, the development of a biomechanical model to simulate a virtual patient performing the knee flex-extension. Second, the identification of ANNIM, and third the identification of NF. After these three steps, the EMC was ready to be evaluated in comparison with other controllers and in response to mechanical disturbances.

3.1 The Biomechanical Model

In order to simulate neuromuscular skeletal features of the lower limb of a paraplegic subject, a complex model, modified from the one developed by Riener [17], was implemented in Simulink®. The plant was constrained to move on the sagittal plane and the knee was assumed to be an ideal hinge joint in accordance with previous studies [18]. Inputs to the plant were the PW of the stimuli delivered to the quadriceps through surface electrodes. The plant output was the knee joint angle.

Five muscle groups were considered: hamstrings (i.e., semimembranosus, semitendinosus, biceps femoris long head), biceps femoris short head, rectus femoris, vastus muscles and lateral and medial gastrocnemius. Each muscle group was characterised by activation parameters (recruitment curve, calcium dynamics, muscular fatigue) and contraction parameters (maximal force, force–length relationship and muscle force–velocity relationship).

The fatigue/recovery model was based on a fitness function fit(t), modified from Riener [17]. This function can be expressed by the following first order relation:

$$\frac{dfit}{dt} = \frac{(fit_{min} - fit)a(t)\lambda(f)}{T_{fat}} + \frac{(1 - fit)(1 - a(t))\lambda(f)}{T_{rec}}, \qquad (1)$$

where $a(t)$ was the activation of the non-fatigued muscle. Minimum fitness was given by fit_{min}. The time constants for fatigue T_{fat} and for recovery T_{rec}, as well as fit_{min}, were estimated from stimulation experiments as proposed by Riener [17]. The term $\lambda(f)$ was

$$\lambda(f) = 1 - \beta + \beta\frac{f}{100} \qquad (2)$$

for $f < 100\,\text{Hz}$; β was a shape factor.

In our simulations, the frequency f was always fixed at $40\,\text{Hz}$.

Finally, the activation of the fatiguing muscle was given by

$$a_{fat}(t) = a(t)fit(t), \qquad (3)$$

where a(t) is the activation before the fatigue model. The fatigue occurrence showed a decrease of the muscle input gain to 50% of its nominal value over 100 s, comparable to [12].

All the not stimulated muscles contributed to limb dynamics by their passive viscous and elastic properties.

3.2 The Feedforward Controller: ANNIM

Following the direct inverse modelling approach [19], the ANNIM was trained to identify the inverse model of the system to be controlled. The ANNIM desired outputs were PW waveforms and its inputs were the knee angle and the knee velocity obtained stimulating the plant not including fatigue effects in response to the chosen PW signals. To capture the system dynamics, the ANNIM inputs were augmented with signals corresponding to past inputs. Therefore, ANNIM inputs were the actual knee angle and velocity and their 4 previous samples $(q(t), q(t-1), \ldots, q(t-4)$ and $dq(t), dq(t-1), \ldots, dq(t-4))$. As suggested by Matsuoka [20], a white noise was added to the input signals (mean 0, standard deviation equal to 5% of the maximum value of angle or velocity) in order to maximise the exploration of the input/output space and to improve the quality of learning. Several multi-layer feedforward perceptrons with ten inputs, one hidden layer and one neuron in the output layer were trained in order to choose the optimal number of neurons in the hidden layer. The activation functions were hyperbolic tangents for the hidden layers and the logarithmic sigmoid function in the output layer in order to map respectively the non-linearity of the system and the bounded and positive PW range. The ANNIM architecture is shown in Fig. 2, it was trained with the Levenberg–Marquardt learning algorithm [21].

Training of ANNIM

One crucial aspect in the design of ANNs is to build the training set (TS), i.e., the collection of input/output examples used to identify the network weights. With the aim of exploring the whole space of the possible system behaviour, the TS was made up of an enlarged and diversified set of stimulation patterns. The waveforms used were rectified sinusoids, with a plateau at the maximum PW value and/or with a plateau at the minimum PW value or without any plateau. The different types of networks were evaluated on the basis of three factors: first, the value of the mean square error (MSE) between the desired

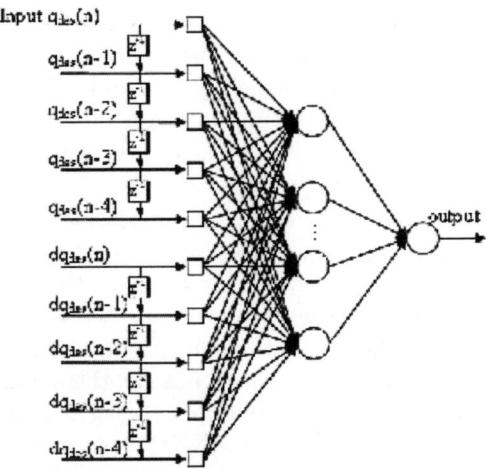

Fig. 2. ANNIM architecture

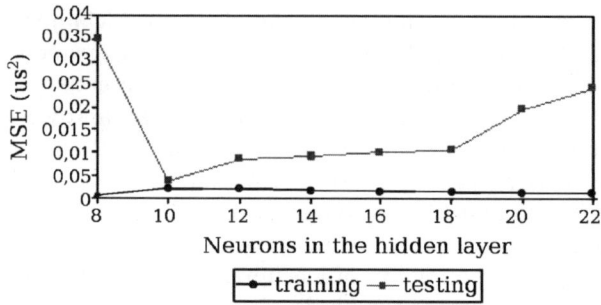

Fig. 3. MSE for training and testing data, calculated for networks with different number of neurons in a hidden layer

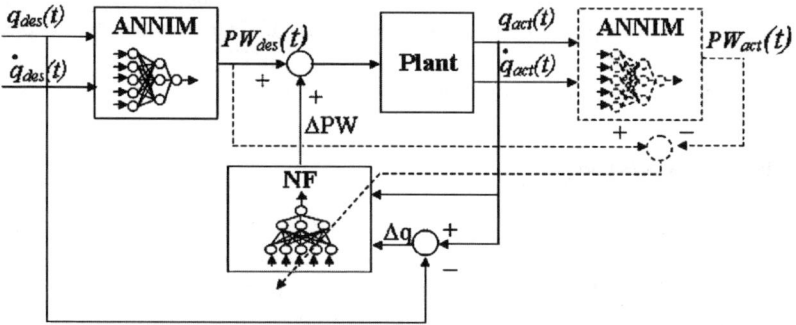

Fig. 4. NF training setup

output and the one produced by the network. Second, the network has to minimise the difference between the MSE computed on training and testing data [22] which would indicate that the network is generalising well over the whole working space and not overfitting the TS. Third, it is important to minimise the complexity of the network, avoiding long computing times. Thus, the smallest network architecture which gave a good MSE and a similar performance between training and testing data was chosen. As shown in Fig. 3, the ANN with ten neurons in the hidden layer produced the best performance and was selected as the ANNIM.

3.3 The Feedback Controller: NF

The rule of NF was to produce the PW correction necessary to compensate the angular error due to the effect of muscular fatigue and to any time variation or disturbance occurred. To obtain the TS of NF the simulation setup shown in Fig. 4 was developed. A detailed description of it is reported in [23].

This scheme included the series of ANNIM, plant and another ANNIM. The desired angle (q_{des}) was the input to the ANNIM that had already been trained, producing the corresponding desired PW (PW_{des}) as an output.

PW_{des} was then given as an input to the plant. The output of the plant was the actual angle (q_{act}), i.e., the angle generated by stimulating the plant and in which the fatigue effect was included. After that, q_{act} was used as an input to the ANNIM, producing PW_{act}. Therefore, the angular error could be translated as PW domain: $\Delta PW = PW_{act} - PW_{des}$. ΔPW, as it was built, was the extra PW which would had been required to produce the desired angle at the current state of the plant. Thus NF was trained to produce ΔPW as an output, when it received the corresponding angle (q_{act}) and angular error Δq as inputs.

Training of NF

Figure 5 shows the clear correspondence between angular errors and PW errors: ΔPW contained a piece of information of the muscular fatigue phenomenon already present in Δq.

The signal used to build the TS of NF (q_{des} in Fig. 5) was a repeated sequence of 15 consecutive flex-extension trajectories lasting 100 s. In the TS different shapes of angular trajectories were included: sinusoids with different amplitudes and durations with and without a plateau during the

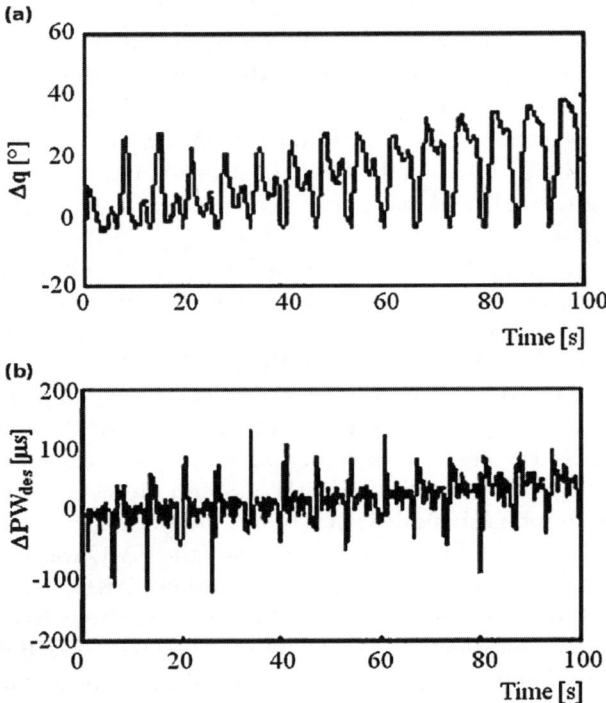

Fig. 5. An example of repeating sequences collected for the NF TS. (**a**) shows the Δq trajectory and (**b**) the ΔPW_{des}

flexion and/or during the extension [23]. NF was again a multi-layer perceptron with ten inputs (four paste samples of the knee angle q_{act} and of the
angular error Δq), eight neurons in the hidden layer and one in the output
layer. The activation functions were hyperbolic tangents on both the layers
allowing positive and negative PW corrections. The training algorithm was
Levenberg–Marquardt [21].

3.4 EMC Evaluation via Simulations

The performance and the behaviour in response to disturbances obtained using
the EMC were compared with traditional controllers already developed in
literature: a PID controller in an anti-windup configuration (PIDAW) and a
NEUROPID controller [7] which included the ANNIM and a PID as a feedback
controller. In both the controllers the PID parameters were identified using
an iterative procedure based on the minimisation of root mean square error
(RMSE) [4], where the initial estimation of the optimisation was derived from
the Ziegler–Nichols rules [8].

EMC Performance in Comparison with Other Controllers

Without fatigue, the tracking capability of the EMC was very similar to the
NEUROPID one, while the PIDAW showed a typical time-lag, as reported in
Fig. 6. The RMSE between the actual trajectory and the desired one in the
first oscillation was about 1.6° for the EMC, 10.9° for the PIDAW and 5.1°
for the NEUROPID. In the second oscillation, when both the PIDs reached
their optimal behaviour, the RMSE was about 1.8° for the EMC, 7.2° for
the PIDAW and 2.9° for the NEUROPID. The comparison of the three controllers in fatigue conditions was performed in terms of the RMSE obtained in
response to simulations of 100 s. Six different angular trajectories were tested.
Each oscillation lasted from 2 to 10 s and had an amplitude ranging from
40° to 70°. Figure 7 shows that the three controllers behaved very differently
when the tracking error increased because of the occurrence of the fatigue.
Differently from PIDAW and NEUROPID, the EMC did not increase the
stimulation PW rapidly to compensate the tracking error and this avoided an
unhelpful over-stressing to the system. In fact, such high stimulation was not
enough to achieve the desired trajectory when fatigue occurred and produced
an increase of the tiring conditions of the system. Moreover, the EMC maintained always an interval of no stimulation in between successive oscillations,
which was fundamental for recovery. This way, the EMC was able to prolong
the exercise for more time with satisfying extensions. After 100 s (10th cycle),
the RMSE with respect to the desired amplitude was about 19° for the EMC,
while it was about 31° for the PIDAW and 25° for the NEUROPID.

In Table 1 the comparison of the three controllers is reported in terms of the
25th, 50th and 75th percentile of the RMSE obtained on the six trajectories
tested during the exercise.

The statistical comparison is detailed in [23].

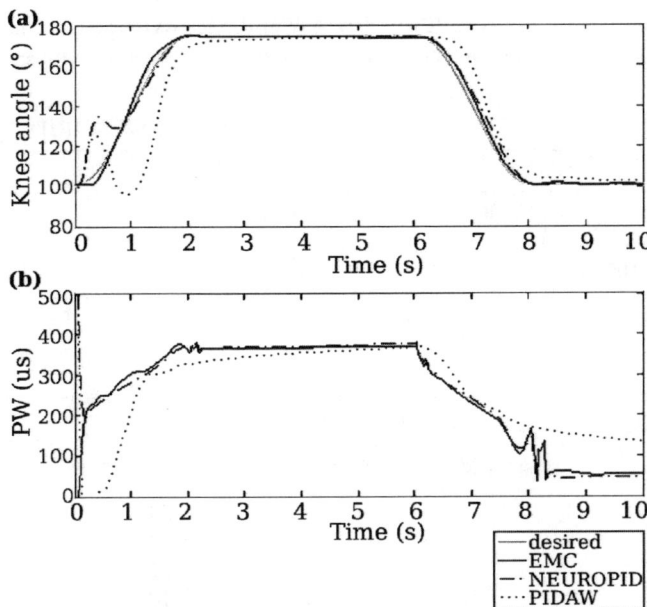

Fig. 6. A comparison of the performance obtained by the three controllers in term of angular trajectories (**a**) and PW (**b**) without considering the muscular fatigue effect

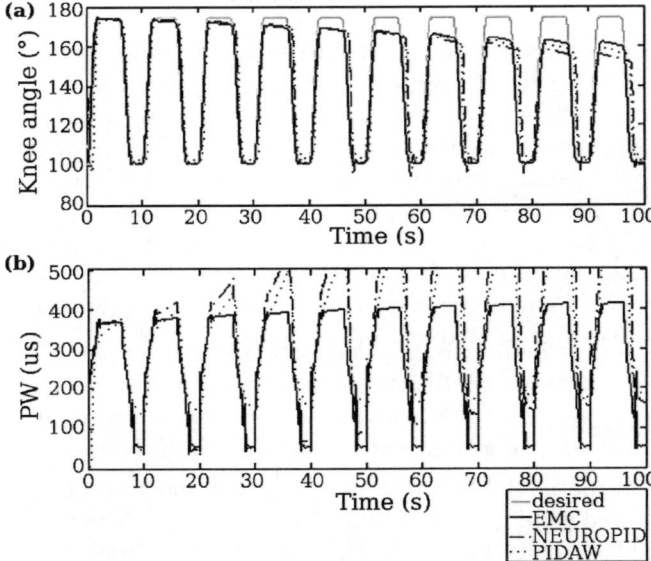

Fig. 7. A comparison of the performance obtained by the three controllers in term of angular trajectories (**a**) and PW (**b**) in presence of the muscular fatigue effect

Table 1. Comparison between EMC, PIDAW and NEUROPID in terms of the 25th, 50th and 75th percentiles of the RMSE obtained on six different testing angular trajectories

	EMC (°)			PIDAW (°)			NEUROPID (°)		
	25th	50th	75th	25th	50th	75th	25th	50th	75th
0–30	1.83	2.26	2.88	7.24	9.44	11.71	2.59	4.75	5.59
30–60	2.19	3.91	6.34	8.32	10.71	13.03	4.52	7.76	12.42
60–100	4.96	8.14	13.4	9.36	16.11	19.3	5.32	15.08	20.06

Such comparison was split in three periods (0–30, 30–60 and 60–100 s)

Table 2. Comparison of the extra angular RMSE due to spasms obtained by the three controllers while delivering the six different spasms

Spasm	EMC (°)	PIDAW (°)	NEUROPID (°)
1	3.12	2.71	3.24
2	4.20	4.59	5.11
3	6.03	6.57	6.91
4	6.29	6.70	7.10
5	6.63	7.31	7.85
6	7.79	6.43	6.78

Capability of Resisting to Mechanical Disturbances

Following [12], a mechanical disturbance such as a spasm was modelled as a square wave lasting for 2 s. Six different spasms were delivered to the plant with the limb in different positions during the simulated movement. The spasm amplitude ranged between 20 and 30% of the maximal total torque of the knee. The RMSE of the extra error between the desired angular trajectory and the obtained one only due to the spasm are reported in Table 2.

The EMC was able to resist well even if such spasm occurrences were not included in the examples used for training showing similar extra errors as PID based controllers. A more detailed analysis on the behaviour of the EMC in response to disturbances and spasms is reported in [23].

EMC Robustness

Robustness in the model parameters was tested and the satisfactory results obtained ensured good generalisation for successive sessions on the same subject. A complete study on the EMC generalisation capabilities is reported in [15, 23].

3.5 Discussion and Further Developments

The applicability and the potentiality of the controller were analysed in simulation and demonstrated very promising results. In fact, the EMC produced a good tracking capability when fatigue was not affecting a lot the performance. In these cases, the EMC was more accurate with respect to the other two controllers implemented. In addition, the EMC strategy showed good potential particularly when fatigue was really affecting the movement. The behaviour of the EMC during the process of fatiguing was completely different from the other two controllers, reducing the error by a third after 100 s. The EMC achieved such different performance because the feedback correction considered the tracking of the desired trajectory as well as the level of fatigue. The training solution of the EMC actually translated the angular error in PW error taking into consideration the differences between the actual fatigue performance with respect to the nominal one. In this way, the EMC corrected the stimulation parameters by giving an extra PW correlated to the level of fatigue. The main effect of this strategy was that the stimulation parameters grew much more slowly during repeated flex-extensions, thereby not saturating and not over-stressing the stimulated muscle. This behaviour was exactly opposite for PID based controllers. The latter stimulated the muscle to a maximum, depending only on the angular error and not evaluating the feasibility of tracking. This solution, once fatigue was too strong to permit correct tracking, provoked an over-stimulation of the muscle, inducing an even more rapid increase of fatigue. Thus, it is crucial to highlight the capability of EMC in lengthening the exercise in time, i.e., in producing the effectiveness of the rehabilitation therapy.

One of the main advantages of a completely neural control approach is the possibility of exploiting the generalisation capability of the ANNs. This property could be essential in the control of FES because it permits the use of the same controller on different patients as well as on the single subject in different moments of the same rehabilitation sessions, which correspond to different muscular conditions. As reported in [23] the EMC generalisation capability was tested in simulation and showed really good results.

Another important aspect to be emphasised is that the use of ANNs can render the training phase needed for tuning the controller parameters, shorter and simpler than the identification phase required by traditional PID controllers. The collection of the training set consisted of single flex-extension movements, which could be positive for the patient muscular tone rehabilitation. This initial phase would even be reduced by using neural networks already trained on data coming from other disable people previously treated; starting from these ANNs it would be advisable to retrain them only on a few experimental data collected from the new patient. In this way it would be possible to obtain a universal ANN able to generalise on different patients.

To confirm the simulation results the controller was designed experimentally and it was tested on a paraplegic patient.

4 Experimental Study

After verifying the feasibility of the EMC controller in simulation, we designed and tested it with experimental trials on both the legs of a paraplegic patient.

4.1 Subject and Experimental Setup

After giving his written consent, a 30 years old complete paraplegic patient with a spinal cord lesion at the T5–T6 level, took part to the experimental trials. He was not muscularly conditioned and at his first experience with FES. The patient sat on a bench, allowing the lower leg to swing freely and placing the hip angle fixed at 60° flexion and the ankle at 0°. Quadricep muscles were stimulated by adhesive rectangular surface electrodes with a cathode placed proximally over the estimated motor point of the rectus femoris and the anode, at approximately 4 cm proximal to the patella. The knee joint angle was measured by an electrogoniometer, which comprised two plastic bars attached to the thigh and the shank and a linear potentiometer fixed on the joint. This device was interfaced with the PC by means of an A/D board with a sample rate of 100 Hz. A computer-driven multi-channel stimulator, MotionStim 8^{TM}, (Krauth and Timmermann, Hamburg, Germany) delivered balanced bipolar rectangular pulses.

Once verified the correct placement of the stimulation electrodes, some preparatory tests were used to define the stimulating current and the range of PW, as reported in [15]. In particular, the current-level was fixed at 60 mA for the left leg and 70 mA for the right one. The PW was selected in order to achieve the required maximal and minimal knee extension. The chosen PW range was from 250 to 400 μs for both the legs. The stimulation frequency was 20 Hz because the patient was at his first experience with FES and a greater frequency value could provoke a too high level of fatigue.

4.2 Training of ANNIM

As in the simulation study, the first step to design EMC is the training of the inverse model of the patient, ANNIM, and therefore, the collection of the TS data. To build the TS the subjects' muscles were stimulated with a PW variable in the chosen range (250–400 μs). This permitted us to span the whole parameter space well, both in the PW and angle (TS input/output). As already done in simulation, the chosen stimulation waves were rectified sinusoids with a plateau in extension and/or in flexion or without any plateau. The duration and the maximum PW are reported in Table 3.

The TS for the patient consisted of more than 10,000 samples. TS data was collected by delivering some PW signals to both the legs in order to increase the generalisation capability of the ANN. To avoid the occurrence of the fatigue effect in the TS data collected for the ANNIM training, we delivered each training signal alternatively to the right and the left leg and

Table 3. Waveform characteristics for building the TS and the testing set of ANNIM

PW profile	Duration (s)	PW (µs)
Flexion plateau	3	250
Extension plateau	4	300
No plateau	5	350
	6	400
	7	

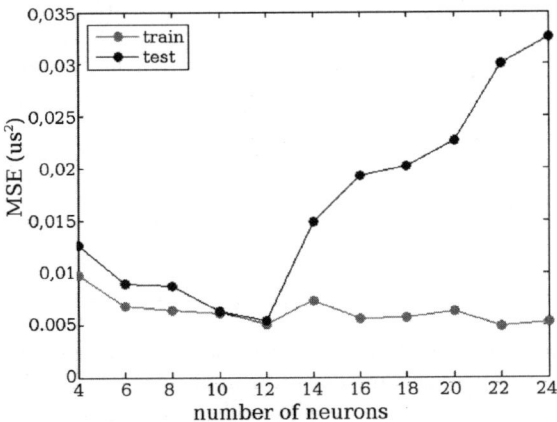

Fig. 8. RMSE for training and testing data, calculated for networks with a different number of neurons in the hidden layer

there was a rest of 2 min in between two successive trials on the same leg. The trained ANNs had the same number of layers and activation functions as those adopted during simulation and described in Sects. 3.2 and 3.3. However, the number of neurons in the hidden layer was chosen so as to optimise the performance of the trained ANNs according to the experimental data. The network architecture for the experimental session was selected following the three factors described in Sect. 3.2. As shown in Fig. 8, the network with 12 neurons in the hidden layer proved to be the best one, showing good performance on both the training and testing sets.

4.3 Training of NF

Analogously to the simulation study, the NF architecture was again a multi-layer perceptron with ten input neurons (four past samples of the knee angle q_{act} and of angular error Δq) and one neuron in the output layer. In order to correctly train the NF, the error information in the angular domain was transformed into a correction in the PW domain. The NF TS was obtained using the same simulation setup which included the series of the ANNIM, the subject and another ANNIM as described in Sect. 3.3. This way, the NF learnt

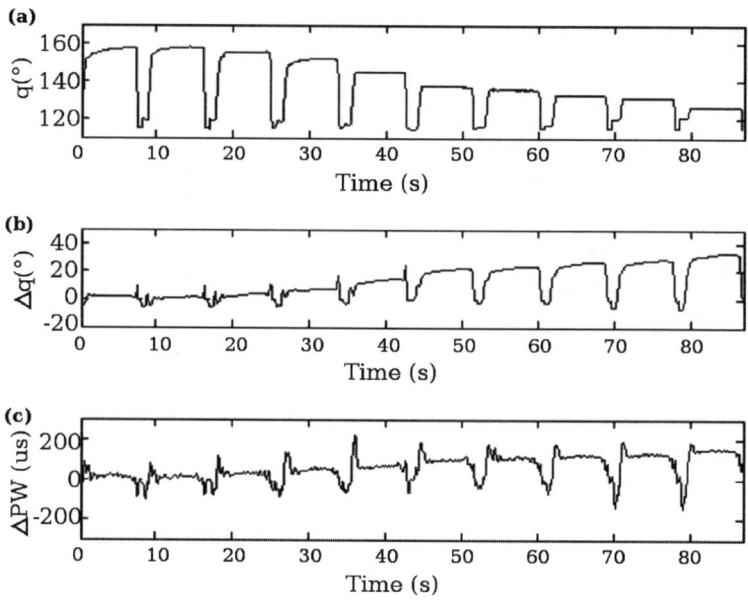

Fig. 9. An example of repeating sequences collected for the NF TS. The angular trajectory is shown in (**a**), the angular error in (**b**) and the ΔPW in (**c**)

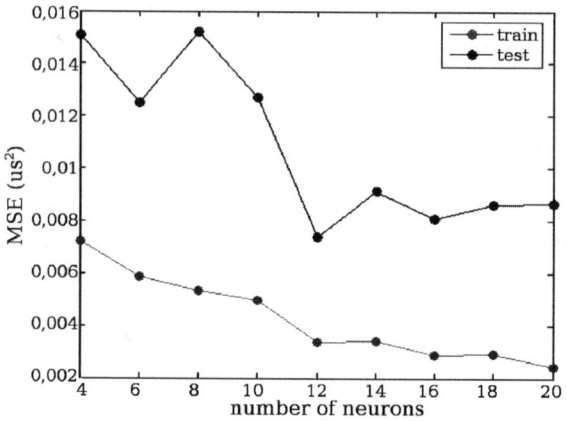

Fig. 10. RMSE for training and testing data, calculated for NF with a different number of neurons in the hidden layer

to predict the correct extra PW depending on the observed angular error. An example of the training signals is reported in Fig. 9.

The TS included signals that were generated by oscillations that lasted about 100 s. The chosen NF had 12 neurons in the hidden layer as reported in Fig. 10.

4.4 Experimental Results

Once designed EMC we had the possibility to test it in comparison with the feedforward ANNIM controller.

ANNIM Performance on Fresh Muscles

The chosen ANNIM was tested on the patient in order to evaluate its performance in terms of the knee angle. Figure 11 shows the comparison between the desired angular trajectory and the one obtained with the ANNIM.

The ANNIM produced good angular trajectories in response to both a training signal (Fig. 11a1,a2 and testing signal (Fig. 11b1,b2) with a similar shape to the training one (sinusoidal shape for PW and the corresponding angular profile obtained from the patient). The mean values and the standard deviations of the tracking error obtained in response to testing signals by the left leg, had a mean value of $3.62°$ and a standard deviation of $0.97°$. Regarding the right leg, the mean value was $4.29°$ and the standard deviation was $0.81°$. The ANNIM was a feedforward controller and therefore it produced the same output, independently of external perturbations. In Fig. 11 the angle obtained in the three successive oscillations is quite similar (particularly in panel a1); this fact demonstrates that the fatigue is not yet affecting a lot the muscles.

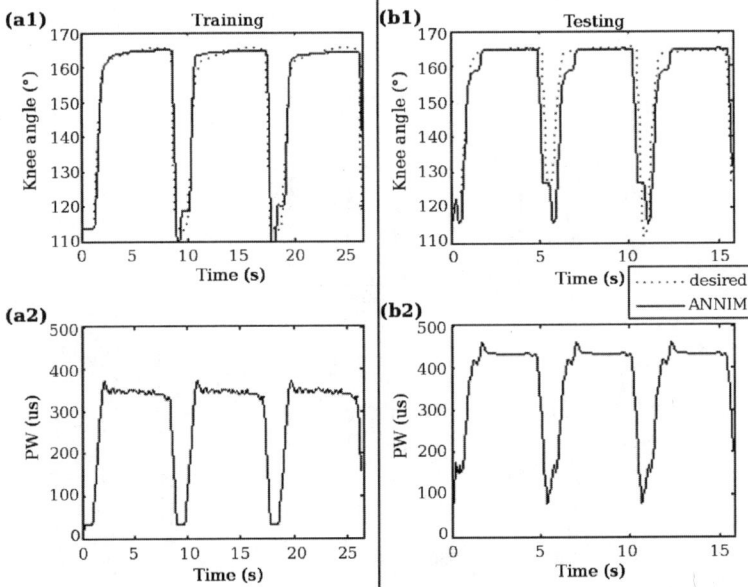

Fig. 11. ANNIM performance obtained in response to a training (**a1**), (**a2**) and to a testing signal (**b1**), (**b2**). (**a1**), (**b1**), show the angular trajectories and (**a2**), (**b2**) the delivered PW

EMC Compensation of Muscular Fatigue

The performance of the developed EMC controller was compared with the one obtained using the ANNIM, in order to evaluate the improvement gained mainly when the muscles were not fresh. Figures 12 and 13 report the performance obtained in response to a signal included in the NF TS and a testing signal never seen from the NF before, respectively. It is clear from the two figures that the desired testing signal reaches the highest extension value and the PW delivered by the EMC reaches the maximum value allowed by a stimulator (500 µs) in the end.

The reported figures refer to two trials performed on the same leg. It is important to underline that the initial peak of the PW delivered by the EMC was due to the delay of the muscle response, which resulted in a huge error between the angle obtained and the one desired. This peak did not affect the results because it was an initial push which permitted the leg to reach a good level of extension anyway, even when it was suffering from fatigue. In Fig. 12, the effort of the controller became relevant in the last oscillations, when the leg tended to fall down because of the force of gravity and the controller gave a rapid PW peak to recover the extension.

The angle obtained during testing conditions (Fig. 13) reached similar performance levels in terms of tracking error but the PW was more jagged and produced more jerky movements in the extension phase.

Figure 14 shows the angular tracking error computed by the EMC and ANNIM.

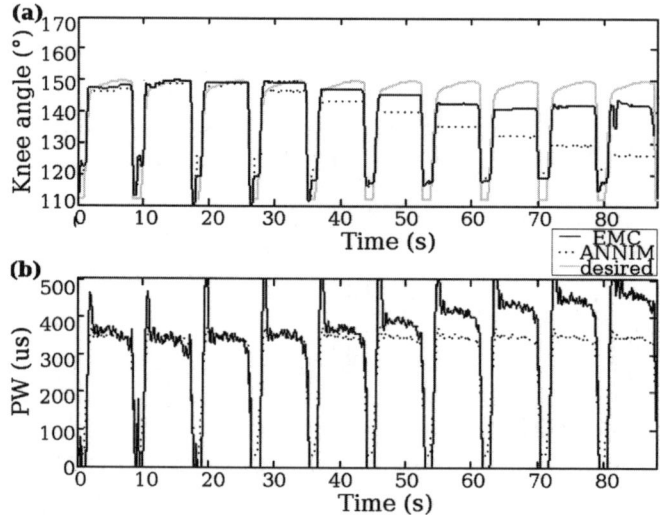

Fig. 12. A comparison of performance obtained with the EMC and ANNIM in response to a signal included in the NF TS. The angle (**a**) and the PW (**b**) are reported

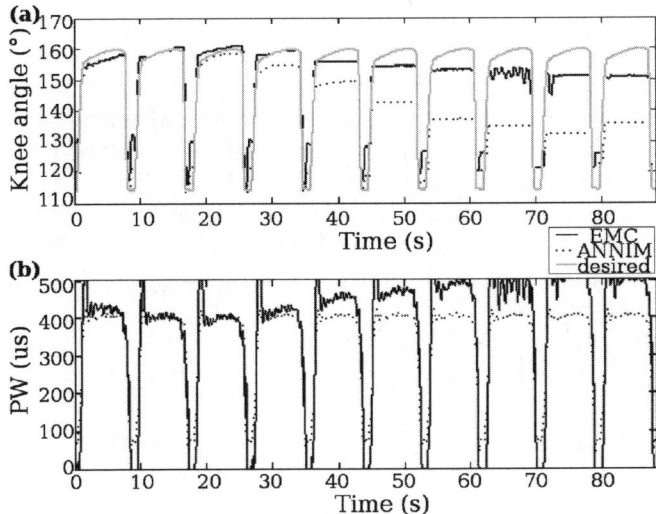

Fig. 13. A comparison of the performances obtained with the EMC and ANNIM in response to a signal that is not included in the NF TS. The angle (**a**) and the PW (**b**) are reported

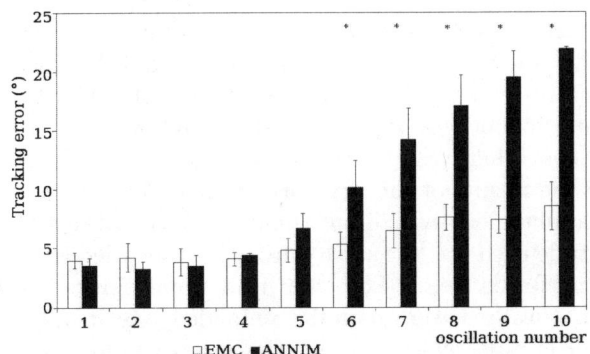

Fig. 14. Mean values and standard deviations of the tracking error obtained with the ANNIM and EMC in each movement. The asterisks indicate that, in each oscillation, a t-test statistical analysis showed a significant difference between the two controllers

The reported values are the mean values and standard deviations computed in response to both training and testing signals on both the legs. It is important to highlight that, in the left leg, several trials were performed whilst only two trials were performed on the right leg, which was very slow to recover. The EMC produced a tracking error which was lower than 8° until the end of the 100 s of movement.

After verifying that all the data were normally distributed according to the Kolmogorov–Smirnov test, a t-test was performed to identify the differences

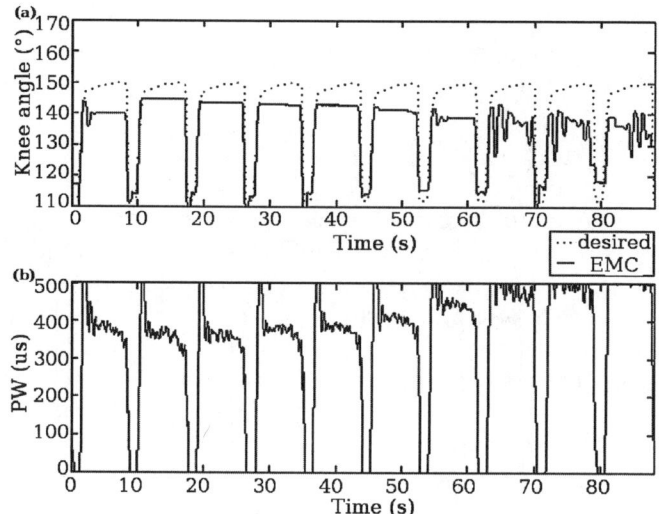

Fig. 15. EMC performance obtained with a load of 0.5 kg fixed to the ankle. In (**a**) the *dotted line* is the desired angle and the *solid line* is the angle obtained by the patient. In (**b**) the PW generated by EMC is reported

between the tracking error generated by EMC and ANNIM in each movement. In the initial five movements, the were not significant differences between the two controllers. Instead, after the fifth oscillation, the ANNIM tracking error was significantly higher ($p < 0.01$) than the EMC one.

The EMC controller was also tested in the presence of a load of 0.5 kg on the ankle and the results of this trial are reported in Fig. 15.

Since the applied load was about the 20% of the total shank load and the current amplitude used was the same during the unloaded trials, the controller showed good performance, as expected from simulations [23]. The extension reached was naturally lower than the unloaded one in all the oscillations. The fatigue effect was really clear in the last four movements when there were several oscillations of the leg during the extension phase. When the angle obtained went down, the controller delivered a PW peak to compensate this flexion. These oscillations also had a greater amplitude with respect to the ones produced during the unloaded conditions, because the inertia of the movement was increased. It is important to highlight that the presence of the load at the ankle does not increase the delay in the muscular response.

4.5 Discussion and Further Developments

The experience achieved during this study allowed us to firmly state that ANNs are optimal techniques to control non-linear and time-varying systems and, in particular, physiological systems. It is interesting to highlight that the choice of using ANNs allowed us to mimic the way in which the central nervous system produces a motor task in order to perform any movement, as proposed by the theory of internal models discussed by [16]. First, we proposed the use of an inverse model to build the nominal motor command (PW) correspondent to a desired motor task (knee angle trajectory). In addition, the use of a feedback controller was analogously based on the existence of an internal "feedback controller" able to translate the error measured by sensors in the sensory domain (angular error) into an error in the motor command domain (PW correction). This point, which is the idea for collecting the NF training set, is probably the most innovative aspect of EMC controller.

A good advantage of the presented EMC controller is that it did not use any extra setup to identify its parameters and therefore, it is easy to use also in a clinical environment. In addition, the generalisation capability showed by EMC both in simulation and in experimental tests is crucial in the FES applications control, being every single session not exactly the same as the previous one because of many variables which could not be controlled in normal clinical practice (i.e., the re-positioning of electrodes is not exactly the same, the position of the sensor is not exactly repeatable as well as the general condition of the patient is always different).

Moreover, this controller has a good translational property over multiple muscles because ANNs can process many inputs and have many outputs. Therefore, they are indeed readily applicable to multivariable systems. The ANNs ability to identify multi input multi output systems would also be an appealing perspective to exploit. In this way, having a major amount of input information to describe the system condition, it would be possible to obtain a control action that is more similar to a natural one, exploiting the simultaneous contractions of different muscles involved in the movement. However, the most important difficulty to face in the translation of EMC to complex motor tasks is due to the intrinsic redundancy of the motor apparatus. In fact, there are a lot of different stimulation strategies able to induce correctly a complex motor task such as cycling or walking (different co-contractions levels and different distribution between agonist muscles). On one side, we could define one of the possible solutions and use it as the stimulation strategy. On the other side, the possibility to switch among different motor strategies is very important to face fatigue, allowing a sort of turn-over of the muscles fibers. How to balance between control simplicity and exploitation of motor redundancy is one of the main open issue in the FES control of complex movements. The use of ANNs could not help in defining one single solution but in controlling multiple and possible solutions simultaneously, but this point needs to be deeply explored in future works.

5 Conclusions

We proposed a controller, called EMC, for neuromuscular stimulation of knee flex-extension which is composed by a feedforward inverse model and a feedback controller, both implemented using neural networks. The training of the networks is conceived to avoid to a therapist and a patient any extra experiment, being the collection of the training set included in the normal conditioning exercises. The EMC philosophy differs from classical feedback controllers because it does not merely react to the error in the tracking of the desired trajectory, but it estimates also the actual level of fatigue of the muscles. The controller was deveoped and evaluated in simulation and then tested on one paraplegic patient showing promising results. A further study on other patients would be advisable in order to stress properly the generalisation capability of the controller developed.

References

1. Jaeger RJ (1996) Principles underlying functional electrical stimulation techniques. J Spinal Cord Med, 19(2):93–96.
2. Hausdorff JM, Durfee WK (1991) Open-loop position control of the knee joint using electrical stimulation of the quadriceps and hamstrings. Med Biol Eng Comput, 29:269–280.
3. Ferrarin M, Palazzo F, Riener R, Quintern J (2001) Model-based control of fes-induced single joint movements. IEEE Trans Neural Syst Rehabil Eng, 9(3):245–257.
4. Quintern J, Riener R, Rupprecht S (1997) Comparison of simulation and experiments of different closed-loop strategies for functional electrical stimulation: experiments in paraplegics. Artif Organs, 21(3):232–235.
5. Veltink P (1991) Control of fes-induced cyclical movements of the lower leg. Med Biol Eng Comput, 29:NS8–NS12.
6. Munih M, Hunt KJ, Donaldson N (2000) Variation of recruitment nonlinearity and dynamic response of ankle plantar flexors. Med Eng Phys, 22(2):97–107.
7. Chang GC, Luh JJ, Liao GD (1997) A neuro-control system for the knee joint position control with quadriceps stimulation. IEEE Trans Rehabil Eng, 5:2–11.
8. Ferrarin M, D'Acquisto E, Mingrino A, Pedotti A (1996) An experimental PID controller for knee movement restoration with closed-loop fes system. Proceedings of the 18th Annual International Conference of the IEEE Engineering in Medicine and Biology Society, Amsterdam, The Netherlands 1:453–454.
9. Hatwell MS, Oderkerk BJ, Sacher CA, Inhar GF The development of a model reference adaptive controller to control the knee joint of paraplegics. IEEE Trans Automatic Control, 36:683–691.
10. Riener R, Edrich T. (1999) Identification of passive elastic joint movements in the lower extremities. J Biomech, 32:539–544.
11. Haykin S (1999) Neural Networks – A Comprehensive Foundation, Second Edition, Prentice Hall, Upper Saddle River, NJ.
12. Abbas J, Chizack H (1995) Neural network control of functional neuromuscular stimulation systems: Computer simulation studies. IEEE Trans Biomed Eng, 42(11):1117–1127.
13. Riess J, Abbas J (2000) Adaptive neural network control of cyclic movements using functional neuromuscular stimulation. IEEE Trans Rehab Eng, 8(1):42–52.

14. Riess J, Abbas J (2001) Adaptive control of cyclic movements as muscles fatigue using functional neuromuscular stimulation. IEEE Trans Neural Syst and Rehab Eng, 9(3):326–330.
15. Ferrante S, Pedrocchi A, Iannò M, De Momi E, Ferrarin M, Ferrigno G (2004) A Functional electrical stimulation controlled by artificial neural networks: pilot experiments with simple movements are promising for rehabilitation applications. Funct Neurol, 19(4):243–252.
16. Wolpert D, Miall R, Kawato M (1998) Internal models in the cerebellum. Trends Cognit Sci, 2(9):338–347.
17. Riener R, Fuhr T (1998) Patient-driven control of fes-supported standing-up: a simulation study. IEEE Trans Rehab Eng, 6:113–124.
18. Stein RB, Zehr EP, Lebiedowska MK, Popovic DB, Scheiner A, Chizeck HJ (1996) Estimating mechanical parameters of leg segments in individuals with and without physical disabilities. IEEE Trans Rehabil Eng, 4:201–211.
19. Kumpaty S, Narendra K, Parthasarathy K (1990) Identification and control of dynamical systems using neural networks. IEEE Trans Neural Networks, 1:4–27.
20. Matsuoka K (1992) Noise injection into inputs in back-propagation learning. IEEE Trans Syst Man Cybernet, 22(3):436–440.
21. Hagan M, Menhaj M, (1994) Training feedforward networks with the marquardt algorithm. IEEE Trans Neural Netw, 5(6):989–993.
22. Principe J, Euliano N, Lefebvre W (2000) Neural and adaptive systems: fundamentals through simulation. Wiley, New York.
23. Pedrocchi A, Ferrante S, De Momi E, Ferrigno G (2006) Error mapping controller: a closed loop neuroprosthesis controlled by artificial neural networks. J NeuroEng Rehabil, 3:25.